Lecture Notes in Computer Science   1212

Edited by G. Goos, J. Hartmanis and J. van Leeuwen

Advisory Board: W. Brauer   D. Gries   J. Stoer

Springer

Berlin
Heidelberg
New York
Barcelona
Budapest
Hong Kong
London
Milan
Paris
Santa Clara
Singapore
Tokyo

Jonathan P. Bowen
Michael G. Hinchey   David Till (Eds.)

ZUM '97: The Z Formal Specification Notation

10th International Conference of Z Users
Reading, UK, April 3-4, 1997
Proceedings

 Springer

Series Editors

Gerhard Goos, Karlsruhe University, Germany

Juris Hartmanis, Cornell University, NY, USA

Jan van Leeuwen, Utrecht University, The Netherlands

Volume Editors

Jonathan P. Bowen
The University of Reading, Department of Computer Science
Whiteknights, PO Box 225, Reading, Berks RG6 6AY, UK
E-mail: J.P.Bowen@reading.ac.uk

Michael G. Hinchey
New Jersey Institute of Technology, Real-Time Computing Laboratory
University Heights, Newark NJ 07102, USA
and
University of Limerick, Dept. of Computer Science and Information Systems
National Technological Park, Castletroy, Limerick, Ireland
E-mail: hinchey@cis.njit.edu

David Till
City University, Department of Computer Science
Northampton Square, London EC1V 0HB, UK
E-mail: till@soi.city.ac.uk

Cataloging-in-Publication data applied for

Die Deutsche Bibliothek - CIP-Einheitsaufnahme

The Z formal specification notation : proceedings / ZUM '97,
10th International Conference of Z Users, Reading, UK, April 3
- 4, 1997. Jonathan P. Bowen ... (ed.). - Berlin ; Heidelberg ;
New York ; Barcelona ; Budapest ; Hong Kong ; London ;
Milan ; Paris ; Santa Clara ; Singapore ; Tokyo : Springer, 1997
  (Lecture notes in computer science ; Vol. 1212)
  ISBN 3-540-62717-0
NE: Bowen, Jonathan P. [Hrsg.]; ZUM <10, 1997, Reading>; GT

CR Subject Classification (1991): D.2, I.1.3, F.3.1, D.1, G.2, F.4.1

ISSN 0302-9743
ISBN 3-540-62717-0 Springer-Verlag Berlin Heidelberg New York

© Springer-Verlag Berlin Heidelberg 1997
Printed in Germany

Typesetting: Camera-ready by author
SPIN 10549454   06/3142 – 5 4 3 2 1 0   Printed on acid-free paper

Preface

This Z User Meeting (ZUM), was the tenth in the series originally started by Ib Sørensen in December 1986 at Oxford, under the auspices of the Oxford University Computing Laboratory. The first five meetings were all held in Oxford, initially at the Department of External Studies in Rewley House, and the last at one of the colleges, Lady Margaret Hall. There is no written record of the first meeting, but an informal proceedings was produced by Jonathan Bowen for the second and third meetings, still available on-line for those interested in the historical development of Z! In 1989 the first formal proceedings were edited by John Nicholls and published in the Spring-Verlag *Workshops in Computing* series. John helped develop and maintain the momentum of the Z User Meetings throughout their early years.

In 1991, the first ZUM outside Oxford was held at the University of York, another centre of research in formal methods including the Z notation. The following year, the first meeting outside the realms of academe was held at the offices of the UK government Department of Trade and Industry (DTI) in London. This was particularly apt since the results of a DTI-funded initiative on Z, the ZIP project, were presented, and all the invited speakers were from industry. The meeting also saw the formal launch of the Z User Group (ZUG), whose main function is to organize the Z User Meetings, and facilitate interaction between users of Z.

1994 saw the largest Z User Meeting ever with around 140 delegates, perhaps encouraged by the elegant surroundings of St. John's College, Cambridge and the change of timing from December to the summer. Subsequently ZUM has been held at approximately 18 month intervals, partly to avoid clashing with the Formal Methods Europe symposium which also runs to an 18 month cycle.

The first meeting outside the United Kingdom was held at Limerick in Ireland, with a change of name to the *International Conference of Z Users* and a change of series for the proceedings to the well-regarded *Lecture Notes in Computer Science*, again published by Springer-Verlag. The meeting recorded in this volume was held back in the UK, at the University of Reading, and we intend to hold ZUM alternatively inside and outside the UK in the future.

Our invited speakers for ZUM'97 were drawn from Italy, the UK, and the USA. Prof. Egon Börger of the University of Pisa is a recognized European expert on formal methods. Dr. Anthony Hall is well-known in Z circles as an industrial user of Z and other formal methods, working for Praxis plc, a company in Bath, UK, which applies formal methods widely, especially for the development of safety-critical systems. Dr. Constance Heitmeyer of the US Naval Research Laboratories comes from outside the Z community, but is widely respected for her work in the application of formal techniques to real-time systems.

Tool demonstrations were organized by Ali Abdallah (University of Reading,UK) throughout the main meeting. The conference also saw the launch of a new International Series in Formal Methods published by Academic Press. As is traditional with ZUM, there were a number of associated activities both before and after the main meeting. Tutorials were held immediately beforehand, organized by Sam Valentine of the University of York. On the day after the conference, the third in a successful series of informal Educational Issues Sessions was held, again organized by their original initiator, Neville Dean of Anglia Polytechnic University.

ZUM'97 was supported by a number of companies and organizations. Praxis continue to give valuable support in the running of the Z mailing list, which lightens the administrative burden on the Z User Group considerably. This year, IBM United Kingdom Laboratories provided generous support for expenses of invited speakers, student bursaries, and the best paper prize. FACS, the Formal Aspects of Computer Science specialist group of the British Computer Society (BCS), continue to support ZUG by providing publicity. The University of Reading provided facilities for the conference; in particular, Roy Briggs has offered the use of the Museum of English Rural Life on the campus for a reception.

During the week after ZUM'97, the fifth EU ESPRIT **ProCoS-WG** Working Group Meeting on *Provably Correct Systems* was held at Reading, enabling group members to attend both events easily if they wished. **ProCoS-WG** (ESPRIT Working Group 8694) gave financial support to cover secretarial assistance for ZUM which helped greatly with the organization of this conference as well as previous meetings. Christina Simmons, the local organizing secretary in the Department of Computer Science at Reading, provided invaluable support in the administrative planning and smooth running of the conference. Joan Arnold, the **ProCoS-WG** secretary based at the Oxford University Computing Laboratory, was an experienced and appreciated helper at many of the Z User Meetings.

On-line information concerning the conference is held under the following URL:

http://www.cs.reading.ac.uk/zum97/

This will be kept up to date after the conference with any relevant information, and provides links to other on-line resource concerning the Z notation such as previous Z User Meetings, and also formal methods in general. We welcome delegates to the Tenth International Conference of Z Users, and look forward to developments for the next ten in the series after this milestone point.

Reading, Limerick and London   Jonathan Bowen (Conference Chair)
January 1997   Mike Hinchey and David Till (Programme Co-Chairs)

Programme Committee

Ali Abdallah, University of Reading, UK
Jonathan Bowen, University of Reading, UK (*Conference Chair*)
Paolo Ciancarini, University of Bologna, Italy
Neville Dean, Anglia Polytechnic University, UK
Andy Evans, University of Bradford, UK
David Garlan, Carnegie Mellon University, USA
Martine Guerlus, France Telecom CNET, France
Jonathan Hammond, Praxis, UK
Howard Haughton, JP Morgan, UK
Ian Hayes, University of Queensland, Australia
Mike Hinchey, NJIT, USA & Univ. of Limerick, Ireland (*Programme Co-Chair*)
Hans-Martin Hörcher, DST GmbH, Germany
Jonathan Jacky, University of Washington, USA
Kevin Lano, Imperial College, London, UK
Shaoying Liu, Hiroshima City University, Japan
Nimal Nissanke, University of Reading, UK
Norah Power, University of Limerick, Ireland
Chris Sennett, DRA Malvern, UK
David Till, City University, UK (*Programme Co-Chair*)
Sam Valentine, University of York, UK
Jim Woodcock, Oxford University, UK
John Wordsworth, IBM UK Laboratories, UK

External Referees

All submitted papers were reviewed by the programme committee and/or a number of external referees. We are very grateful to these people, and apologize in advance for any names omitted from this list:

Rob Arthan	Michael Butler	Stelvio Cimato
Roger Duke	Cecilia Mascolo	Paul Mukherjee
Larry Paulson	Gordon Rose	Margaret West

Sponsors

The 10th International Conference of Z Users greatly benefited from the support and financial assistance of the following:

IBM United Kingdom Laboratories
Praxis plc
ProCoS-WG ESPRIT Working Group
University of Reading

Tutorial Programme

Three tutorials were presented on the day prior to the main sessions (2nd April); they were:

How to Read the (draft) Z Standard
Stephen Brien, Andrew Martin, and Jim Woodcock

Larch: Theory and Practice
John Wordsworth

The Z/EVES System
Mark Saaltink

Contents

Formal Methods: A Panacea or Academic Poppycock?*

Constance Heitmeyer

Naval Research Laboratory
Washington, DC 20375
heitmeyer@itd.nrl.navy.mil

1 Introduction

Much has been written in the past decade about the usefulness of formal methods for developing computer systems. In this talk, I describe first what I mean by a formal method, discuss some systems to which formal methods have been successfully applied recently, and present several guidelines that I and my colleagues have found useful in applying formal methods in the development of practical systems and software. I conclude with a summary of what has been accomplished to date and why some skepticism about the utility of formal methods in computer system development remains well-founded.

2 What is a formal method?

In a lecture in 1934 [6], Gödel provides the following definition of a formal method:

> A *formal mathematical system* is a set of symbols along with rules for employing them...[such that] for each rule of inference there shall be a finite procedure for determining whether a given formula B is an immediate consequence (by that rule) of given formulas $A_1, \ldots, A_n$.

Thus, for Gödel, a formal method allows one to reason about a system using the rules of logical deduction.

In computer science, we usually assign a broader meaning to the term. By a *formal method*, we mean any mathematically-based technique useful in the development of computer systems. In developing computer systems and software, two kinds of formal methods are valuable: formal specification notations and formal verification techniques.

A *formal specification* is a computer system (or a computer system part) that is machine-processable; that is, one can tell whether the description is well-formed. A formal specification can reduce errors by reducing ambiguity and imprecision and by making some instances of inconsistency and incompleteness obvious.

The product of *formal verification* is a formal proof, often based on mechanical support, that a computer system (or a computer system part) satisfies certain properties of interest. The formal proof can be based on deductive reasoning as described by Gödel, or it can be developed with a model-based approach, such as model checking [5]. In a model-based approach, one typically represents the system of interest as a finite state machine, enumerates all the possible states of the machine, and checks that the properties of interest are satisfied in each machine state.

* NRL's research in formal methods is supported by ONR and SPAWAR.

3 Recent Applications of Formal Methods

We can identify two broad classes of problems in computer system development where formal methods have proven useful. First, formal methods have had utility in proving the correctness of algorithms used in software applications. Second, formal methods can provide convincing evidence that a specification of a computer system or a computer system part satisfies certain properties of interest, or that a description (for example, a design or an implementation) of a system or system part satisfies a given specification. Below, I provide examples of practical applications of formal methods, some in each class.

3.1 Applying Formal Methods to Verify Algorithms

Recently, Rushby and his colleagues have used the mechanical verification system PVS to formally specify and mechanically verify fault-tolerant algorithms useful in digital flight-control systems [11]. One of these algorithms was developed by C. S. Draper Laboratory. Its purpose is to achieve interactive consistency in Draper's "Fault Tolerant Processor" (FTP).

A fundamental result in interactive consistency is that at least $3n + 1$ processors are required to withstand n simultaneous Byzantine faults; for example, four processors are required to withstand a single fault. The FTP architecture uses another approach: it requires only three main processors plus three much simpler "interstages" whose sole function is to relay messages between the main processors. This architecture is attractive because it is both cheaper and more reliable than the conventional four-processor architecture. Using PVS, Lincoln and Rushby not only verified the algorithm used in FTP. In addition, they developed and verified an improved algorithm that allows some processors to lack interstages and that uses a hybrid fault model to withstand a wider range of faults.

Another example recently reported by Kaufmann and Moore uses ACL2, a new version of the Boyer-Moore theorem prover, to verify the correctness of a new algorithm for doing division [10]. Recently, a hardware company developed a new microprocessor with a nonstandard algorithm for doing division. To increase confidence in the algorithm's correctness, the company hired Kaufmann and Moore to mechanically verify the algorithm. After three weeks, Kaufmann and Moore, using ACL2, successfully completed the verification. After receiving this evidence of the algorithm's correctness, the company will have higher confidence when it places the new microprocessor on the market.

These two applications of formal methods are significant because each provides important evidence that a given algorithm performs as intended. Given this evidence, the algorithms can be confidently used in the systems for which they were originally designed (i.e., the FTP architecture and the new microprocessor) as well as in new future systems. An additional benefit in using formal methods is that the availability of the original formal specifications and proofs should dramatically reduce the amount of effort needed to specify and verify new revised versions of the original algorithms.

3.2 Applying Formal Methods to Specifications

Since 1992, our research group at NRL has been developing software tools for detecting errors, often automatically, in requirements specifications represented in the SCR tabular format [9, 8, 2, 3]. In 1993, Lockheed-Martin used a version of the SCR format to specify the requirements of the Operational Flight Program (OFP) for their C-130J aircraft. Currently, we are using one of our tools, the consistency and completeness checker [9], to automatically check the Lockheed specifications for various errors, such as type errors, missing cases, nondeterminism, and circular definitions.

To date, we have used the tool to analyze 80 of the 1100 tables that comprise Lockheed's OFP specification and, in the process, have detected a number of errors — numerous type errors as well as six cases of undesired nondeterminism. Left uncorrected, such errors could lead to serious (or even catastrophic) problems in the operation of the C-130J.

Recently, we began applying our tools to components of various systems under development by the U.S. Navy. A current effort is to analyze the specification of an operator panel for torpedo control. To date, our tools have detected numerous syntactic errors in the panel's specification. Once we have discussed these errors with the developers of the specification and have made the needed corrections, we will apply the tools to the improved specification and report any additional errors that are detected. Subjecting the panel specifications to formal analysis should significantly improve the correctness of the specifications and thus reduce the likelihood of problems in the software controlling the torpedos.

Leveson and her colleagues have used mechanically-supported formal methods, similar to the NRL methods described above, to analyze large portions of the requirements specification for TCAS II, a collision avoidance system for commercial aircraft [7]. Like us, they found that their automated methods detected significant errors not caught by an extensive manual review.

4 Guidelines for Applying Formal Methods

In applying formal methods to practical systems, I and my NRL colleagues have developed a number of guidelines. Several of these are described below.

1. *Describe the computer system requirements formally as a relation between entities in the system environment.* More specifically, the required computer system behavior should be expressed as a mathematical relation between quantities in the environment that the system monitors (i.e., *monitored quantities*) and environment quantities that the system controls (i.e., *controlled quantities*). By describing the required system behavior in terms of the environment, the requirements specification provides a black-box description of the system that avoids design and implementation details. The existence of this formal statement of the required system behavior is extremely valuable: it can serve as the formal foundation for the design, implementation, testing, and maintenance of the computer system. The Parnas-Madey Four Variable Model [11] provides a formal framework for specifying the system and software requirements.

2. *Represent the requirements in a notation that produces readable, precise, and concise specifications.* Some (see, e.g., [4]) advocate the use of a logic language to capture the system and software requirements. In contrast, we advocate a more user-friendly notation, such as a tabular notation, but require an explicit formal semantics for the notation [2]. Although logic languages can be highly expressive, using them requires training and mathematical sophistication that most software developers lack. Without this training, creating high-quality specifications in a logic language is very difficult. Moreover, we find that the expressive power of logic languages is seldom needed in specifying practical systems and finally that logic languages usually lack structures to support readability.

   In contrast to logic languages, the SCR tabular notation sacrifices generality for both ease of use and improved support for analysis. Moreover, tables provide a natural organization which permits independent construction, review, modification, and analysis of smaller parts of a large specification. Finally, in our experience, tables, unlike alternative notations, scale very well to large problems, such as the C-130J OFP.

3. *Use mechanically-supported formal analysis to detect errors in both the specification and hand proofs.* Although both the tables used in the OFP specifications and the Statechart-like specifications of TCAS II were checked manually, software tools detected many errors overlooked by the reviewers. Reference [1] describes a recent example in which a software tool we built on top of the mechanical prover PVS helped us detect more than a dozen errors (including two major errors) in a formal specification and proofs of the Steam Boiler Controller problem. Not only are software tools better than people for detecting certain classes of errors (e.g., type errors, missing cases, etc.). In addition, they can reduce the difficulty of manual inspection, a labor-intensive task that humans find tedious and boring. The tools also liberate people to do more creative work.

4. *Use simulation to validate the specifications.* Even though a specification has been checked mechanically and no errors have been detected, the specification may still be wrong. For example, some required behavior may have been omitted unintentionally, or some requirements may be stated incorrectly. There is no guarantee that formal analysis will detect these kinds of errors. To uncover such errors, we advocate the use of a "simulator", which symbolically executes the system based on the formal specifications. The user can use the simulator to execute selected scenarios and study the simulated system behavior to make sure that the simulated behavior matches the intended behavior.

5. *Use incremental development.* One approach to describing the required system behavior is to first specify the *ideal* system behavior, which abstracts away timing delays and imprecision, and to later specify the *allowable* system behavior, which bounds the timing delays and imprecision. Typically, a function specifies ideal system behavior, whereas a relation specifies the allowable system behavior. The allowable system behavior is a relation rather than a function because it may associate a monitored quantity with more than a single value of a controlled quantity.

The required system behavior is easier to specify and to reason about if the ideal behavior is defined first. Then, the required precision and timing can be specified separately. This is standard engineering practice. Moreover, this approach provides an appropriate separation of concerns, since the required system timing and accuracy can change independently of the ideal behavior.

6. *Apply formal analysis techniques in the appropriate order.* I envision the following process for developing computer system specifications using formal methods. Although such a process is an idealization of a real-world process, it demonstrates how tools such as our consistency checker can be used to produce high quality specifications. After a formal notation is used to specify the required system behavior, an automated consistency checker is used to check the specification for syntax and type correctness, coverage, determinism, and other application-independent properties. Then, the specification is executed symbolically using a simulator to ensure that it captures the customers' intent.

In the later stages of the specification phase, mechanical support can be used to analyze the specification for application properties. Initially, a small subset with fixed parameters and only a few states is extracted from the specification and a tool, such as a model checker, is used to detect violations of the properties. This may be repeated, each time with a different or larger subset. Once there is sufficient confidence in the specification, a mechanical proof system may be used to verify the complete specification or, more likely, safety-critical components.

5   Concluding Remarks

Above, I have given a number of examples that demonstrate that formal methods can be useful in developing computer systems. They can

- detect errors that people miss,

- scale to handle real-world problems, and

- provide added confidence in the correctness of a system or a system component.

However, software developers still have some reason to be skeptical about formal methods:

- Generally, the specification languages and tools associated with formal methods are difficult to use, lack user-friendliness, and give minimal feedback when errors are detected.

- Sometimes, the effort required to use a formal method outweighs the added confidence that the formal method gives.

- Although formal methods allow developers to perform some valuable analysis of a specification, they give developers little help in validating that the specification is the *right* specification.

I expect these problems to subside in the next five years. The cost of using formal methods is going down. As an example, consistency checking as described by [9, 7] is very cheap, yet it detects errors that, left uncorrected, can have serious consequences. Moreover, the difficulty and effort required to use mechanical verifiers has decreased significantly. Both the mechanical verification of the division algorithm and the detection of errors in the Steamer Boiler Controller (SBC) problem each required only three weeks. Moreover, mechanically checking a corrected version of the SBC proofs required only a few hours. In many cases, producing more readable specifications and better feedback when errors are detected is an engineering problem rather than a research problem. Finally, specifications can be validated by using simulation as described above.

I expect the use of formal methods in the many ways described above to produce high-quality specifications. Such specifications should produce systems that are more likely to perform as required and less likely to lead to accidents. They should also lead to significant reductions in software development costs.

References

1. M. Archer, C. L. Heitmeyer. Verifying hybrid systems modeled as timed automata: A case study. *Proceedings, Hybrid and Real-Time Systems Workshop (HART '97)*, Grenoble, France, March, 1997.
2. R. Bharadwaj, C. L. Heitmeyer. Applying the SCR requirements specification method to practical systems: A case study. *Proceedings, Twenty-First Annual Software Engineering Workshop*, Greenbelt, MD, December 1996.
3. R. Bharadwaj, C. L. Heitmeyer. Verifying SCR requirements specifications using state exploration. *Proceedings, First ACM SIGPLAN Workshop on the Automated Analysis of Software*, Paris, France, January 1997.
4. Ricky W. Butler. *An Introduction to Requirements Capture Using PVS: Specification of a Simple Autopilot*. NASA Technical Memorandum 110255. NASA Langley Research Center, Hampton VA 23681.
5. E. M. Clarke, E. A. Emerson and A. P. Sistla. Automatic verification of finite-state concurrent systems using temporal logic specifications. *ACM Transactions on Programming Languages and Systems* 8(2):244–263, 1986.
6. K. Gödel. On Decidable Propositions of Formal Mathematical Systems. Notes by Kleene and Rosser on Gödel's 1934 Lecture to the IAS. Reprinted in *The Undecidable*, ed., Martin Davis.
7. M. Heimdahl, N. Leveson. Completeness and consistency analysis of state-based requirements. *Proceedings, 17th International Conference on Software Engineering (ICSE '95)*. Seattle, WA, April 1995.
8. Constance Heitmeyer, et al. SCR*: A toolset for specifying and analyzing requirements. *Proceedings, 10th Annual Conference on Computer Assurance(COMPASS '95)*, Gaithersburg MD, June 1995.
9. C. L. Heitmeyer, R. D. Jeffords, and B. G. Labaw. Automated consistency checking of requirements specifications. *ACM Trans. on Software Engg. and Methodology*, 5(3)231–261, July 1996.
10. Matt Kaufmann and J Strother Moore. ACL2: An industrial-strength version of Nqthm. *Proceedings, 11th Annual Conference on Computer Assurance (COMPASS '96)*. Gaithersburg, MD, June 17-21, 1996.

An Introduction to the Event Calculus

Bill Stoddart

School of Computing and Mathematics, University of Teesside, U.K.

Abstract. The Event Calculus is a model of communicating state machines which is defined in Z. The machines have a number of behavioural states, which can be represented in diagrammatic form, and have data states which can be described with Z data base schemas. Machines change state on the occurrence of an "event". A diagrammatic notation is used in which each transition may be labelled with an event together with a Z operation schema which describes any change in the machines data state. Communication between machines is modelled by means of shared events, associated with a simultaneous change of state of two or more machines. In this paper the key concepts of the calculus are introduced through tutorial examples based on vending machines. This is followed by a case study of a distributed seat booking system.

1 Introduction

Information systems may be described in terms of *data*, *process* and *behaviour* [1]. In this terminology *data* refers to the static aspects of the data held by a system; usually described in Z by means of a "data base schema". By *process* we mean the dynamic aspects of the data; usually described in Z by means of "operation schemas". By *behaviour* we mean how a system responds to its environment, and how component subsystems communicate and respond to one another.

The classical style of Z specification provides a powerful and expressive means of describing *data* and *process* aspects of a system. It is less obvious how Z should be used to describe *behaviour*.

The calculus described in this article allows *behaviour* to be described in Z, whilst still allowing *data* and *process* to be expressed in the usual way.

1.1 A simple vending machine and customer

The Event Calculus is based on a simple formulation of the idea of a "state machine". A machine has a number of states, and changes state on the occurrence of an "event". Only one event can occur at a time. Communication between machines, including the passing of data, is expressed in terms of shared events that simultaneously change the state of two or more machines

As a first example we consider a vending machine V that sells chocolate bars at two pounds each, and a customer C. Their next state relations, which are denoted by $\psi\, V$ and $\psi\, C$, are shown in Fig. 1.

First consider the vending machine V without reference to the customer. Suppose it starts in initial state V_0. The machine can accept one pound and

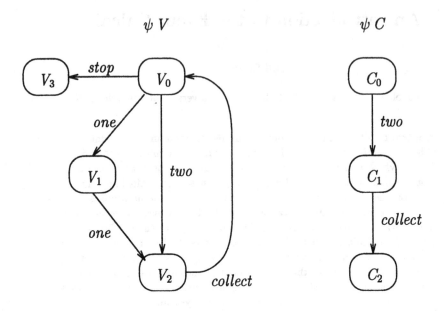

$$\psi\,V \qquad\qquad\qquad \psi\,C$$

$$one \in \mathrm{dom}\ C$$

Fig. 1. Unreliable Vending Machine and Customer

two pound coins, and after insertion of two pounds allows a customer to collect his chocolate bar. The machine may also stop accepting coins and dispensing chocolate bars, perhaps because it is empty.

Next consider the vending machine with its customer, starting in the initial composite state $\{V_0, C_0\}$. We frame the rules of our calculus so that the presence of the customer will constrain the behaviour of the vending machine, which can now only sell at most a single chocolate bar in exchange for a two pound coin. Equally the behaviour of the vending machine may constrain the behaviour of the customer, as the vending machine may *stop* before the customer has made his purchase.

From initial state $\{V_0, C_0\}$ the possible events that may occur are *two* or *stop*.

The event *two* is a shared event: the vending machine has a two pound coin inserted, and the customer inserts the coin. This results in a simultaneous change of state in both machines, from $\{V_0, C_0\}$ to $\{V_2, C_1\}$.

The event *stop* concerns machine V only, and results in a change of state from $\{V_0, C_0\}$ to $\{V_3, C_0\}$.

We do not want the rules of our calculus to allow the event *one* to occur from state $\{V_0, C_0\}$. But how should we frame these rules to differentiate between *one* and *stop*, since these are both events which the customer is not prepared to take part in? We associate with each machine a set of events known as its *repertoire*.[1] This set will contain all the events evident in the machines behaviour (i.e. all events used in forming its next state function) but may also include additional events. We include *one* in the repertoire of C.

We can now give an informal set of rules for event calculus transitions (a formal version of these rules is given in [7])

- For an event to occur in a given composite state there must be a machine that is ready to take part in that event.

- An event can only occur when all machines that have that event in their repertoire are ready to take part in it.

- When an event occurs each machine that has that event in its repertoire changes according to one of the possibilities offered by its next state relation. Other machines are unaffected.

These rules allow us to derive the behaviour of a system (a set of machines) from the behaviour of its constituent machines.

We now give a translation of our example into Z. We first introduce the types $MACHINE$, $STATE$ and $EVENT$.

$MACHINE ::= V \mid C$
$STATE ::= V_0 \mid V_1 \mid V_2 \mid V_3 \mid C_0 \mid C_1 \mid C_2$
$EVENT ::= one \mid two \mid collect \mid stop$

The function ψ gives the behaviour of individual machines.

$\psi : MACHINE \leftrightarrow (STATE \times EVENT \leftrightarrow STATE)$

$\psi V = \{(V_0, one) \mapsto V_1, (V_0, two) \mapsto V_2, (V_0, stop) \mapsto V3,$
$\quad (V_1, one) \mapsto V_2, (V_2, collect) \mapsto V_0\}$
$\psi C = \{(C_0, two) \mapsto C_1, (C_1, collect) \mapsto C_2\}$

We define the repertoire of each machine.

$repertoire : MACHINE \longrightarrow \mathbf{P}\, EVENT$

$repertoire\ V = \mathrm{ran}(\mathrm{dom}(\psi\ V))$
$repertoire\ C = \mathrm{ran}(\mathrm{dom}(\psi\ C)) \cup \{one\}$

To these definitions would be added the formal rules for deriving the composite behaviour of the system. These rules define a function $\mathcal{X}_0$, which takes a composite state and returns the behaviour of a system starting from that composite state. The declaration of $\mathcal{X}_0$ is

$\mathcal{X}_0 : \mathbf{P}\, STATE \nrightarrow (\mathbf{P}\, STATE \times EVENT \leftrightarrow \mathbf{P}\, STATE)$

---

[1] This idea is borrowed from CSP, where the set of events relevant to a process is referred to as its alphabet.

The event calculus rules, together with the definitions of ψ and *repertoire*, give the behaviour of the system from initial state $\{V_0, C_0\}$ as:

$$X_0\{V_0, C_0\} = \{((\{V_0, C_0\}, stop) \mapsto \{V_3, C_0\}, (\{V_0, C_0\}, two) \mapsto \{V_2, C_1\},$$
$$(\{V_2, C_1\}, collect) \mapsto \{V_0, C_2\}, (\{V_0, C_2\}, stop) \mapsto \{V_3, C_2\}\}$$

1.2 A vending machine with internal state

We extend our model to one in which the vending machine has an internal state.

We specify the number of chocolate bars the machine can hold, the price of one bar, and the set of coins that can be used with the vending machine.

> $maxbars == 50$
> $price == 2$
> $coin == \{1, 2\}$

We will need to refer to the customers of the vending machine.

> $cust : \mathbf{P}\ MACHINE$

The internal state of our system consists of the number of bars currently held, the cash taken from previous transactions, and the value of the coins that have been input for the current purchase. The invariant records the fact that as bars are sold the cash held increases. We also give an initialisation schema that establishes the invariant.

___ VM _____
$chocs : \mathbf{N}$
$cash\_held : \mathbf{N}$
$coins\_in : 0 .. price$
_____
$chocs \leq maxbars$
$(maxbars - chocs) * price =$
$\qquad cash\_held$
_____

___ Init _____
VM
_____
$chocs = maxbars$
$cash\_held = 0$
$coins\_in = 0$
_____

Coins may only be inserted when the vending machine has at least one chocolate bar left. The value of the input coin is added to the running total for the current transaction.

___ InCoin _____
ΔVM
$coin? : coin$
_____
$chocs > 0$
$chocs' = chocs$
$cash\_held' = cash\_held$
$coins\_in' = coins\_in + coin?$
_____

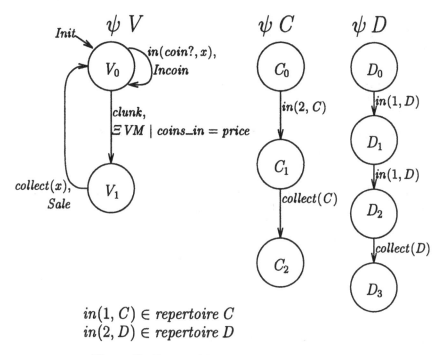

$$in(1, C) \in repertoire\ C$$
$$in(2, D) \in repertoire\ D$$

Fig. 2. Vending machine with internal state

A sale can be made when sufficient coins have been inserted. When a sale occurs the number of bars held is reduced by one and the coins input for that transaction are added to the cash held.

Sale
ΔVM

$coins\_in = price$
$chocs' = chocs - 1$
$cash\_held' = cash\_held + coins\_in$
$coins\_in' = 0$

The Event Calculus diagram for this model is given in fig2. Our vending machine has two customers, C and D.

$$cust = \{C, D\}$$

We use descriptive functions to refer to events. For example the expression $in(2, C)$ is used to refer to the event that customer C inserts a two pound coin. We need the following declarations of the functions used in the diagram. Note that by specifying that these functions are injections we ensure (for instance) that $collect(C)$ and $collect(D)$ are *different* events.

$$in : coin \times cust \rightarrowtail EVENT$$
$$collect : cust \rightarrowtail EVENT$$

We want to think of the state of the vending machine as being made up of two components. The first of these components is its *behavioural* state (V_0 or V_1). The second is its *data* state, which will be some binding allowed by the vending machine database schema *VM*. A convenient way to achieve this is for V_0 and V_1 to be injective functions with the following declarations.

$$V_0, V_1 : VM \rightarrowtail STATE$$

When the Event Calculus "model" (diagrams plus supporting formal text) is translated into Z, these declarations of paramatrized states and events are augmented by declarations of individual states and events, along with the restrictions necessary to ensure that states and events denoted by means of descriptive functions are different from one another. The repertoires of each machine are defined as the events associated with each machine plus any additional events given in the model, and the next state relations for each machine are obtained by interpreting each labelled transition from the Event Calculus diagram as a set of primitive transitions. Full details are given in [7].

The initial data state as given by the initialisation schema is the binding:

$$(\!| chocs \rightsquigarrow 50, cash\_held \rightsquigarrow 0, coins\_in \rightsquigarrow 0 |\!)$$

So that if we assume the initial behavioural state of V is V_0, then the initial state of machine V is:

$$V_0 \, (\!| chocs \rightsquigarrow 50, cash\_held \rightsquigarrow 0, coins\_in \rightsquigarrow 0 |\!)$$

and assuming the initial states of C and D are C_0 and D_0 respectively, the initial composite state of the system is:

$$\{ V_0 \, (\!| chocs \rightsquigarrow 50, cash\_held \rightsquigarrow 0, coins\_in \rightsquigarrow 0 |\!), C_0, D_0 \}$$

This suggests a token semantics in which we represent the state of a system by placing tokens on an Event Calculus diagram. To represent the initial state of our system we place tokens on states V_0, C_0 and D_0, and we label the token on V_0 with the binding $((\!| chocs \rightsquigarrow 50, cash\_held \rightsquigarrow 0, coins\_in \rightsquigarrow 0 |\!)$ to show its associated data state. We are now in a position to look at a possible event sequence for our system, and to see how the effect of each event is represented in our token semantics.

– The possible events that can occur from the initial composite state given above are $in(2, C)$ and $in(1, D)$. Suppose that the event that does occur is $in(2, C)$. To represent the effect of this event we relabel the counter on state V_0 according to the specification given in the schema *InCoin*, and we move the the counter for machine C to state C_1.

- The next event to occur must be *clunk*. This event is internal to the vending machine and we represent its effect by moving the counter for machine V to state V_1. The data state of machine V is not changed, so we do not change the label on the counter.

- Customer C can now collect his chocolate bar. To represent this event we move the token for machine V to V_0 and relabel it according to the specification given in *Sale*, and we move the token for machine C to state C_2.

Full details of the above sequence are given in the following trace:

Event	State
	$\{\,V_0\ (\!]\,chocs \rightsquigarrow 50,\, cash\_held \rightsquigarrow 0,\, coins\_in \rightsquigarrow 0[\!),\, C_0, D_0\}$
$in(2, C)$	$\{\,V_0\ (\!]\,chocs \rightsquigarrow 50,\, cash\_held \rightsquigarrow 0,\, coins\_in \rightsquigarrow 2[\!),\, C_1, D_0\}$
$clunk$	$\{\,V_1\ (\!]\,chocs \rightsquigarrow 50,\, cash\_held \rightsquigarrow 0,\, coins\_in \rightsquigarrow 2[\!),\, C_1, D_0\}$
$collect(C)$	$\{\,V_0\ (\!]\,chocs \rightsquigarrow 49,\, cash\_held \rightsquigarrow 2,\, coins\_in \rightsquigarrow 0[\!),\, C_2, D_0\}$

1.3 Summary

The Event Calculus is based on a particularly simple form of state machine, in which machines change state on the occurrence of events. Communication is modelled by shared events, which cause a simultaneous change of state in two or more machines. In Section 1.1 we have presented a simple vending machine example in this basic calculus and given its translation into Z.

The calculus obtains its full expressive power when we use descriptive functions to refer to states and events. This allows us to formulate the idea of state in terms of two components, which are a behavioural state, and an internal data state. It also allows us to model parameter passing between machines. In Section 1.2 we have illustrated these features by the example of a vending machine with internal state. The internal state is described with a Z data base schema, and changes to the state are described by Z operation schemas. Note however, that the syntactic pre-conditions of these operation schemas describe enabling conditions rather than true pre-conditions: this obviously has consequences for refinement.

In the following Section we present a distributed booking system case study. The only new concept to be modelled (in a very simple way) is that of a *class* of machines. The case study will include a treatment of time, but this will not require any extension to the theory.

2 Case Study

2.1 System Outline

We provide the specification of a reservation system that maintains a database of flight details, and can handle multiple concurrent reservations from different users. Reservation queries which cannot be handled immediately because the system is too busy or because the flight they refer to is currently receiving bookings from a different source, are queued and processed as soon as possible.

The booking system is made up of four *classes* of machine.

- There is a central database machine S which holds the bookings for each flight and an enquiry queue.

- There are a number of *booking_clerk* machines C^0, C^1, ..., which take responsibility for making the flight bookings.

- There are a number of *booking_agent* machines denoted by A^0, A^1, ... These provide an interface between the system and its users.

- Associated with each booking agent A^j is a watchdog timer T^j, used to abort the interaction between a booking agent and a user if the user fails to respond within a given time.

Our specification allows for new flights to be added to the database, for flights to be removed from the database, and for seat bookings to be made for any flight held in the database. The seat booking process drives the design of the system due to the need for concurrent booking sessions. We consider a "user" of the system to be an entity (man or machine) who knows which flight he is interested in, and intends to book seats on that flight. We outline below the events that take place when a user interacts with the system.

- A user wishing to make a reservation communicates a flight number (supposed to uniquely identify a flight) to a *booking_agent* machine which is dedicated to him until his reservation has been completed.

- On obtaining the flight number from the user, the booking agent sends an enquiry to the central database machine, which places it in a queue.

- When a booking clerk is available and the flight in question is free for booking (i.e. is not accepting reservations from another booking agent) the central database machine passes the flight reservations record and booking agent i.d. to the booking clerk.

- The clerk then sends the booking agent the list of free seats on the flight.

- The booking agent reports the list of free seats to the user.

- The booking agent inputs the user's bookings. If the user fails to respond within a given time the booking agent times out and assumes a nil response.

- The booking agent sends the new bookings to the reservation clerk who updates the flight reservation record.

- The booking clerk returns the flight reservation record to the central database.

The maximum number of booking clerks that it is useful to have is equal to the number of flights. It is then possible to have every flight being booked simultaneously. The number of booking agents should be at least as great as the number of booking clerks. If there are more agents than clerks and all clerks are busy, free agents can be used to input flight numbers and send enquiries to the database manager.

We tend to use long descriptive names for operation schemas, e.g. *Pass Flight-Details To Booking Clerk*. These names give an idea of how the schema will be *used* at the behavioural level of the specification.

2.2 Preliminaries

The Event Calculus Formal Model is to be considered as a pre-amble to any Event Calculus specification. As well as giving a formal representation of the rules for the underlying calculus, this introduces the given types *MACHINE*, *STATE* and *EVENT*.

We also require given types for flight numbers, seat numbers, names of passengers, and users.

$$[FLIGHT\_NO, SEAT, NAME, USER]$$

We need to identify certain machines as booking agents and booking clerks.

$$booking\_agent, booking\_clerk : \mathbf{P}\ MACHINE$$

The details held for each flight are the flight number, the seats, the free seats, and the current bookings. No seat may be double booked.

$$
\begin{array}{l}
\hline
\textit{Flight} \\
\hline
flight\_no : FLIGHT\_NO \\
seats, free : \mathbf{P}\ SEAT \\
booked : SEAT \nrightarrow NAME \\
\hline
seats \setminus \mathrm{dom}\ booked = free \\
\hline
\end{array}
$$

2.3 The Central Data Base Machine

Our central database holds reservation details for multiple flights. We distinguish flights which are being booked (and are hence unavailable for other agents to book) from those which are not being booked at the moment (and are hence, in general, available for booking).

Flights1

flights, being_booked, not_being_booked : FLIGHT_NO ⤖ Flight
flight_nos, available, unavailable : **P** FLIGHT_NO

$(\forall f : \text{dom } flights \bullet (flights\, f).flight\_no = f)$
being_booked ∩ not_being_booked = {}
being_booked ∪ not_being_booked = flights
flight_nos = dom flights
available = dom not_being_booked
unavailable = dom being_booked

We keep note of any flights which are to be closed to further bookings.

Flights2

Flights1
closing : **P** FLIGHT_NO

---

closing ⊆ flight_nos

We hold a queue of pending requests. Each request identifies a flight number and the agent which is handling the booking. The queue contains at most one entry for each booking agent.

Request

flight_no : FLIGHT_NO
agent : booking_agent

Flights

Flights2
queue : seq Request

$\forall i,j : \text{dom } queue \mid i \neq j \bullet$
$\quad (queue\, i).agent \neq (queue\, j).agent$

The initial state of the database has no flights recorded and an empty request queue.

Init

Flights

---

flights = {} ∧ closing = {} ∧ queue = ⟨⟩

We specify an operation to add a new flight to the database.

```
┌─ AddFlight ────────────────────────────────────
│ ΔFlights
│ new_flight? : Flight
├────────────────────────────────────────────────
│ new_flight?.flight_no ∉ flight_nos
│ flights' = flights ∪ {new_flight?.flight_no ↦ new_flight?}
│ being_booked' = being_booked
│ closing' = closing
│ queue' = queue
└────────────────────────────────────────────────
```

We specify an operation to add an agent's transaction request to the request queue.

```
┌─ QueueRequestFromAgent ─────────────────────────
│ ΔFlights
│ agent? : booking_agent
│ flight_no? : FLIGHT_NO
├────────────────────────────────────────────────
│ (∃ r : Request | r.flight_no = flight_no? ∧ r.agent = agent? •
│     queue' = queue ⌢ ⟨r⟩)
│ flights' = flights
│ being_booked' = being_booked
│ closing' = closing
└────────────────────────────────────────────────
```

We define a function to return the queue position of the next request to be processed, or a zero if no suitable request is waiting. The next request to be processed may not be the one at the head of the queue, as another request may already be being processed for that flight. The arguments to the function are a queue of requests, and a set of flight numbers. The set of flight numbers to be thought of as representing *unavailable* flights.

```
│ next_enquiry : seq Request × P FLIGHT_NO → N
├────────────────────────────────────────────────
│ ∀ q : seq Request; unavailable : P FLIGHT_NO •
│     ∃ s : P N •
│         s = {n : dom q | (q n).flight_no ∉ unavailable} ∧
│             next_enquiry(q, unavailable) = if s = {} then 0 else min s
```

We now use this function in the specification of an operation to select the next flight to be booked from the request queue, and output the flight details and the identity of the agent wishing to book the flight. The flight becomes unavailable for booking. (It will become available for booking again when the current bookings have been completed). When this schema is used in the Event Calculus model, the output *clerk!* specifies the the *destination* of the other output data. The value of *clerk!* is loosely specified, indicating that the operation can output to any booking clerk that is ready to receive the data.

```
┌─ PassFlightDetailsToBookingClerk ──────────────────────────────
│ ΔFlights
│ flight! : Flight
│ agent! : booking_agent
│ clerk! : booking_clerk
├────────────────────────────────────────────────────────────────
│ ∃ n : N₁ • n = next_enquiry(queue, unavailable) ∧
│     (queue n).flight_no ∈ available ∧
│     flight!.flight_no = (queue n).flight_no ∧
│     agent! = (queue n).agent
│ ∧
│     queue' = squash({n} ⊲ queue)
│
│ flights' = flights
│ being_booked' = being_booked ∪ {flight!.flight_no ↦ flight!}
│ closing' = closing
└────────────────────────────────────────────────────────────────
```

We will allow for the possibility that the request queue can contain requests
with invalid flight numbers. This may occur because a bad flight number is input
by a booking agent. It may also occur because a flight has closed to further bookings, and even been deleted from the system, whilst a request for a transaction
on that flight has been waiting in the queue.

An important point is that *unavailable* flights are valid; they are just unavailable at the moment because they are actually being booked. Now if the first
request in the queue that does not refer to an unavailable flight refers to an invalid flight, the pre-condition of the above operation is false, and at the Event
Calculus level this will mean that the event associated with the operation cannot
fire. We need a further operation to discard such a request from the queue, and
to inform the booking agent that placed it there that it is invalid. The booking
agent that will receive the signal is identified by the output *agent!* in the following
schema.

```
┌─ ReportDuffRequestToBookingAgent ──────────────────────────────
│ ΔFlights
│ agent! : booking_agent
├────────────────────────────────────────────────────────────────
│ ∃ n : N₁ • n = next_enquiry(queue, unavailable) ∧
│     (queue n).flight_no ∉ available ∧
│     agent! = (queue n).agent
│ ∧
│     queue' = squash({n} ⊲ queue)
│
│ being_booked' = being_booked
│ not_being_booked' = not_being_booked
│ closing' = closing
└────────────────────────────────────────────────────────────────
```

We specify the operation that accepts the updated flight details returned by
a booking clerk, and note that this flight is now available for booking. When this

schema is used in the Event Calculus model, the input *clerk?* specifies the *source* from which the other data is received. The value of *clerk?* is loosely specified, indicating that any booking clerk that has updated flight details to communicate is allowed to send those details.

___ *ReceiveUpdatedFlightDetailsFromBookingClerk* _____
$\Delta Flights$
flight? : *Flight*
clerk? : *booking_clerk*
___
flight?.flight_no $\in$ *unavailable*
being_booked' = {*flight?.flight_no*} $\lhd$ *being_booked*
not_being_booked' = *not_being_booked* $\cup$ {*flight?.flight_no* $\mapsto$ *flight?*}
queue' = *queue*
closing' = *closing*

An Event Calculus model that incorporates the operations defined so far is shown in figure 2.3.

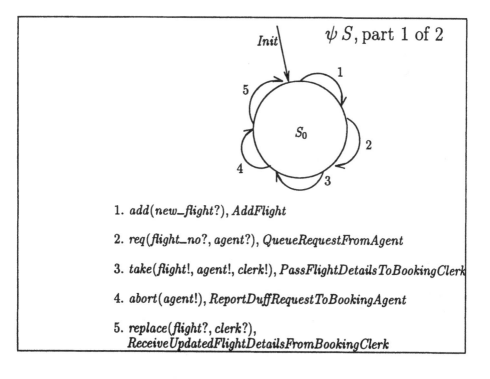

1. *add(new_flight?)*, *AddFlight*

2. *req(flight_no?, agent?)*, *QueueRequestFromAgent*

3. *take(flight!, agent!, clerk!)*, *PassFlightDetailsToBookingClerk*

4. *abort(agent!)*, *ReportDuffRequestToBookingAgent*

5. *replace(flight?, clerk?)*,
   ReceiveUpdatedFlightDetailsFromBookingClerk

Fig. 3. Flight Bookings, Central Database Machine, Part 1 of 2

Note that the transition arcs are labelled with numbers, which refer to the full labels that are given below the diagram. Supporting type declarations for the parametrized states and events are:

$S_0 : Flights \rightarrowtail STATE$
$add : Flight \rightarrowtail EVENT$
$req : FLIGHT\_NO \times booking\_agent \rightarrowtail EVENT$
$take : Flight \times booking\_agent \times booking\_clerk \rightarrowtail EVENT$
$abort : booking\_agent \rightarrowtail EVENT$
$replace : Flight \times booking\_clerk \rightarrowtail EVENT$

Our model of the central database machine is not yet complete, as we have still to consider the deletion of flights from the database. We must bear in mind that a flight may at any time be in the process of being booked, and that in this case the central database will not have an up to date record of the flight's reservation details. Our solution will be to remove a flight from the database in two stages. We first request that the flight be closed to further bookings: this does not affect any bookings that are currently taking place.

__ Close _____

$\Delta Flights$
$flight\_no? : FLIGHT\_NO$

$closing' = closing \cup \{flight\_no?\}$
$being\_booked' = being\_booked$
$not\_being\_booked' = not\_being\_booked$
$queue' = queue$

The database machine can delete a flight that has been closed and is not being booked. A final report of the flight reservation details is given.

__ DeleteFlightAndReport _____

$\Delta Flights$
$flight! : Flight$

$flight!.flight\_no \in available \cap closing$
$closing' = closing \setminus \{flight!.flight\_no\}$
$flights' = \{flight!.flight\_no\} \lhd flights$
$being\_booked' = being\_booked$
$queue' = queue$

The behaviour of the central database machine with respect to the closure and deletion of flights is shown in fig 4.

Supporting type declarations for the parametrized states and events are:

$\psi\,S$ part two of two

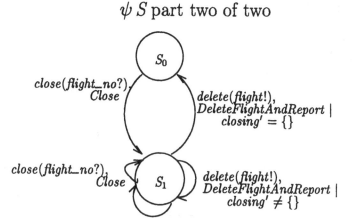

Fig. 4. Flight Bookings, Central Database Machine, Part 2 of 2

$S_1 : Flights \rightarrowtail STATE$
$close : FLIGHT\_NO \rightarrowtail EVENT$
$delete : Flight \rightarrowtail EVENT$

We use behavioural specification to prioritise the deletion of flights over other activities. This ensures that a closed flight not only *can* be deleted, but that it *will* be deleted, rather than chosen for further bookings.

Now this prioritisation could have been expressed in a specification that used a single behavioural state, but bringing out the behavioural dimension surely makes the specification clearer. Also note this: to achieve a correct specification with a single behavioural state we would have to concern ourselves with the deletion of flights whilst specifying their selection for booking. The use of behaviour in our specification allows us to consider one problem at a time.

Our specification ensures that deletion of flights is prioritised by introducing a behavioural state S_1 in which the only permitted activities are receiving updated flight details, receiving requests to close flights, and deleting flights. We only return to behavioural state S_0 when all closing flights have been deleted. This achieves the aim of ensuring that no further bookings can take place on a flight that is closing, but we do not need to be so severe. In state S_1 it is perfectly safe to engage in all activities, except the booking of flights which are closing. These changes to the specification are easily made at the level of the Event Calculus diagrams, and are left as an exercise for the reader.

2.4 Booking Clerks

Each active booking clerk has a database which holds the reservation details of a flight and the identity of the agent dealing with the reservations.

```
 ┌─ BookingClerkDB ─────────────────────────────────────
 │ flight_no : FLIGHT_NO
 │ seats, free : P SEAT
 │ booked : SEAT ⇸ NAME
 │ agent : booking_agent
 ├───────────────────────────────────────────────────
 │ seats \ dom booked = free
 └───────────────────────────────────────────────────
```

We need an operation to input the above date. This resembles an initialisation schema, but rather than having only a "before state", it has only an "after state". At the event calculus level we will use it to describe an event which creates an internal state in a machine which previously did not have one.

```
 ┌─ TakeFlightDetailsFromCentralDB ─────────────────────
 │ BookingClerkDB'
 │ flight? : Flight
 │ agent? : booking_agent
 ├───────────────────────────────────────────────────
 │ flight_no' = flight?.flight_no
 │ seats' = flight?.seats
 │ free' = flight?.free
 │ booked' = flight?.booked
 │ agent' = agent?
 └───────────────────────────────────────────────────
```

We need an operation to report the free seats of the flight to the booking agent. At the Event Calculus level the output $agent!$ will determine the destination of the other output data.

```
 ┌─ ReportFreeSeatsToBookingAgent ──────────────────────
 │ Ξ BookingClerkDB
 │ seats! : P SEAT
 │ agent! : booking_agent
 ├───────────────────────────────────────────────────
 │ seats! = free
 │ agent! = agent
 └───────────────────────────────────────────────────
```

We specify an operation to make some seat reservations.

```
 ┌─ MakeBookingsRequestedByAgent ───────────────────────
 │ ΔBookingClerkDB
 │ new_bookings? : SEAT ⇸ NAME
 │ agent? : booking_agent
 ├───────────────────────────────────────────────────
 │ dom new_bookings? ⊆ free
 │ booked' = booked ∪ new_bookings?
 │ seats' = seats
 │ flight_no' = flight_no
 └───────────────────────────────────────────────────
```

Note that the specification of these last two operations is greatly simplified by the fact that the database under consideration contains only a single flight. *It is the behavioural nature of our specification that provides this simplification,* whilst allowing the system as a whole to handle details of multiple flights.

The final duty of a booking clerk is to return the updated flight details to the central database. After doing this, a booking clerk machine does not retain any data, and this is reflected in the following schema which has a before state and no after state.

```
┌─ ReturnUpdatedFlightDetailsToCentralDB ─────────────────────
│ BookingClerkDB
│ flight! : Flight
├─────────────────────────────────────────
│ flight!.flight_no = flight_no
│ flight!.seats = seats
│ flight!.free = free
│ flight!.booked = booked
└─────────────────────────────────────────
```

A description of the *class* of booking clerk machines is given in fig. 5. We need some supporting Z to declare the functions used to describe events:

$$free : \mathbf{P}\, SEAT \times booking\_agent \times booking\_clerk \rightarrowtail EVENT$$
$$book : (SEAT \rightarrowtail NAME) \times booking\_agent \times booking\_clerk \rightarrowtail EVENT$$
$$return : Flight \times booking\_clerk \rightarrowtail EVENT$$

We also need some "pseudo Z" to define the number of booking class machines, and to declare their parametrized states.

let $m = 4$ in:
  $\#booking\_clerk = m$
  for $i = 1$ to m assert:
  $C^i \in booking\_clerk$

  $\quad C_1^i, C_2^i, C_3^i : BookingClerkDB \rightarrowtail STATE$

  end
end
This "pseudo Z" is processed in an obvious way during the translation of the Event Calculus specification into standard Z.

2.5 Booking Agents

A booking agent machine communicates directly with a user who wishes to make seat reservations. The booking agent inputs a flight number, and then lodges a request regarding this flight. The agent is then contacted by a booking clerk, which supplies details of free seats on the flight. The booking agent transmits

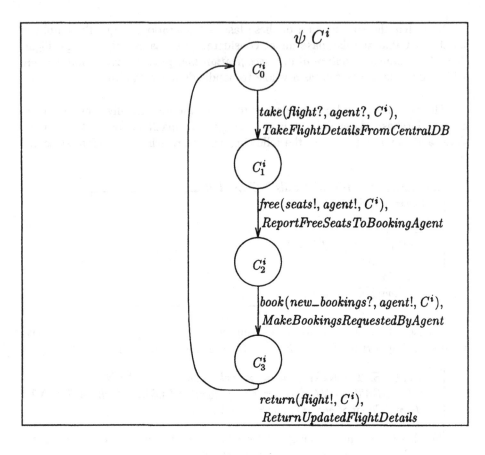

$$\psi \ C^i$$

$$C_0^i$$

$take(flight?, agent?, C^i),$
$TakeFlightDetailsFromCentralDB$

$$C_1^i$$

$free(seats!, agent!, C^i),$
$ReportFreeSeatsToBookingAgent$

$$C_2^i$$

$book(new\_bookings?, agent!, C^i),$
$MakeBookingsRequestedByAgent$

$$C_3^i$$

$return(flight!, C^i),$
$ReturnUpdatedFlightDetails$

Fig. 5. Booking Clerks

these details to the user and then inputs the users valid reservations, which are transmitted back to the booking clerk. If the user fails to respond within a given time after being provided with the flight details, a timeout occurs and the booking agent reports that no reservations are to be made.

A request lodged by a booking agent may be an invalid request, either because an invalid flight number was originally entered, or because a flight number has become invalid since being entered (the flight has closed between the time the flight number was entered and the time the central database began to process the enquiry). In this case the transaction is aborted.

We require the following database schema to describe the internal state of active booking agents (when inactive between transactions they will have no internal state).

```
┌─ AgentDB ────────────────────────────────────────────────
│ user : USER
│ clerk : booking_clerk
│ flight_no : FLIGHT_NO
│ freeseats : P SEAT
│ new_bookings : SEAT ⇸ NAME
├───────────────────────────────────────────────────────────
│ dom new_bookings ⊆ freeseats
└───────────────────────────────────────────────────────────
```

At the start of a transaction a booking agent inputs and records a flight number and a user i.d. Although "users" are not part of the system, we can think of the input *user?* as indicating where (i.e. which user) these inputs are coming from. Next the agent will post a request for service to the central database. At the level of the operation schema the only data that needs to be output for this purpose is the flight number. At the event calculus level the agent also sends its own identity.

```
┌─ Start ──────────────────────        ┌─ SendRequestToCentralDB ─────
│ AgentDB'                             │ Ξ AgentDB
│ flight_no? : FLIGHT_NO               │ flight_no! : FLIGHT_NO
│ user? : USER                         ├──────────────────────────────
├──────────────────────────            │ flight_no! = flight_no
│ flight_no' = flight_no?              └──────────────────────────────
│ user' = user?
└──────────────────────────
```

If the request is invalid when checked by the central database machine, due to a bad flight number being input or the given flight having closed to further bookings since the request was queued, the booking agents activity will be aborted at this point. We do not need a schema for this, as it is dealt with at the Event Calculus level of the specification.

If the request is valid when checked by the central database machine, the booking agent will be sent details of the free seats on this flight and the identity of the booking clerk who is dealing with this transaction. At the Event Calculus level the latter input tells us where the information is coming from.

```
┌─ InputFlightDetailsFromBookingClerk ────────────────────────
│ Δ AgentDB
│ seats? : P SEAT
│ clerk? : booking_clerk
├──────────────────────────────────────────────────────────
│ freeseats = seats?
│ clerk = clerk?
│ flight_no' = flight_no
│ user' = user
└──────────────────────────────────────────────────────────
```

The booking agent then reports the flight details to the user. Although "users" are not part of the system, we can think of the output *user!* as specifying which user these outputs are being sent to.

```
┌─ ReportFreeSeatsToUser ──────────────────────────────
│  Ξ AgentDB
│  user! : USER
│  free! : P SEAT
│ ─────────────────────────────────────────────────────
│  user! = user
│  free! = freeseats
└───────────────────────────────────────────────────────
```

Normally the user should reply with details of the bookings he wishes to make. The booking agent only accepts bookings for free seats.

The input *user?* in the following schema may seem pointless as it appears to provide information already known to the system. However, we can think of it as indicating that these inputs can only come from the particular user with whom we are negotiating the transaction.

```
┌─ InputNewBookingsFromUser ──────────────────────────────
│  Δ AgentDB
│  user? : USER
│  bookings? : SEAT ↦ NAME
│ ─────────────────────────────────────────────────────
│  user? = user
│  dom bookings? ⊆ freeseats
│  clerk' = clerk
│  new_bookings' = bookings?
└───────────────────────────────────────────────────────
```

If the user fails to reply within a certain time limit the transaction is timed out and there are no new bookings. The details of this time limit and of the timer mechanism are given later.

```
┌─ Timout ──────────────────────────────────────────────
│  Δ AgentDB
│ ─────────────────────────────────────────────────────
│  clerk' = clerk
│  new_bookings' = {}
└───────────────────────────────────────────────────────
```

Finally the booking agent reports the new bookings to the booking clerk. After it has performed this duty it no longer needs a data state: hence the following schema does not specify an after state for the operation.

$\_\_$ *SendNewBookingsToBookingClerk* $_____$
AgentDB
bookings! : *SEAT* $\rightarrowtail$ *NAME*
clerk! : *booking_clerk*
$_____$
bookings! = *new_bookings*
clerk! = *clerk*

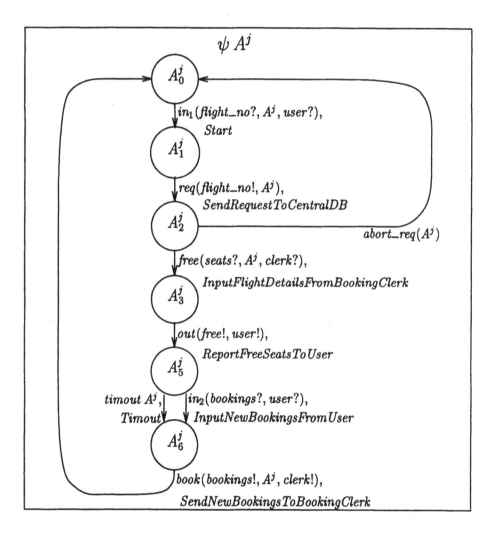

Fig. 6. Booking Agents

A description of the *class* of booking agent machines is given in fig. 6. We need some supporting Z to declare the functions used to describe events:

$$in_1 : FLIGHT\_NO \times booking\_agent \times USER \rightarrowtail EVENT$$
$$out : \mathbb{P}\, SEAT \times booking\_agent \times USER \rightarrowtail EVENT$$
$$in_2 : (SEAT \rightarrowtail NAME) \times USER \rightarrowtail EVENT$$
$$timout : booking\_agent \rightarrowtail EVENT$$

We also need some "pseudo Z" to define the number of booking agent machines and to declare their parametrized states.

let $n = 6$ **in:**

$\#booking\_agent = n$

for $j = 1$ **to** n **assert:**

$A^j \in booking\_agent$

$$A_1^j, A_2^j, A_3^j, A_4^j, A_5^j : BookingAjentDB \rightarrowtail STATE$$

end

end

2.6 Watchdog Timers

Time in our specification is measured in terms of clock ticks. It is not part of this formal specification to say how often these occur, but we will say informally that for this specification the clock rate is one tick per minute.

Each booking agent has an associated watchdog timer which is set running when the agent reports the flight details to the user. The timer is reset when the agent receives the user's bookings. If the user fails to respond within a given time limit, the user is timed our and the agent assumes no bookings are to be made.

We define the time out limit.

$maxwait == 10$

The database of a timer consists a count of ticks. When the timer is set, the count of clock ticks is zero. A clock tick increments the count.

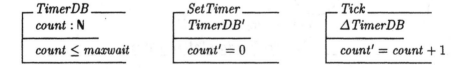

A description of the *class* of watchdog timer machines is given in fig. 7. We also need some "pseudo Z" to define the number of timer machines and to declare their parametrized states.

let $n = 6$ **in:**

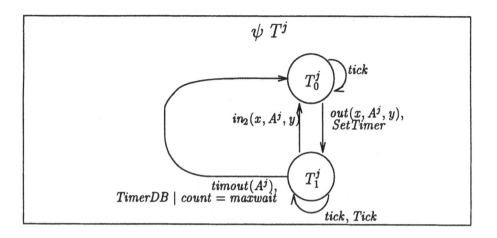

Fig. 7. Watchdog Timers

$\#timer = n$
for $j = 1$ **to** n **assert:**
$T^j \in timer$

$\quad\big|\quad T_1^j : TimerDB \rightarrowtail STATE$

   end
end

A timer T^j has two behavioural states. In state T_0^j, which is a primitive state, the timer allows *tick* events to occur, but they have no effect on it. The timer moves to its second behavioural state when a booking agent outputs flight details. On entering this state, the timer's counter is set to zero. In this state, the event *tick* has the effect of incrementing the timer's counter.

When the counter value equals the maximum waiting time, the constraint given in the timer database schema prevents any further ticks occurring. This does not mean that time is stopped (although it could if we make a mistake in our specification). Rather it means that the event $timout(T^j)$ must occur before another clock tick.

The effect of a time out is to reset the timer to its initial state. The timer will also be reset if booking agent A^j inputs a user response before the waiting period is terminated. This is what should happen during normal operation.

3 Conclusions

We have presented a calculus of communicating state machines in which communication is modelled by means of shared "events" which cause a simultaneous change of state in two or more machines. The primitive idea of "state" in this basic calculus has been developed into the notions of "data state" and "behavioural state". A machine's data state can be described using a Z data base schema, and changes to that state can be described using Z operation schemas. The use of behavioural states allows us to say *in what circumstances* and *in what order* operations will be performed, and *where* inputs come from and outputs go to. The calculus allows a machine to have its own private data state, and only allows other machines to see components of that data state that are communicated to them.

The Event Calculus was first described in [5]. In this version internal data states were modelled by tuples. A second version, in which internal data states are modelled by objects which have a schema type, is described in [6]. An extension of this version to model hybrid systems was reported in [9]. Further details of the current version, in which internal data states are modelled by bindings, can be found in [7] and [8].

Other formalisms that model data, process and behaviour use a complex multi-paradigm approach. LOTOS has process algebra and algebraic components [3], CCZ (an integration of CCS and Z) has process algebra and model based components [2], and RAISE has process algebra, model based *and* algebraic components [4]. In the Event Calculus however, everything is done in Z.

Acknowledgements

The author wishes to express his thanks to the referees for their comments and corrections to the draft text, and also his particular gratitude to Steve Dunne and Andy Galloway for interesting and non-terminating discussions on the wonders and limitations of Z.

References

1. S E Dunne and R Hodgson. Process, data and behaviour. *Information and Software Technology, Vol 32, no 8*, 1990.
2. Andy Galloway. *Integrated Formal Methods*. PhD thesis, University of Teesside, School of Computing and Mathematics, 1996.
3. ISO. IS8807: Information Processing Systems – Open System Interconnection – LOTOS. Technical report, International Standards Organisation, Geneva, Switzerland, 1988.
4. R. Milne. RAISE technical reports REM/11 and REM/12. Technical report, STC Technology, 1990.
5. Bill Stoddart and Peter Knaggs. The Event Calculus, (formal specification of real time systems by means of Z and diagrams). In H Habrias, editor, *5th International Conference on putting into practice methods and tools for information system design*. University of Nantes, 1992.
6. W J Stoddart. The Event Calculus, vsn 2. Technical Report SCM-95-1, University of Teesside, UK, 1995.
7. W J Stoddart. The Event Calculus, vsn 3. Technical Report SCM-96-1, University of Teesside, UK, 1996.

8. W J Stoddart. The Event Calculus. In Micheal G Hinchey and C Neville Dean, editors, *A Handbook of Z*. Academic Press International Series in Formal Methods, to appear.

9. W J Stoddart, S E Dunne, and P C Fencott. Modelling Hybrid Systems in Z. In H Habrias, editor, *Z Twenty Years on: What is its Future*. Univ Nantes, France, 1995.

Experiences with PiZA, an Animator for Z*

M A Hewitt, C M O'Halloran and C T Sennett

Defence Research Agency, Malvern, UK

Abstract. Large Z specifications require testing if they are to be at all credible and translation to Prolog provides one means of doing this. Experience with PiZA (Prolog Z Animator), a system designed to carry out this translation automatically, is described. The paper gives a brief description of the principles of the translation with some practical details of using PiZA. Experience in animating large specifications of about the size of a small compiler is discussed in terms of the performance, the problems presented by the form of the Prolog translation, the processing of input and output values and the use of diagnostics. The paper concludes with a discussion of the use of Z as a design language.

1 Introduction

A large Z specification is a daunting object to read or to check. Tools are available to ensure that a specification is correct as far as the syntax and types are concerned, but there are still many slips which can be present in a specification which passes these checks. For example, dropping a decoration or making a mistake in the priority of operators are errors which are frequently made but difficult to spot and which will not be detected by syntax and type checkers. The source text of a program is a similarly complicated object, but programs do have the advantage that they can be run and tested. Lacking the ability to test a specification, one must have substantial doubts as to whether it is correct, which rather destroys the whole point of using a formal method at all.

A solution is to animate the specification. Because of its ability to handle logical expressions Prolog provides a natural medium for this and it is possible to animate a Z specification by translating to Prolog and running the result. PiZA is a system for carrying out this translation automatically: in effect it provides a means for generating a rapid prototype from a Z specification. This paper is a report of our experiences using the animator PiZA, which was developed by one of us (MAH). We hope the report is a dispassionate account of both the benefits and disadvantages of using the PiZA system.

The next section of this paper gives an overview of the animation process while subsequent sections give the experiences and conclusions.

---

2 Overview of the PiZA animator

2.1 The representation in Prolog

The purpose of the animation is to evaluate a schema. Typically, the schema describes some operation and what is required is to print output values as a result of some given inputs. Prolog can carry out this evaluation by solving for the unknowns in a logical expression in terms of the knowns by using the process of resolution. Consequently, animation consists in translating Z logical expressions into Prolog goals in such a way that the resolution process can solve for the unknowns[7, 8, 1].

The variables in the goals stand for values of the Z identifiers and expressions for the particular evaluation being carried out: they are not the same identifier names, which makes for some difficulties in diagnosing errors. For example the following schema taken from Spivey's birthday book example in [6]

```
┌─ AddBirthday ─────────────────────────────────
│ ΔBirthdayBook
│ name? : NAME
│ date? : DATE
├────────────────────────────────────────────────
│ name? ∉ known
│ birthday' = birthday ∪ {name? ↦ date?}
└────────────────────────────────────────────────
```

is translated into the Prolog goal

```
schema_def('AddBirthday',[A,B,C,D,E,F]) :-
  negate(
    element(F,D)
  ),
  dom(A,D),
  maplet(F,C,G),
  construct_singleton_set(G,H),
  set_union(A,H,B),
  dom(B,E).
```

where the following mapping associates Prolog variable names with the Z variables they represent: A = *birthday*, B = *birthday'*, C = *date?*, D = *known*, E = *known'*, F = *name?*. The difficulty in mapping Z variable names into Prolog variable names arrises from the different naming semantics adopted by the different languages. The schema inclusion of *ΔBirthdayBook* within *AddBirthday* is achieved by copying the property of *ΔBirthdayBook* into *AddBirthday* before it is translated into Prolog.

The translation from *AddBirthday* into Prolog uses predicated definitions from the PiZA *run time library*. This is a collection of Prolog predicates that implement the Spivey mathematical toolkit given in chapter 4 of [6].

The Prolog representations of the Z elements used by the run time library usually involve Prolog lists. Thus a set is represented by a list, a tuple is represented by a list, a schema is represented by a list of pairs of the component names and values and a sequence is represented by a list. This last representation is an optimisation as a Z sequence is really a relation: the PiZA run time library has to catch all the possible uses of sequences to ensure that the correct semantics are obeyed.

To make changes to these representations possible without breaking existing code, constructor functions are used to create instances of data types. In the code excerpt given previously `maplet/3` and `construct_singleton_set/2` are constructor functions. Note that, following the Prolog convention, the constructor functions may also be used as destructor functions by reversing the mode of the arguments. For example the following Prolog session illustrates the use of `maplet/3` as both a constructor and a destructor:

```
| ?- maplet( 1, 2, X ).
X = rtuple([1,2])

| ?- maplet( X, Y, rtuple([1,2]) ).
X = 1,
Y = 2
```

Complex Z expressions can give rise to the creation of temporary variables in the Prolog translation, as each part of the expression is evaluated separately. For example the schema $test3$

```
┌─ test3 ─────────────────────────────────────
│ a?, b?, c?, d?, e! : \mathbb{F}\, Z
├─────────────────────────────────────────────
│ a? ∪ b? ∪ c? ∪ d? = e!
└─────────────────────────────────────────────
```

is converted to the following Prolog code

```
schema_def(test3,[A,B,C,D,E]) :-
    set_union(A,B,F),
    set_union(F,C,G),
    set_union(G,D,E).
```

in which the Prolog variables F and G represent intermediate results produced in evaluating the expression $a? \cup b? \cup c? \cup d?$.

A few of the run time library predicates are meta predicates. A meta predicate takes as its argument one or more predicates. The `negate/1` goal in the Prolog translation of *AddBirthday* is an example of a meta predicate which succeeds if its argument is false and fails if its argument is true. Use of the `negate/1` meta predicate is only safe when all the Prolog variables, with the exception of temporary variables, in its argument have been instantiated with values.

PiZA is unable to represent infinite sets and any expression which produces an infinite set will lead to difficulties. Furthermore, PiZA, at present, is unable to detect at translate time if a Z expression evaluates to an infinite set. For example, upon calling the schema *test4*

```
┌─ test4 ────────────────────────────────────────────────────
│ a! : ℙ Z
├──────────────────────────────────────────────────────────
│ a! = { x : Z | x mod 2 = 0 }
└──────────────────────────────────────────────────────────
```

PiZA appears to freeze while calculating an output for *a!*. The tool continually generates elements of *a!* until the computer memory becomes exhausted whereupon PiZA reports an error and resets itself.

2.2 Relationship to other work

Prolog is a popular choice for animating Z since it is based upon predicate logic. Some previous attempts at using Prolog to animate Z are detailed in Knott et al[11], Dick et al[1] and West et al[9]. This section compares the approaches taken in [1] and [9] with the approach taken by PiZA.

The approach taken in [1] is classified as a *generate and test* approach. Prolog is generated from the signature of the schema to enumerate every possible solution to the schema. The predicate part of the schema is translated into Prolog goals which filter the candidate solutions produced from the signature, discarding the solutions which do not satisfy the schema. Except for trivial examples the number of candidate solutions becomes so high that the pure generate and test paradigm becomes infeasible.

To try to improve matters the authors discuss *filter promotion* where the combinatorial explosion of candidate solutions is reduced by interleaving the generator predicates produced from the schema signature with the filter predicates produced from the schema property. Although this helps, the generate and test paradigm still does not scale up to specifications of a realistic size.

In contrast the PiZA animator tries to instantiate the output variables of the schema from the input variables. If the Prolog code generated from the schema cannot instantiate the output variables using only the input variables then it is discarded and an error message reported. However generate and test can sometimes still occur within PiZA generated Prolog. For example with the following schema

```
┌─ test1 ────────────────────────────────────────────────────
│ a? : Z ↔ Z
│ b!, c! : Z
├──────────────────────────────────────────────────────────
│ (b!, c!) ∈ a?
│ b! = c! + 1
└──────────────────────────────────────────────────────────
```

PiZA would create Prolog that would successively instantiate $(b!, c!)$ to every element of *a?* until it found a set of instantiations that satisfied the $b! = c! + 1$

constraint. When PiZA generates non determinate code such as this it issues a warning.

The approach taken in [9] differs from that in [1] in that the overall goal of the animation is not to instantiate output variables. In this animation technique the user provides values for every Z variable and the Prolog interpreter will return true if the variables satisfy the specification and false otherwise. PiZA can be used in this fashion if desired, but it does not compel its user to know in advance what the output from the specification should be.

The disadvantage of PiZA with respect to the techniques adopted by [9] and [1] is that the signature of a schema is not converted into Prolog so implicit predicates in the signature do not constrain the output of the animator. For example in the schema *test2* shown here

```
┌─ test2 ─────────────────────────────────
│ a : Z
│ b : 1 .. 10
└─────────────────────────────────────────
```

the variable b has the same type as the variable a but the signature declares an implicit predicate that b is in the range one to ten.

Because the current version of PiZA does not generate Prolog from declarations any important implicit predicates should be removed from the signature and placed in the property of the schema as in this version of *test2*.

```
┌─ test2 ─────────────────────────────────
│ a, b : Z
│─────────────────────────────────────────
│ b ∈ 1 .. 10
└─────────────────────────────────────────
```

Finally both [9] and [1] were reports on techniques adopted to animate Z using Prolog. The Prolog itself was written by hand although [1] mentioned an ongoing effort to provide tool support. PiZA is different in that it is a tool which automatically generates Prolog from Z without user intervention. Provided the translation used by PiZA is sound this should reduce the errors that are associated with hand translations.

2.3 Practical details

The specification is written in the PiZA input language rather than using LaTeX directly as in the popular type checking systems available for Z. There are a number of reasons for this, the most pressing being that the animation needs more information than is provided by a standard LaTeX description of Z. An example is that a global constraint may provide a definition and the animator needs to know which identifier is being defined. The language uses reserved words rather than the LaTeX use of the special \ symbol and this results in an immense improvement in the readability of the input text. However, like a LaTeX Z specification, the input can include general descriptive text and this is delimited by the characters {_ _}.

The *AddBirthday* schema given earlier, is represented in the PiZA language by the following text

```
{sb AddBirthday
  delta BirthdayBook ;
  name? : NAME ;
  date? : DATE
----
  name? not_elem known ;
  birthday' = birthday union { name? :--> date? }
}
```

This shows how Z symbols such as ∪ are represented by reserved words such as union.

Specifications are animated in batch mode using a transaction file. This calls a succession of schemas and provides them with input values. The output values can be printed out and are piped on to the input values of succeeding schema calls. For example, the birthday book specification could be queried with the following transaction file:

```
specification file( "bb" ).
begin.
{cs init_BirthdayBook}

{cs AddBirthday
  name? = helen;
  date? = nov_20th
}

{cs AddBirthday
  name? = jon;
  date? = jun_5th
}

{cs Remind
  today? = nov_20th
}
end.
```

The PiZA analysis provides only enough analysis to carry out the translation into Prolog. Type checking needs to be done additionally to this and is carried out by translating the PiZA representation into the appropriate form of LaTeX for the widely available type checkers ƒuzz[5], CADiZ[2, 3] and ZTC[4].

Type setting of the input language is also done by translation into LaTeX: to assist with this the PiZA input language contains some special symbols to allow new lines and spaces to be inserted and to cope with the possibility of needing an identifier identical to a reserved word. During the process of translation, PiZA can also be instructed to add LATEX index commands, to build up an index of definitions and uses.

To cope with an existing Z specification written directly in LATEX a UNIX *sed* (stream editor) script is provided which is able to do about 85% of the modifications required to convert a LATEX format file to the PiZA input syntax. This solution is far from ideal however and ultimately a parser for the LATEX representation of Z may be written.

3 Experiences with the animator

3.1 Performance

Two specifications have been animated. The first was for an algorithm to analyse usage of pointers and unions in a language such as C. The input to the language is not C itself but output from a C compiler in the OSF's Architecture Neutral Distribution Format (ANDF)[10]. The details of the algorithm are not relevant to this paper but some size statistics are. The specification document is 300 pages long: it contains 934 schema definitions, 57 abbreviation definitions, 62 free types, 2 given sets and 191 axiomatically defined identifiers. This large specification takes about 2 minutes to syntax analyse and 4 minutes to translate to Prolog, one minute to compile and 2 minutes to analyse the output from a 20 line C program.

The specification is obviously very complex. The abstract syntax of the basic input contains well over a 100 constructors and the algorithm itself involves 2 passes, one for pointer analysis and one for union analysis. The algorithm is experimental and has been subject to a considerable amount of development, but the problem as a whole is of a similar order of complexity to the code generation part of a compiler.

The second specification was concerned with translating to an intermediate language, in this case the input language for the Malpas[12] software analysis suite. This is a recurring problem as it is often necessary to apply software analysis to subsystems written in various obscure languages. In this case, the target is fairly well defined but the source language may have peculiarities needing careful modelling. Specifying the translation does therefore have a number of advantages. For this fairly infrequently used process, the animation is sufficiently fast for it to be an adequate implementation. Typically, a small module of a 100 lines or so takes a minute to translate. Note that this does depend upon the input: one particularly unfriendly module took 3 hours.

3.2 PiZA as a text preparation tool

The PiZA input language is very much easier to use than LATEX. This is easily illustrated by one of the simplest schema declarations in the analysis specification which is written in PiZA as

```
{sb Aligned_Shape
   sh : Shape;
   al : Alignment
}
```

and converted by PiZA into the LaTeX form:

```
\begin{schema}{Aligned\_Shape}
\index{Aligned\_Shape@$Aligned\_Shape$}
sh\index{sh@$sh$} : Shape\index{Shape@$Shape$} ;
al\index{al@$al$} : Alignment\index{Alignment@$Alignment$}
\end{schema}
```

The complexity of the LaTeX arises from the use of keywords introduced by the \
character and the heavy use of index commands. Clearly a general purpose type
setting language has to have this complexity and with a large scale, constantly
evolving document, this level of indexing is essential. However maintenance of
the LaTeX document would be almost impossible as it is very difficult to see the
wood for the trees. The PiZA language, because it uses reserved words specialised
to the Z application and because it can insert index commands automatically,
makes for documents which are very much easier to maintain. PiZA is worth
using on these grounds alone.

3.3   Ability to handle Z

Any animation in Prolog is going to have limitations. If a specification is inde-
terminate then the Prolog translation of it will also be, and it is possible that
the Prolog resolution engine will loop forever while searching for a solution, even
for expressions which can be given a value.

An example[1] is a specification for Pythagorean numbers, that is, x, y, z
such that $x^2 + y^2 = z^2$. The Prolog interpreter will assign x, y and z the value
one and test the equality which will fail. It will then loop trying different values
of x. Since with y and z equal to one there is no solution to the equation this
loop will not terminate with a solution. This has not proved a problem for us,
probably because the specifications we have animated are for algorithms and
given in a very constructive fashion, so indeterminacy does not often arise.

Apart from this, there are some minor limitations in PiZA's ability to translate Z which have not caused severe problems. An example is that an axiomatic
box can define only one identifier. For example the following axiomatic box defines both *is_even* and *is_odd*:

$is\_even \_ : \mathbb{P}\,\mathbb{Z}$
$is\_odd \_ : \mathbb{P}\,\mathbb{Z}$

$\forall i : \mathbb{Z} \bullet is\_even\ i \Leftrightarrow i \bmod 2 = 0$
$\forall i : \mathbb{Z} \bullet is\_odd\ i \Leftrightarrow i \bmod 2 = 1$

This single axiomatic box would need to be separated into two for PiZA to
be able to generate Prolog code to implement *is_even* and *is_odd*. A related

issue is that the definition of an identifier must be completed within one global constraint. Global constraints which do not define identifiers are not translated.

Writing in a constructive style is something which is generally to be encouraged (although not at the expense of clarity) and is particularly important for the specification of an algorithm. A constructive style can be summarised as rule definitions being preferred to constraints. A predicate should, where possible, be expressed as a conjunction of equalities, each one defining one or more variables. This style is more easily understandable as well as having fewer problems in animation and is more likely to be consistent.

A global definition can often be given as a syntactic abbreviation using the == symbol. Except for particularly simple cases we have tended to use axiomatic boxes instead of abbreviations as it is helpful when reading a specification to see the signature of the identifier being defined. Abbreviations also have the disadvantage that PiZA simply substitutes the definition of the abbreviation wherever it is referenced. If an abbreviation is frequently referenced the size of the Prolog translation from the specification will increase and debugging will become more difficult.

The PiZA animator uses the definition of the Z language given in Spivey's reference manual[6].

3.4 Treatment of errors

An error in the specification may appear either as an inability to translate the Z into Prolog, or as a failure of a Prolog goal in execution. Obviously, false predicates can appear almost anywhere without there being a fault in the specification, so it is necessary to have some criterion for an error other than the simple failure of a Prolog goal. The criterion adopted is that a specification is deemed to have failed when a user defined function application fails.

Faults appearing as an inability to carry out the translation from Z to Prolog are usually concerned with the appearance of an unbound variable in the Prolog. A schema value created by a theta expression, for example, must have all of its fields bound. (It is arguable whether this is a limitation in the tool or a sensible restriction: in our experience it has been a sensible restriction and found many examples of unwittingly unset identifiers.)

PiZA also outputs a warning message when an identifier is bound in one arm of a disjunction, but not another. This frequently arises in situations such as the definition of the following function to sum the integers in a sequence.

$$
\begin{array}{|l}
\hline
sum : \text{seq}\,\mathbf{Z} \nrightarrow \mathbf{Z} \\
\hline
sum = \\
\quad (\lambda\, sz : \text{seq}\,\mathbf{Z} \\
\quad \bullet\ (\mu\, n, n_1 : \mathbf{Z}\ | \\
\qquad sz = \langle\rangle \wedge n = 0 \\
\qquad \vee \\
\qquad \#sz = 1 \wedge n = head\ sz \\
\qquad \vee \\
\qquad \#sz > 1 \wedge \\
\qquad n_1 = head\ sz \wedge n = n_1 + sum(tail\ sz) \\
\qquad \bullet\ n)) \\
\end{array}
$$

In this axiomatic description, n_1 is local to the μ clause as a whole, but only defined in the non-termination case.

Binding an identifier down only one arm of a disjunction is an error in fewer than one in ten cases. To prevent these genuine cases being swamped by needless warnings we have adopted a style in which the scope of local variables is reduced as much as possible. In the case of sum, n_1 is removed from the μ declaration and the third arm of the disjunction replaced by an existential quantification:

$$
\begin{array}{|l}
\hline
sum : \text{seq}\,\mathbf{Z} \nrightarrow \mathbf{Z} \\
\hline
sum = \\
\quad (\lambda\, sz : \text{seq}\,\mathbf{Z} \\
\quad \bullet\ (\mu\, n : \mathbf{Z}\ | \\
\qquad sz = \langle\rangle \wedge n = 0 \\
\qquad \vee \\
\qquad \#sz = 1 \wedge n = head\ sz \\
\qquad \vee \\
\qquad \#sz > 1 \wedge \\
\qquad (\exists\, n_1 : \mathbf{Z} \bullet n_1 = head\ sz \wedge n = n_1 + sum(tail\ sz)) \\
\qquad \bullet\ n)) \\
\end{array}
$$

3.5   Order of evaluation

The analysed PiZA source text is translated into Prolog in a way which attempts to ensure that Prolog variables are instantiated before they are used. As a trivial example, the expression $x > 0 \wedge x = 1$ needs to be translated with the second arm of the conjunction as the first Prolog goal. As a result, the ordering of the Prolog is not the same as the ordering of the predicates in the Z, a source of considerable confusion. The reordering of predicates to achieve a satisfactory Prolog translation is to some extent a question of heuristics. If PiZA cannot achieve a satisfactory ordering it reports an error. In most case where this has occurred, it has been been the result of an error in the specification.

The ordering issue is probably the main drawback with Prolog animation. The first problem is that when there is an error, the diagnostics report the execution of the Prolog goals which led up to the error, which will appear in a different order from the corresponding Z predicates. (They also appear in a different language, but this causes surprisingly few difficulties.) This order problem is compounded by the fact that when a schema is included as a predicate, the effect on the translation is simply to include the appropriate Prolog goals from that schema. Consequently, when a function is defined using several schemas, it can be hard to find which schema has failed.

Another rather irritating aspect of this problem is that in a disjunction such as $x > 0 \land y = f(x) \lor x = 0 \land y = 0$, the animator could in principle start tackling the predicates in any order, for example $y = f(x)$, even if, as is quite likely with this type of construction, $x > 0$ is a guard for the call of f (that is, $f(x)$ would fail if $x = 0$). In this particular expression, the goal reordering heuristics in PiZA are such that this would not happen, but in the algorithm there have been occasional cases of error reporting functions being obeyed during the course of a subsequently failed goal.

The solution to this problem is to recast the disjunction into a conjunction of implications, e.g. $(x > 0 \Rightarrow y = f(x)) \land (x = 0 \Rightarrow y = 0)$, because the translation is such that the right hand side of an implication is only evaluated if the left hand side succeeds. A more elegant solution would be the use of an 'andthen' command in PiZA, or an 'if then else' predicate in Z.

3.6 Input to the animation

Testing a specification requires input values. Numbers present no problems and values from a given set can simply be represented by Prolog atoms, as illustrated in the birthday book example given above. This is adequate for simple specifications, but for a specification of any complexity, and any sort of realistic testing, the input values will also be complex and will need to be constructed. Input values may be constructed within the specification, but this has the disadvantage that the specification is changed every time the input data is changed and in addition, the construction of the input value in Z is rather tedious. In our case, input values are drawn from the abstract syntax of the input language and can contain hundreds of terms.

An alternative approach is to write a little compiler in Prolog which takes the input language and converts it to the appropriate representation in Z: this requires a knowledge of how the animator represents free types, sets, sequences and schemas in Prolog, so that the appropriate Prolog term can be constructed. This is fairly straightforward, although not exactly trivial.

The analysis algorithm used the compilation approach whereas the translation algorithm used input in Z. Both required supplementary programs. For the analysis algorithm, a program to transform the ANDF representation into Prolog, followed by a Prolog program to transform this into the Prolog representation of the corresponding Z were required, whereas for the translation algorithm a single program was required to convert the input language into the PiZA representation of the corresponding Z term.

The two approaches both have advantages and disadvantages. Compiling to Prolog directly is more efficient but requires knowledge of the Prolog representation of Z. Compiling to PiZA is independent of the representation (and hence of changes to PiZA) but is less efficient. The specification does change with each change of input, but this is not a severe problem in practice because PiZA has an import directive which allows a specification to be spread over a number of files.

3.7 Diagnostics and other output

This is the converse problem to input. On successful completion of a transaction the output values are displayed. On an error, a trace of the goals leading up to the failure in a function call is displayed, together with their input and output values. The animator has a default way of printing Z values, but for large structures this can be too bulky, since every field of a schema is printed out. Usually it is possible to simplify the output considerably. The animator itself has features which allow it to suppress the output from selected Prolog goals or not print some or all of the fields of selected schemas. This is a great help, but it takes time to decide how to do the selection and inevitably some bug always requires a little extra to be printed out.

It is usually also possible to simplify the print out, particularly in cases dealing with terms in an abstract syntax. This is because it is frequently possible to tell the type of the term from its appearance, so the constructor can often be omitted without construction. For example, a Z term like $make\_value(3)$ can usually be printed as 3 without confusion. To do this, the Prolog diagnostic output can be intercepted with user supplied printing instructions.

As with input, supplementary programs may be required to convert the output for further processing. In the case of the analysis algorithm, this was not necessary but for the Malpas translation a program was required to convert from the abstract syntax output by the animation to the concrete syntax required by Malpas.

3.8 Interface to Prolog

For both input and diagnostics it is necessary to have some interface to the underlying Prolog. This is provided by the initial parts of the specification and transaction files which are treated as Prolog until a begin. statement is read. These preludes have been getting steadily bigger as testing has progressed and a certain amount of Prolog programming is probably inevitable for any realistic animation.

In addition to these preludes, an interface to Prolog is provided from within the PiZA language. This is provided by two pseudo Z predicates:

1. The construction prolog(" ... ") results in the Prolog within the quotation marks being executed.
2. The construction call(*Word, expression,...*) enables a Prolog predicate named *Word* to be called with parameters provided by the Prolog representation of the Z expressions.

For our particular purpose, the uses made of this facility were to provide a set of instance values (rather tedious to do in Z) and to carry out error reporting. The latter in particular is fairly essential for any large scale specification.

4   Conclusions

It should be clear that the PiZA animator is a tool capable of dealing with substantial specifications. At its simplest level of use as a text preparation system it is very easy to use and substantially improves on LaTeX, although not having the graphical interface capabilities of CADiZ. Animation of simple specifications is also easy, but the difficulty of using it increases with the size of the specification. For substantial specifications, prior knowledge of Prolog is desirable, but a relatively superficial knowledge is all that is required. Like most free software, user documentation is fairly primitive, but improving.

For both specifications, the object of the exercise was to design an algorithm: correctness was not an issue. Consequently, the experiences reported here are concerned with Z as a design language, not with its use for verification. For the Malpas translator the animation *was* the implementation, but the ANDF analyser will ultimately be implemented in a conventional language, probably C. We expect that the implementation will be considerably more reliable than would be the case had the algorithm been implemented in C directly. In the mean time, the animation is sufficiently good that we can spend time improving the algorithm before starting the implementation.

A key requirement for design languages, and one facilitated by the use of Z, is the ability to evolve a design as the complexity of the problem is gradually understood. This is particularly the case with specifications concerned with the processing of a language, which have unique problems related to the complexity and expressive power of the language. Malpas, as a language independent static analysis system, requires the semantics of the language under consideration to be explicitly incorporated into its input language. The semantics of this language are simpler than the programming language being modelled and various mathematical functions need to be used to model a program accurately. Unfortunately, the appropriateness of the model often depends upon the particular implementation which means that to construct the appropriate translator a translation needs to be made by hand. If the analysis of the result is difficult, the translation needs to be modified and augmented with mathematical functions to make it easier. Even if the language is relatively simple this process is time consuming and therefore expensive. If the language is large and expressive the process becomes very difficult because of the number of ways language constructs could be composed to give different meanings.

PiZA has allowed us to specify the translation (or modelling) process precisely and animate the specification to determine its worth. This is an important point which is often missed by people involved with formal methods. This is an example of a class of problem for which it is not appropriate to abstract away the complexity. It is the complexity which needs to be tackled head on and this seems to be quite common where the analysis of real world programming languages is involved. For this type of problem, animation is essential to provide control over the complexity.

The advantage of specifying the translation process in Z and then animating it is twofold. The translation is peer reviewable, like any other Z specification, in a way which a program implementing a translator is not; this is obviously an advantage when the analysis is part of a safety assessment. The second advantage is that the ability to rapidly prototype a translator, try it out, and then change the specification in the light of experimentation means that the correct specification can be arrived at within a reasonable timescale. The team at DRA have used this novel approach on a translator development which experience has shown would take 1 person year; it in fact took half a person year saving more than half the allocated budget and allowing unexpected technical difficulties to be overcome.

Using Z as a design language has required a substantial effort, but not incommensurate with the size of the problems. In this type of situation the alternatives are either a specification language with animation, or direct implementation in code. Design languages which do not have possibilities for animation are much less satisfactory as it is necessary to implement the design in order to carry out experimental testing. Once the design is implemented the code becomes the working description and the design itself will usually rapidly become out of date.

Using Z as a design language requires total familiarity with the language and with the use of Z to express software designs. In order that alternatives can be explored, the effort of writing a design has to be small. Formalisation rather works against this as there is a tendency to formalise at too detailed a level. However, given experienced Z users, the advantages stem from the clarity of the design and the power of set theory to express designs at a high level.

The PiZA Z animator is distributed under the GNU general public licence and is available without charge from the URL

```
http://www.noodles.demon.co.uk/PiZA/PiZAHome.html
```

References

1. Dick AJJ, Krause PJ, and Cozens J. Computer aided transformation of Z into Prolog. Workshops in Computing, pages 71–85. Springer-Verlag, 1990.
2. Jordan D, McDermid JA, and Toyn I. CADiZ – computer aided design in Z. Workshops in Computing, pages 93–104. Springer-Verlag, 1991.
3. Toyn I. *CADiZ Quick Reference Guide.* York Software Engineering Ltd, University of York, York YO1 5DD, UK, 1990.
4. Xiaoping J. *ZTC: A Type Checker for the Z Notation.* School of Computer Science, DePaul University, Chicago, Illinois, USA, 1995.
5. Spivey JM. *The fUZZ Manual.* Computing Science Consultancy, 2 Willow Close, Garsington, Oxford OX9 9AN, UK, 2nd edition, 1992.
6. Spivey JM. *The Z Notation: A Reference Manual.* Series in Computer Science. Prentice Hall International, 2nd edition, 1992.

7. Hewitt MA. Automated animation of Z using Prolog. B.S.c. Project Report, Department of Computing, Lancaster University, 1991.
8. Hewitt MA. Optimization of Prolog generated from Z specifications. Master's thesis, Department of Computing Science, University of Aberdeen, 1992. Available from http://www.noodles.demon.co.uk/PiZA/PiZAOldDocs.html.
9. West MM and Eaglestone BM. Software development: Two approaches to animation of Z specifications using Prolog. *IEE/BCS Software Engineering Journal*, 7(4):264–276, July 1992.
10. OSF. ANDF, application portability and open systems. Technical report, Open Software Foundation, 11 Cambridge Center, Cambridge, MA 02142, USA, 1991.
11. Knott RD and Krause PJ. The implementation of Z specifications using program transformation systems: The SuZan project. In C. Rattray and R. G. Clark, editors, *The Unified Computation Laboratory*, volume 35 of *IMA Conference Series*, pages 207–220, Oxford, UK, 1992. Clarendon Press.
12. TA Consultancy Services Ltd, 'The Barbican' East Street, Farnham, Surrey, UK, GU9 7TB. *Intermediate Language Manual for MALPAS*, 6th edition, 1996.

Automating Test Case Generation from Z Specifications with Isabelle

Steffen Helke, Thomas Neustupny, Thomas Santen

GMD FIRST
Rudower Chaussee 5, D-12489 Berlin, Germany
email:{steffen | thomas | santen}@first.gmd.de

Abstract We use a structure preserving encoding of Z in the higher-order logic instance of the generic theorem prover Isabelle to derive test cases from Z specifications. This work shows how advanced theorem provers can be used with little effort to provide tool support for Z beyond mere type-checking. Experience with a non-trivial example shows that modular reasoning according to the structure of a specification is crucial to keep the proof-load manageable in practical applications. Support for modular reasoning can be based on higher-order equational reasoning as implemented in Isabelle.

1 Introduction

Experience has shown that application of formal methods in early phases of the software life-cycle can help avoid specification errors and thereby enhance quality and reduce cost of software products. The precision of specifications written in a formal specification language like Z forces specifiers to express requirements and design decisions in an unambiguous way, bringing unclear points in a specification into view and avoiding misconceptions.

Still, developing a formal specification can be a time consuming and intellectually demanding task, so that to set up a formal specification as a means of documentation alone may not be justified. Taking into account that testing efforts can amount to 75% of the total cost of a piece of software, developing support to exploit formal specifications in the later development phases seems to be a particularly attractive research goal from the perspective of utility to industry.

Such support need not mean the formal verification of a program against a specification. Today, considering cost-effectiveness, a formal proof of correctness can only be the exception rather than the rule. Recently, more pragmatic approaches to support verification and validation (V&V), in particular testing, based on formal specifications have been proposed [5, 6, 8, 16]. These approaches also acknowledge that testing will always be necessary because even a formal proof of correctness cannot ensure that the actual code executed on the target platform exhibits the desired behavior.

While type checkers like fUZZ [14] are useful to validate specifications at the level of language, more powerful tools are needed to support working with specifications at the level of semantics. Since Z specifications are not "executable", this kind of support in general leads to logical deduction problems.

The contribution of this paper is to show how modern theorem proving technology can support V&V activities based on Z specifications. We argue that advanced logical frameworks like the generic theorem prover Isabelle [12] can be adapted with little effort to support Z in a way that is trustworthy, flexible, easily maintainable, and able to support non-trivial, medium-sized specifications.

The paper is organized as follows: In the next section, we describe a technique to derive test cases from Z operation schemas that we use as an example V&V activity. Section 3 introduces Isabelle and proof support for Z based on an encoding in higher-order logic. How this encoding can be used to implement the generation of test cases is described in Section 4. Results for a non-trivial example are shown in Section 5. We put our work into context in Section 6 before drawing some conclusions in Section 7.

2 Testing Based on Z

We first give a short presentation of the general approach to testing underlying our work. For more information on testing based on formal methods we refer to [5], [8] and [6]. As an example throughout the paper we use a medium-sized Z specification based on a problem originally posed by Abrial. Then we describe how to generate test cases from a specification using a transformation to disjunctive normal form.

2.1 An Overview

The prerequisites for any testing activities are a Z specification and an implementation which shall be tested for conformance to the specification. Furthermore, we need a set of test inputs for the implementation which is a subset of all possible inputs, because we usually cannot perform a complete test. Testing activities can be divided into four steps:

1. **test case generation:** For each operation on the state space the pre-/post-operation relation is divided into sub-relations, for which the *uniformity hypothesis* seems justified. It assumes that all members of a sub-relation lead to similar behavior of the implementation. These sub-relations are the *test cases*. (Ideally they are disjoint, but they need not be.)

2. **test input extraction:** For each test case (at least) one representative is chosen, so that coverage of all test cases is provable. From the resulting test data, test input data processible by the implementation is extracted.

3. **test execution:** The implementation is run with the selected test input data in order to determine the test output data.

4. **test result verification:** The pair of input/output data is checked against the specification of the test case.

Since Z specifications in general are not executable, all steps except the third lead to deduction tasks. In this paper we concentrate on step one: the use of advanced theorem proving techniques to automating test case generation.

2.2 The Specification of a Steam Boiler Control

Abrial posed a specification problem as a benchmark for formal methods. It consists of an informal description of control software for a steam boiler. We illustrate test case generation with the Z specification of Büssow and Weber [3] for Abrial's problem.

The steam boiler is a physical unit for producing steam by boiling water (cf. Fig. 1). The boiler is constantly heated while new water is pumped into the boiler by four pumps. Other components of the steam boiler need not be considered here.

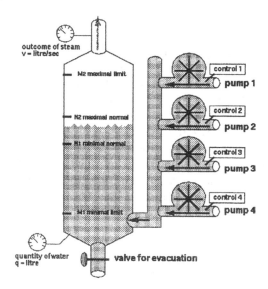

Figure 1: The steam boiler

For safe operation, the water level must be kept in a normal range. During normal operation, the water level can only be controlled by the four pumps: no pumping decreases and maximum pumping increases the water level. If some critical condition arises, the steam boiler changes its mode from *running* to *stopped*, and an *alarm* is raised. This is specified as follows:

PhysModes ::= *waiting* | *adjusting* | *ready* | *running* | *stopped*

Alarm ::= *ON* | *OFF*

```
┌─ Modes ─────────────────────────┐    ┌─ UnitModels ──────────────┐
│ st : PhysModes; alarm : Alarm   │    │ ActorModels; qa1, qa2 : N │
├─────────────────────────────────┤    ├───────────────────────────┤
│ st = stopped ⇒ alarm = ON       │    │ 0 ≤ qa1 ≤ qa2 ≤ C         │
└─────────────────────────────────┘    └───────────────────────────┘
```

The state of the steam boiler depends on models of the physical units like sensors and actuators, *UnitModels*, and on *Modes*:

```
┌─ SteamBoiler ────────────────────────────────────────────────┐
│ UnitModels; Modes                                             │
├───────────────────────────────────────────────────────────────┤
│ st ∈ {waiting, adjusting, ready} ⇒                            │
│     (alarm = OFF ⇔ NoDefects ∧ SteamZero)                     │
│ st = running ⇒                                                │
│     (alarm = OFF ⇔ WaterTolerable ∧ TolerableDefects)         │
│ (st = running ∨ PumpsOpen) ⇒ ValveClosed                      │
└───────────────────────────────────────────────────────────────┘
```

These schemas serve us to illustrate test case generation in the following.

2.3 Test case generation with the DNF method

For extracting test cases from a Z specification we choose the *disjunctive normal form* (DNF) approach of Dick and Faivre [5]. We generate DNFs for single operations only. How these results can be used to construct a separation of the whole specification state space into test cases is described in [5].

The schema predicate of the operation to be tested is transformed into a disjunctive normal form, where each disjunction represents one test case. During this step all predicates of the operation are transformed into $\wedge$, $\vee$ and $\neg$ combinations. For example, the predicate $A \wedge (B \Rightarrow C)$ has the DNF representation $A \wedge \neg B \vee A \wedge B \wedge C$. For the transformation, we deal only with propositional connectives. Quantified formulas, equations, etc. are regarded as atomic.

As an example, let us consider the schema *SteamBoiler* from above. It is not an operation schema, but it is imported by each operation and its predicate is part of the operation's predicate. We restrict our attention to the last predicate. By expansion of the included schema *Modes* we get

```
┌─ SteamBoiler ────────────────────────────────────────────────┐
│ UnitModels; st : PhysModes; alarm : Alarm                     │
├───────────────────────────────────────────────────────────────┤
│ ...                                                           │
│ st = stopped ⇒ alarm = ON                                     │
│ (st = running ∨ PumpsOpen) ⇒ ValveClosed                      │
└───────────────────────────────────────────────────────────────┘
```

We can now syntactically calculate the DNF. The result is:

__*SteamBoiler*_____
| *UnitModels*; st : *PhysModes*; *alarm* : *Alarm*

...

$(st = stopped \wedge alarm = ON \wedge st = running \wedge ValveClosed) \vee$
$(st = stopped \wedge alarm = ON \wedge PumpsOpen \wedge ValveClosed) \vee$
$(st = stopped \wedge alarm = ON \wedge st \neq running \wedge \neg\, PumpsOpen) \vee$
$(st \neq stopped \wedge st = running \wedge ValveClosed) \vee$
$(st \neq stopped \wedge PumpsOpen \wedge ValveClosed) \vee$
$(st \neq stopped \wedge st \neq running \wedge \neg\, PumpsOpen)$

Without any further information about the steam boiler, two things can be noted. First, the first case (i.e. disjunction term) contains the contradiction $st = stopped \wedge st = running$ and can therefore be removed. Second, in the third and forth case the terms $st \neq ...$ are redundant because of the $st = ...$ terms. These two observations allow us to simplify the DNF of *SteamBoiler*:

__*SteamBoiler*_____
| *UnitModels*; st : *PhysModes*; *alarm* : *Alarm*

...

$(st = stopped \wedge alarm = ON \wedge PumpsOpen \wedge ValveClosed) \vee$
$(st = stopped \wedge alarm = ON \wedge \neg\, PumpsOpen) \vee$
$(st = running \wedge ValveClosed) \vee$
$(st \neq stopped \wedge PumpsOpen \wedge ValveClosed) \vee$
$(st \neq stopped \wedge st \neq running \wedge \neg\, PumpsOpen)$

Finally, the DNF transformation including simplifications leads to five cases. We have only expanded the definition of *Modes* while the other included schema *UnitModels* is left untouched. Expanding *UnitModels*, too, would have two effects: first, more cases will be generated, because of splitting each $\leq$-term in two cases $<$ and $=$, and second, at least the cases with $0 = qa1 = qa2 = C$ turn out to be unsatisfiable (because of $0 < C$) and thus will be eliminated by simplification.

The schema *SteamBoiler* specifies the data state of the steam boiler. It therefore appears in many operation schemas, for which the DNF must be calculated. For these, the precalculated DNF version of *SteamBoiler* can be used.

We conclude, that the following aspects become important for the DNF method:

- deciding about replacing terms by their definition

- finding and removing unsatisfiable test cases

- finding and removing redundant terms

The rest of the paper describes how this process can be automated using Isabelle.

3 Proof Support for Z Using Isabelle

The technique to generate test cases from operation schemas, especially the detection of unsatisfiable test cases, essentially amounts to a number of logical deduction problems. Hence, a mechanical theorem prover for Z seems suitable to effectively support this technique.

In principle, there are two ways to come up with deduction support for Z: we can implement a prover tailor-made for Z, or we can use an existing prover that is flexible enough to be adapted to the semantics of Z. For the latter choice, generic theorem provers, often called *logical frameworks*, are particularly attractive, because they are implemented so as to support various logics.

We prefer to embed Z into a logical framework for several reasons: first of all, we can re-use an existing implementation and do not have to deal with basic algorithms like substitution and unification that are needed in virtually every implementation of a logical calculus. But these systems also provide us with sophisticated implementations of advanced theorem proving techniques like higher-order rewriting that we would have neither the expertise nor the resources to implement ourselves. Furthermore, systems that have a large community of users are more trustworthy and often come with a large library of theorems and proof procedures that we can use for free.

In this section, we show how proof support for Z can be obtained by semantically embedding the language into an implementation of higher-order logic that is provided by the generic prover Isabelle [12]. We first give a brief introduction to Isabelle, and describe the key aspects of our encoding of Z afterwards. A detailed description of that encoding and a justification why it conforms to the Z draft standard [11] is given elsewhere [9].

3.1 Isabelle

Logical frameworks implement a calculus for a simple meta-logic that is nevertheless expressive enough to encode many other logics, so called object logics, in it. The meta-logic of Isabelle only consists of equality $==$, implication $\Longrightarrow$, universal quantification $\bigwedge$, and functional abstraction λ. A number of object-logics like first-order predicate logic, Zermelo-Fraenkel set theory, constructive type-theory, and higher-order logic have been encoded in Isabelle and are part of its standard distribution. Our work on Z is based on Isabelle's higher-order logic (Isabelle/HOL) which is similar to the one implemented in the HOL system [7].

Logical rules like

$$\frac{P \quad Q}{P \wedge Q} \ \text{(conjI)} \qquad \frac{P \wedge Q}{P} \ \text{(conjunct1)}$$

are represented in Isabelle by meta-implications

$$[\![?P;\,?Q]\!] \Longrightarrow ?P \wedge ?Q \qquad ?P \wedge ?Q \Longrightarrow ?P$$

The variables prefixed by a question mark are so called *meta-variables*. Substitutions for these variables are computed when applying rules by *resolution* which essentially is the application of *modus ponens* ($A \implies B$ and $B \implies C$ implies $A \implies C$) modulo unification.

Proofs are usually constructed in a goal directed manner starting with a goal to prove and reducing it by backward resolution to sufficient subgoals. Whenever a subgoal unifies with a known theorem, a branch of the proof is closed, and, eventually, when no unproven subgoals remain, the proof is complete and a new theorem is created replacing all free variables in the proven formula by meta-variables.

3.1.1 Tactical Theorem Proving The *system* Isabelle is a tactical theorem prover in the tradition of LCF. It is implemented in Standard ML which is a strongly typed functional programming language. The type thm of Isabelle theorems plays a crucial role in the implementation. Objects of this type can only be generated by a small number of functions that implement Isabelle's meta-logic.

Tactics are ML functions that map theorems to theorems. Basic tactics implement various kinds of resolution used in forward or backward proofs. They can be combined by *tacticals* that are higher-order functions combining tactics to new tactics. Typical examples are sequentialization, repetition, or choice of tactics. This yields an elegant way to implement complex proof procedures in a trustworthy way: the logical soundness of these proofs depends solely on the correct implementation of the basic tactics and theorem constructors.

There are two very powerful "parameterized tactics" in Isabelle: the *classical reasoner* and the *simplifier*. The former simulates a classical Gentzen style proof system and is parameterized by sets of introduction and elimination rules.

The simplifier implements conditional higher-order rewriting. It is parameterized by a *simplifier set* whose most important components are a set of *conditional equations* and a set of *congruence rules*. The conditional equations have the form[1]

$$C \implies s = t$$

When the simplifier is applied to a subgoal, it replaces subterms of the goal matching s by t if it can prove the corresponding instance of C. Congruence rules can be used to construct a "local context" under which simplification of subterms takes place. A congruence rule for conjunction is

$$\llbracket P = P'; P' \implies Q = Q' \rrbracket \implies P \land Q = P' \land Q' \qquad \text{(conj\_cong)}$$

This theorem, interpreted as a congruence rule, tells the simplifier how to deal with a conjunction $P \land Q$: first, simplify P to P'; then assume P' while simplifying Q to Q'. When the simplifier is (recursively) presented with $P' \implies Q = Q'$, it adds rewrite rules derived from P' to its simplifier set, and then simplifies Q.

---

[1]The simplifier actually works on the meta-equality ==. *Reflexion* rules relate the meta-equality to object equalities = or equivalences ⇔. We therefore do not distinguish between these in the presentation.

Thus it can make use of the information contained in P' to simplify Q to Q'. As we show in detail later, this mechanism is crucial to find unsatisfiable test cases.

3.1.2 Calculational Proofs For our application, it is important that one can start a proof with a goal containing meta-variables. These can be instantiated during proof by unification, giving us a means to *calculate* results during proofs.

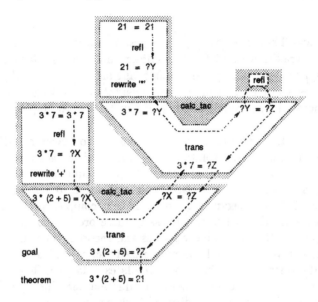

Figure 2: Calculational Proofs

Consider the calculation[2] in Figure 2, and suppose we wish to calculate the result of an arithmetical expression like $3 * (2 + 5)$. The first step is to set up the goal $3 * (2 + 5) = ?Z$ as an equation that contains the metavariable $?Z$ at its right-hand side. The following tactic accomplishes one step of a calculation using Isabelle's simplifier:

```
fun calc_tac S i = (resolve_tac [trans] i) THEN (simp_tac S i)
```

It first resolves subgoal i with transitivity and rewrites the first generated subgoal with the simplifier set S afterwards. In Figure 2, calls to calc_tac are depicted by gray, v-shaped boxes. In the example, we first use a set of rewrite rules for addition to rewrite $2 + 5$ to 7. By default, the simplifier tries to close a subgoal by resolving with the reflexivity rule of equality. This instantiates the metavariable $?X$ that has been introduced by resolving with transitivity before calling the simplifier. The value for $?X$ is propagated into the second subgoal of the transitivity rule, and we are left with the goal $3 * 7 = ?Z$, which again

---

[2]The example is trivial and could be handled by Isabelle in a single step.

is an equality whose right-hand side is a metavariable. This allows us to apply calc_tac repeatedly. In the example, we use rewrite rules about multiplication to compute the final result 21. An explicit application of reflexivity instantiates $?Z$ and finishes the proof. The proven theorem is the concrete, meta-variable free equation $3 * (2 + 5) = 21$.

3.2 Z in Isabelle/HOL

Let us now briefly explain how Z can be encoded in the implementation of higher-order logic in Isabelle. This encoding is described in detail elsewhere [9].

Two major questions arise when we wish to encode Z: how to deal with the set theory of Z and the mathematical tool-kit, and how to represent schemas.

3.2.1 Set Theory and Mathematical Tool-Kit
The set theories of Z and higher-order logic are very similar: both are strongly typed and have some mechanism of parameterization, generic constructs in Z and shallow polymorphism in Isabelle/HOL. Thus, it is easy to define the constants of the mathematical tool-kit in HOL. To date, we have proven most of the theorems about the tool-kit mentioned in [15] and [11] which gives us confidence that our definitions conform to the Z standard. Many other theorems about the basic set operations, Cartesian products, and predicate logic in general are contained in the library of Isabelle/HOL and can directly be used in proofs about Z.

3.2.2 Representation of Schemas
The most crucial part of every Z encoding is to find an appropriate way to deal with schemas. They can be used in different contexts with slightly different semantics. The most notable uses are

schemas as predicates when they are referenced in the predicate part or imported in the declaration part of another schema

schemas as sets when referenced at the "right-hand side of the colon" in declarations

schemas in schema calculus when defining a new schema by conjoining others.

Since we do not wish to use our encoding for reasoning about the Z language but about concrete specifications, we decided to encode just the "logical content" of schemas and leave the "syntactical" side of schema semantics to be dealt with by a parser. This means the parser has to deal with signatures of schemas and the implicit binding of variables in schema references where variables of the schema's signature are identified with variables with the same name in the context of the reference.

From the logical point of view, schemas are most often used with their predicate semantics, e.g. when referencing a state schema. Logical predicates are also best supported by theorem provers like Isabelle. We therefore decided to base our encoding on the "schemas as predicates" semantics.

Basically, a schema is represented by a function from a Cartesian product of the types of its signature components to the type bool of truth values of higher-order logic[3]. Thus, for each schema S, we define a function of the same name mapping a tuple of the schema's signature to the truth value of the schema's predicate. Consider the examples shown in Figure 3.

$$A \widehat{=} \lambda (f, x). [\, x : \mathsf{N} \wedge f : \mathsf{Z} \rightarrow \mathsf{Z} \mid f\widehat{\,}x = 0 \,]$$

$$B \widehat{=} \lambda(f, x). [\, x : \mathsf{N} \wedge f : \mathsf{Z} \rightarrow \mathsf{Z} \mid A(f, x) \wedge x = 5 \,]$$

$$C \widehat{=} \lambda\, y. [\, y : \{(f, x) \mid A(f, x)\} \mid \ldots \,]$$

Figure 3: Schema Encoding

Schema A is represented by a function mapping pairs (f, x) to the predicate of A. In schema B, A is referenced as a predicate. According to the semantics of Z, the members x and f of A's signature are identified with the corresponding variables in B. This is made explicit in the encoding, where the function representing A is applied to the pair (f, x) of parameters of B. In schema C, A is referenced in the declaration of y. The corresponding encoding explicitly builds the extension of predicate A as a set of which y is a member.

4 Test Case Generation with Isabelle

With the encoding of Z at hand, all deduction support offered by Isabelle is immediately available to implement reasoning about Z specifications. We show how to implement the test case generation approach described in Sect. 2. We first give a general overview of the algorithm for a simple schema. The three major steps of that algorithm are discussed in detail afterwards. Finally, we describe how the structure preserving encoding of Z helps us to reduce complexity when dealing with structured specifications.

---

[3]Since the order of declarations is irrelevant in a schema, the parameter tuples are lexically sorted on the names of the declared variables in the Isabelle/HOL encoding.

4.1 Calculating Test Cases

Consider the schema

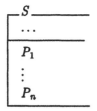

whose predicate[4] is a conjunction of n formulas P_1 through P_n.

We call a predicate P_i *disjunctive* if it – implicitly or explicitly – contains a disjunction, we call it *non-disjunctive* otherwise.

The approach to test case generation of Sect. 2 requires to transform the predicate of an operation schema to disjunctive normal form – leading to a disjunction of operation schemas each of which describes one test case or *sub-operation*. Thus, we conceptually wish to establish an equality like

$$
\begin{array}{c}
\underline{S}\\
\cdots\\
\hline
P_1\\
\vdots\\
P_n
\end{array}
\quad = \quad
\bigvee_{i=1}^{k}
\begin{array}{c}
\underline{S_i}\\
\cdots\\
\hline
R\\
Q_i
\end{array}
\tag{1}
$$

In this transformation, we consider quantified formulas, equations, etc. as atomic, working on the propositional sublanguage of the logic only. An approach to weaken this restriction is discussed in Sect. 7.

The right-hand side of equation (1) contains much redundancy because the conjunction R of the non-disjunctive predicates P_i of S appears in the predicates of all S_i. A straightforward implementation of this transformation would considerably add to the effort required to simplify the predicates of sub-operations and to find unsatisfiable test cases.

We avoid this redundancy by proving an equation like

$$
\begin{array}{c}
\underline{S}\\
\cdots\\
\hline
P_1\\
\vdots\\
P_n
\end{array}
\quad = \quad
\begin{array}{c}
\underline{Sdnf}\\
\cdots\\
\hline
R\\
\bigvee_{i=1}^{k} Q_i
\end{array}
\tag{2}
$$

---

[4]The schemas shown here are meant to sketch the idea of the procedure. Implicit predicates contained in declaration parts of schemas have, of course, to be considered in the process too.

Here, the non-disjunctive P_i are not distributed into the disjunction and only disjunctive P_i are transformed into DNF.

The algorithm to construct *Sdnf* employs the calculational proof style with meta-variables explained in Sect. 3. We start with the goal

$$S = ?X \qquad (3)$$

where S is the name of the constant representing the schema to transform. Thus there is a definition

$$S == \lambda(\ldots). \bigwedge_{i=1}^{n} P_i \qquad (4)$$

in the Isabelle representation of the given Z specification. The theorem eventually proved is

$$S = \lambda(\ldots). R \wedge \bigvee_{i=1}^{k} Q_i \qquad (5)$$

where $?X$ of the goal (3) has been replaced by an HOL function encoding the schema *Sdnf* of (2).

It is important to note that we prove an equality of the constant S, i.e. of the *name*, not of the expanded definition. This allows us to unfold a reference to S somewhere else in the specification not only with the definition (4) but also with the disjunctive normal form (5) of S. We use this feature when dealing with complex specifications in Sect. 4.5 below.

The algorithm computing (5) from the goal (3) consists of three major steps:

1. separate disjunctive predicates

2. compute the disjunctive normal form

3. eliminate unsatisfiable disjuncts and simplify test cases

Steps 1 and 2 are interleaved so that sub-formulas containing disjunctive predicates can easily be determined by examining their outermost connective. Each of the steps is described in detail in the following.

4.2   Step 1: Separate Disjunctive Predicates

The first step after unfolding the definition of S in (3) is to separate disjunctive conjuncts of P from non-disjunctive ones. P must be transformed into a conjunction

$$P = R \wedge Q$$

where R is a conjunction of non-disjunctive formulas and Q is a conjunction of disjunctive ones.

This transformation consists of formula manipulations at a "syntactic" level only: we do not reason about semantic equalities or equivalences but about permutations of sub-formulas of a conjunction.

The basic idea of the transformation is to maintain an equality

$$P = \overline{R} \wedge \overline{P} \wedge \overline{Q}$$

where $\overline{R}$ is a conjunction of non-disjunctive goals, $\overline{Q}$ is a conjunction of disjunctive goals and $\overline{P}$ contains the conjuncts of P not yet sorted into $\overline{R}$ or $\overline{Q}$.

We use equalities like

$$R \wedge ((P_1 \vee P_2) \wedge P) \wedge Q \ = \ R \wedge P \wedge ((P_1 \vee P_2) \wedge Q) \tag{6}$$
$$R \wedge (P_1 \wedge P) \wedge Q \ = \ (P_1 \wedge R) \wedge P \wedge Q \tag{7}$$

to sort the conjuncts of $\overline{P}$ into $\overline{Q}$ or $\overline{R}$. The process terminates when $\overline{P} = true$.

Note that the left-hand side of (6) is an instance of the left-hand side of (7), because it cannot schematically be expressed that P_1 in (7) should match only non-disjunctive formulas. We therefore have to give the application of (6) priority over (7).

Implementing precedences on tactic applications is easily achieved using the tactical ORELSE:

```
tac1 ORELSE tac2
```

tries to apply tac1. Only if this tactic fails, is tac2 applied.

In this way, Isabelle's tactic language allows us to implement the separation of disjunctive predicates without having to manipulate the abstract syntax of formulas directly. This not only results in very concise, abstract code but it also increases confidence in the soundness of our transformations: they solely consist of resolutions against theorems.

4.3 Step 2: Compute the Disjunctive Normal Form

The second major step of the algorithm is to transform the conjunction Q of disjunctive predicates into disjunctive normal form. This is achieved by the usual algorithm that first transforms implications and equivalences to disjunctions, then builds a negation normal form, and finally distributes conjunctions over disjunctions.

We transform implications and equivalences using the rules proposed by Dick and Faivre [5], which lead to mutually exclusive disjuncts.

$$A \Rightarrow B \ = \ \neg A \vee (A \wedge B)$$
$$A \Leftrightarrow B \ = \ (\neg A \wedge \neg B) \vee (A \wedge B)$$

Isabelle's simplifier makes it very easy to implement the three-step transformation to DNF. We just need to construct three simplifier sets with the appropriate equalities, and call the simplifier with each of the sets in turn. The tactic implementing the transformation (on subgoal i) in calculational proof style looks as follows (cf. Sect. 3.1.2):

```
           (calc_tac imp_iff_ss i)
     THEN (calc_tac not_ss i)
     THEN (calc_tac dnf_ss i)
```

The simplifier sets `imp_iff_ss`, `not_ss` and `dnf_ss` contain the equalities needed in each step, e.g. the first is defined by

```
imp_iff_ss = empty_ss addsimps [ imp_to_disj, iff_to_disj ];
```

where `empty_ss` contains the standard parameters for simplification of higher-
order logic formulas but no conditional equations.

Using the simplifier in this way, we can implement the transformation in
a very abstract, declarative way. We do not need to work on the ML data
structure representing the abstract syntax of Isabelle formulas, for example, to
code a recursive decent into sub-formulas let alone deal with variable binding
structures or substitution.

4.4   Step 3: Eliminate Unsatisfiable Disjuncts

As we have seen in Sect. 2.3, the syntactic transformation of a schema's predicate
to DNF can lead to unsatisfiable cases of the resulting disjunction. These are
undesirable not only because they may result in unnecessary testing overhead,
but they can also cause unjustified rejection of tests if test input data is generated
from them and test result verification – necessarily – fails. As we show in Section
5, a large number of cases generated in Step 2 can be contradictory. It is therefore
mandatory to try and eliminate as many of these as possible.

To do so, we use Isabelle's simplifier with the standard simplifier set for
higher-order logic, containing many theorems about predicate logic and set theory, augmented by rewrite rules for the Z Mathematical Tool-Kit and equalities
derived from the Z specification under consideration. A free type definition of Z,
for example, is translated to a `datatype` definition in Isabelle, e.g. the definition
of *PhysModes* in Sect. 2.3, is reflected in Isabelle by

```
datatype PhysModes = waiting | adjusting | ready | running | stopped
```

This construct not only generates the axioms associated with a free type, but
in particular includes the inequalities asserting distinctness of all introduced
constants in the standard simplifier set automatically. This gives the simplifier
the information to rewrite conjunctions like

$$st = waiting \land \ldots \land st = running$$

to *false* without explicit intervention of the users.

For the simplifier to actually find this contradiction, it has to use the equality
$st = waiting$ to rewrite $st = running$ to $waiting = running$; only then can
the distinctness axioms of *PhysModes* be applied. Thus, the simplifier has to
consider the "local context" of the equality $st = running$ in the conjunction. This
is achieved augmenting the simplifier set with the congruence rule `conj_cong`
as described in Sect. 3.1.

As a side effect of simplification to find contradictions, possible redundancies in all Q_i are reduced. A less desirable effect is that new disjunctions may be introduced, for example, by unfolding finite sets as in

$$st \in \{waiting, adjusting, ready\} \Leftrightarrow st = waiting \lor st = adjusting \lor st = ready \tag{8}$$

This is necessary to be able to find contradictions to an equation like $st = running$ as above, but if the case is satisfiable, leaving the disjunction on the right-hand side of (8) in a case Q_i leads to a further case distinction when the DNF is used in a hierarchical DNF computation as described in the next section. This may not be desirable because the three values of st are not distinguished in the specification and the operation to test probably behaves uniformly for all three values.

We therefore post-process all Q_i after elimination of unsatisfiable cases by folding equations like (8). This is again achieved using the simplifier with a specialized simplifier set.

4.5 Hierarchical DNF Computation

Since distributing conjunction over disjunction duplicates formulas, the size of a formula can grow exponentially if it is transformed into DNF. In the worst case, the transformation results in a complete case distinction over the truth values of all atoms. When generating test cases from Z, most of these cases are contradictory and need to be eliminated, because the "atoms" are not propositional variables but are structured. A way to drastically reduce simplification work is to eliminate unsatisfiable conjunctions early and prevent cases depending on these to be generated in the first place.

To this effect, we decompose the computation of DNFs according to the structure of schema references. Wherever a schema that must be unfolded is referenced, e.g. the pre-state schema in an operation schema, we first compute the DNFs of both the referenced and the referencing schema, then unfold the DNF of the referenced schema into the DNF of the latter, and finally compute the DNF of the resulting schema. In this way, many unsatisfiable cases can be eliminated before unfolding schema references.

Two features of our tool are crucial to implement this algorithm. First, our encoding of Z preserves the structure of specifications, i.e. we can actually reason about schemas where references are *not* unfolded. Second, Isabelle allows substituting a schema reference S in arbitrary context using a higher-order equation like (5).

5 Empirical Results

We have used our test case generation tactic in several case studies. The largest is the steam boiler specification of Büssow and Weber [3]. From this specification, which uses a combined notation of Z and Statecharts, we have extracted a pure Z specification consisting of 37 schemas totaling 600 lines of Z.

The operation *SteamBoilerWaiting* specifies a part of the initialization phase before heating of the boiler has begun. If the water level is in normal range the steam boiler state changes to *ready*. Otherwise it changes to *adjusting*, because the water level needs to be adjusted using the valve or the pumps.

```
┌─ SteamBoilerWaiting ─────────────────────────────────
│ ΔSteamBoiler; ΞSensorModels
├──────────────────────────────────────────────────────
│ alarm = OFF ∧ st = waiting
│
│ WaterAboveNormal
│     ⇒ st' = adjusting ∧ ValveOpen' ∧ PumpsClosed'
│
│ WaterBelowNormal
│     ⇒ st' = adjusting ∧ ValveClosed' ∧ PumpsOpen'
│
│ WaterNormal ⇒ st' = ready ∧ ΞActorModels
└──────────────────────────────────────────────────────
```

The results of generating test cases for *SteamBoilerWaiting* on a Sun Ultra-Sparc are summarized in Table 1.

	Schema	Number of Cases before Simplification	after	DNF	Time (sec) Simplification
1.	*SteamBoiler*	48	14	1.4	101.2
2.	*SteamBoilerWaiting*	8	3	0.9	5.0
3.	*SteamBoilerWaiting* DNFs unfolded	42	6	6.6	96.3
4.	*SteamBoilerWaiting* direct	384		14	22h

Table 1: Test case generation for *SteamBoilerWaiting*

Lines 1 and 2 show figures for the state and operation schemas without unfolding schema references. The figures in line 3 refer to the computation for the schema where the DNF of *SteamBoiler* is unfolded into the DNF of *SteamBoilerWaiting* yielding the test cases for this schema.

The figures show that the major part of the cases generated in Step 2 of our the algorithm are not satisfiable, e.g. only 14 out of 48 of the cases produced for *SteamBoiler* remain after simplification. One can also see that the effort of syntactically generating disjunctive normal forms is negligible compared to the work needed to simplify them and eliminate unsatisfiable cases.

The last line of the table illustrates the use of exploiting the structure of the specification: unfolding *SteamBoiler* in *SteamBoilerWaiting* and trying to compute test cases in a single step generates 384 cases in 14 seconds. Their simplification takes approximately 22 hours.

6 Related Work

The generation of disjunctive normal forms from operation specifications described in Sect. 2 is part of a more encompassing approach to testing against model-based specifications proposed by Dick and Faivre [5]. They use DNFs to generate a finite state automaton from a VDM specification, and derive test sequences from this automaton. They also suggest to consider case distinctions derived from data structures and quantifications during test case generation. Unlike them, we do not attempt to ensure disjointness of test cases by transforming disjunctions according to the rule

$$A \lor B \Leftrightarrow A \land B \lor \neg A \land B \lor A \land \neg B$$

We refrained from doing so because of the extreme explosion of cases this rule entails. Since specifiers tend to make disjoint case distinctions, test cases generated by the usual DNF transformation are mostly disjoint anyway.

Dick and Faivre have implemented a 9000 line Prolog tool supporting their approach. It contains a hand-coded simplifier with about 200 inference rules. They advocate to have an extensible set of rules to compensate for the – necessary – incompleteness of the tool in eliminating unsatisfiable test cases. We agree with the latter point but believe that such a tool must be easily and safely extensible. All rules used in our implementation are proven correct in Isabelle giving us confidence that all of our transformations are sound.

Hörcher [8] gives an overview of testing based on Z. He also proposes transforming operation schemas to DNF to generate test cases, but emphasizes the need to automate test evaluation because of the vast amount of data that has to be processed. A test evaluation tool to support this process is described by Mikk [10]. It transforms predicates of schemas into "executable" forms which are then compiled to Boolean-valued C functions. These serve to evaluate test data. Similar to eliminating unsatisfiable cases, the set of transformation rules needs to be extensible to compensate for incompleteness. In addition, predicates of Z operations are in general not executable. This suggests combining theorem proving and compiling schemas to evaluate test results to ensure sound transformations and to deal with non-executable specifications.

Stepney [16] advocates to systematically build abstractions of operation specifications for testing. This approach is based on ZEST [4], an object-oriented extension of Z. A tool to support these activities leaves the deduction steps of simplification and weakening of predicates to the person setting up the tests. It just uses a structured editor to support this process. A tool similar to ours could be used to increase the degree of automation of these steps.

A theory of testing based on algebraic specifications has been proposed by Gaudel [6]. She in particular defines notions such as exhaustive testing, testability, validity and unbias. The counterpart of DNF generation in the algebraic context is unfolding of axioms. Applications of the approach use specially designed axioms to define data types in order to support testing well. For an overview over theorem proving support for Z, we refer to [9].

7 Conclusions

We have described an implementation of an approach to generate test cases from Z specifications. The implementation is based on an encoding of Z in Isabelle/HOL and heavily relies on the theorem proving techniques supplied by that prover.

Neither the approach to generate test cases nor the deduction techniques we have used are new. The contributions of our work, however, lie in showing the feasibility and usefulness of implementing verification and validation techniques on top of a modern state-of-the-art theorem prover.

The major part of the algorithm is realized using Isabelle's simplifier. Here, we only needed to prove the required theorems – which are mostly trivial – and supply the appropriate simplifier sets. Only the separation of disjunctive predicates forced us to implement a specialized tactic. Still, thanks to the powerful tacticals of Isabelle, this tactic consists of only 50 lines of highly abstract code.

Since Isabelle handles all complex formula manipulations, the implementation is flexible and easy to maintain. This is particularly important because most V&V tasks are in general undecidable or have a very high computational complexity. Experiments must show which techniques are suitable to support V&V of practical applications. Therefore, a tool must allow for rapid and safe modifications.

Trustworthiness in general is an important requirement on V&V systems. A major design goal of tactical theorem provers is to maintain soundness of logical deductions regardless of how tactics are combined and theorems are used. Soundness of proofs only depends on the correct implementation of the core deduction functions. This makes such provers a good basis for V&V tools.

Using a state-of-the-art prover like Isabelle which has a large community of users provides additional confidence. It is continuously enhanced and – inevitably present – bugs are found and corrected. Furthermore, our encoding of Z can directly profit from a large library of theorems on higher-order logic.

Tactical theorem provers are undeniably less efficient than hand-optimized special purpose systems supporting just one particular task. Nevertheless, as we have shown in Sect. 5, the major work in test case generation lies in eliminating unsatisfiable test cases. This task needs Isabelle's full theorem proving power, using a large set of rules about predicate logic, set theory, and the Z mathematical tool-kit. Hand-coding this step with comparable deductive power would most probably result in implementing a new theorem prover for Z. Since Isabelle already uses optimizations to control rule applications, we doubt that such a hand-coded tool would significantly improve performance. In contrast, our case studies show that Isabelle can cope with non-trivial specifications in reasonable time.

Future Work. The test case generation tool to date does not consider case distinctions motivated by inductive data type definitions or quantifications to generate test cases. In principle, it is easy to implement such a feature, but it is not obvious how to best control the unfolding process in order to come to sensible test cases. One approach to this problem is to combine DNF generation with methods from classical testing theory, and have the users supply information on important case distinctions, as is proposed in [13].

To date, we have considered test case generation for single operations only. We plan to extend the tool to generate test *sequences* and incorporate information from behavioral specifications. Interestingly, test sequencing as described in [5] involves precondition analyses of operations. A tactic supporting these could also be used in validation activities not directly related to testing.

We believe that the present work can serve as a basis for a tool-kit supporting a broad range of verification and validation activities based on Z specifications.

Acknowledgments. The encoding of Z in Isabelle/HOL has been developed in cooperation with Kolyang and Burkhart Wolff. Thanks to Burkhart and Sadegh Sadeghipour for discussions about test case generation. Maritta Heisel's comments on drafts of this paper helped improve the presentation. So does the picture of the steam boiler provided by Sergio Montenegro.

This work has been supported by the German Federal Ministry of Education, Science, Research and Technology (BMBF), project ESPRESS, grant number 01 IS 509 C6.

References

[1] J. P. Bowen and J. A. Hall, editors. *Z User Workshop*, Workshops in Computing. Springer Verlag, 1994.

[2] J. P. Bowen and M. G. Hinchey, editors. *ZUM'95: The Z Formal Specification Notation*, LNCS 967. Springer Verlag, 1995.

[3] R. Büssow and M. Weber. A Steam-Boiler Control Specification using Statecharts and Z. In J. R. Abrial, editor, *Formal methods for industrial applications: specifying and programming the Steam Boiler Control*, LNCS 1165. Springer Verlag, 1996.

[4] E. Cusack and G. H. B. Rafsanjani. ZEST. In S. Stepney, R. Barden, and D. Cooper, editors, *Object-Orientation in Z*, Workshops in Computing. Springer Verlag, 1992.

[5] J. Dick and A. Faivre. Automating the generation and sequencing of test cases from model-based specifications. In Woodcock and Larsen [17], pages 268–284.

[6] M.-C. Gaudel. Testing can be formal, too. In P. D. Mosses, M. Nielsen, and M. I. Schwartzbach, editors, *TAPSOFT '95: Theory and Practice of Software Development*, LNCS 915, pages 82–96. Springer Verlag, 1995.

[7] M. J. C. Gordon and T. M. Melham. *Introduction to HOL: A theorem proving environment for higher order logics*. Cambridge University Press, 1993.

[8] H.-M. Hörcher. Improving software tests using Z specifications. In Bowen and Hinchey [2], pages 152–166.

[9] Kolyang, T. Santen, and B. Wolff. A structure preserving encoding of Z in Isabelle/HOL. In J. von Wright, J. Grundy, and J. Harrison, editors, *Theorem Proving in Higher-Order Logics*, LNCS 1125, pages 283–298. Springer Verlag, 1996.

[10] E. Mikk. Compilation of Z specifications into C for automatic test result evaluation. In Bowen and Hinchey [2], pages 167–180.

[11] J. Nicholls. Z Notation – version 1.2. Draft ISO standard, 1995.

[12] L. C. Paulson. *Isabelle – A Generic Theorem Prover*. LNCS 828. Springer Verlag, 1994.

[13] H. Singh, M. Conrad, G. Egger, and S. Sadeghipour. Tool-supported test case design based on Z and the classification-tree method, 1996. Proc. Second Workshop on Systems for Computer-Aided Specification, Development and Verification, to appear.

[14] J. M. Spivey. The *fuzz* manual. Computing Science Consultancy, Oxford, UK, 1992.

[15] J. M. Spivey. *The Z Notation – A Reference Manual*. Prentice Hall, 2nd edition, 1992.

[16] S. Stepney. Testing as abstraction. In Bowen and Hinchey [2], pages 137–151.

[17] J. C. P. Woodcock and P. G. Larsen, editors. *FME '93: Industrial-Strength Formal Methods*, LNCS 670. Springer Verlag, 1993.

The Z/EVES System

Mark Saaltink (mark@ora.on.ca)

ORA Canada
267 Richmond Road
Ottawa, Ontario K1Z 6X3
Canada

Abstract. We describe the Z/EVES system, which allows Z specifications to be analysed in a number of different ways. Among the significant features of Z/EVES are *domain checking*, which ensures that a specification is meaningful, and a theorem prover that includes a decision procedure for simple arithmetic and a heuristic rewriting mechanism that recognizes "obvious" facts.

1 Introduction

Technology transfer of formal methods into the university curriculum and into industrial practice is difficult. Among the hurdles to be overcome are resistance to change; a perception that formal methods add to the cost of development; a lack of scientific evidence that these methods are effective; and the dearth of support tools.

The EVES system [2, 4] was developed over the past ten years and, while a technical success, it has not been widely adopted. Like most provers, EVES requires a good deal of expertise to use. EVES also has its own specification language (Verdi) that, while based on ordinary predicate calculus and ZF set theory, has a syntax that is unfamiliar (and sometimes repellent) to many potential users.

The Z language has been widely used in the twenty years since its inception, and its use is growing. There are many books available that introduce Z or present case studies, Z is included in the curriculum of several universities, it has been applied in several industrial efforts, and ISO standardisation is in progress. Thus, Z is achieving some measure of technical transfer.

The Z/EVES project is an effort to join the technical power of the EVES system with the Z notation. The Z notation adds considerable appeal to EVES, and adds some capabilities that were not strongly supported in Verdi. EVES, in turn, provides some powerful analytical capabilities that can be applied to Z specifications in several ways:

- syntax and type checking,
- domain checking,
- schema expansion,
- precondition calculations, and
- general theorem proving.

These ways of analysing a specification also form a technology transfer path. The early kinds of analyses can be carried out with little effort and with a limited understanding of the proof system. More advanced analyses may require the addition of labels and theorems to a specification and a good working knowledge of the Z/EVES prover. We hope that users will be able to start with the simpler analyses and obtain some benefits, then will be drawn into more sophisticated uses of the system.

2 Z/EVES Summary

Z/EVES uses a translation from a slightly extended version of the Z notation, as defined in Spivey's *The Z Notation: A Reference Manual (Second Edition)* [8], to the first-order predicate calculus supported by EVES. The translation is reversible, so the Z/EVES user need not be aware of the details, and works entirely in the Z notation. Such translations have been described before [1, 3], and we will not dwell on the details in this paper. Significantly, however, the translation is structure-preserving so that formulas can retain their compact form during proofs (until the user requests expansion).

The Z/EVES system supports almost the full Z notation; only the unique existence quantifier in schema expressions is not supported in the current version (1.3 at the time of writing).

Interaction with the Z/EVES system uses the "fuzz" syntax [9], which is based on LaTeX and is compatible with the freely available "zed" style for LaTeX. This syntax has been extended with a means for attaching labels to predicates in an axiomatic or generic box; a paragraph that expresses a theorem; and prover commands. Z/EVES can read files that have been prepared for the fuzz typechecker.[1]

Z/EVES supports the Mathematical Toolkit, and has many theorems about the concepts introduced there. We have spent considerable effort in formulating the theorems of the Toolkit in a form that can be used automatically by the theorem prover, so that Z/EVES users need not work too hard to prove simple, obvious facts. Theorems are also formulated in a format more convenient for readers, for cases when the proof needs to be guided.

3 Proving Theorems with Z/EVES

Except for syntax and type checking, all the analyses mentioned above are based on the Z/EVES theorem proving capability. We briefly describe this capability before illustrating the analyses possible in Z/EVES.

---

[1] So long as they use only documented features! We have encountered some users who have used undocumented fuzz features. We may support these features in a future release.

3.1 Starting proofs

Proofs in Z/EVES can be started in four different ways: the `try` command can be used to state a predicate to try proving; a `theorem` paragraph states a theorem and begins its proof; the `try lemma` command can be used to continue the proof of some previously defined theorem; or entering a paragraph can begin a proof of an associated *domain condition*, which is a formula that guarantees that the expressions in that paragraph are well-defined (avoiding, for example, domain errors such as division by zero. See Section 4.1).

A goal is just a Z predicate. Theorems and goals are allowed to refer to free variables—that is, variables that are not bound by any quantifiers.

3.2 The Structure of Proofs

A proof in Z/EVES is a sequence of steps, each of which transforms the current goal predicate into a logically equivalent predicate. When the goal has been transformed to the predicate *true*, the proof is complete.

Z/EVES provides a spectrum of commands, ranging from small manipulations of the formula to powerful heuristically driven proof procedures. These procedures can use theorems that the user has marked as rules (for rewriting), frules (for forward chaining inferences), or grules (for triggered assumptions). The details of the proof commands and proof mechanisms are described in the Z/EVES reference manual [5], and briefly below.

The Z/EVES "linear" proof format is not always easy to use, or the most natural. We are investigating the addition of a proof sketching mechanism to Z/EVES, which would allow for proofs to be presented in a natural deduction style (but allowing justifications like "trivially" or "by rewriting"). These proof sketches can (we think) be converted into a series of lemmas and proofs in the Z/EVES linear style, with the prover being used to fill in the missing parts of the proofs (i.e., where the justification for a step is nontrivial).

3.3 Prover Mechanisms

The Z/EVES prover offers several different mechanisms that can be applied during proofs. We have found fully automatic provers to be very difficult to use; it is hard to know what action to take when a proof fails. Fully user-directed provers can be more easily understood and applied; however providing all the details of a proof can be extremely tedious. In Z/EVES, therefore, we have tried to provide a spectrum of commands ranging from highly automatic to closely user-directed, and a variety of ways to control the automatic capabilities.

Simplification Z/EVES has a decision procedure for quantifier-free linear arithmetic and for equality, using the Nelson-Oppen method [6]. Thus, trivial facts like $1 < 2$ are proved automatically, as well as more complex facts such as $1 \leq x \land 2 + 3 * x \leq y \Rightarrow 5 \leq y$ and $a = b \land b = c \Rightarrow a = c$. The simplifier also has an effect even when presented with a predicate that is not a theorem; for example, $x \in S \Rightarrow (x \in S \wedge x \in T)$ is simplified to $x \in S \Rightarrow x \in T$.

The simplifier implements the one-point rule for quantifiers. Alternatively, there is an **instantiate** command that can be used to specify instantiations of bound variables, and a **prenex** command that removes quantifiers that can be brought to the head of the formula as universal quantifications.

Z/EVES provides a way of extending the power of the simplifier, by adding forward chaining rules (frules) or assumption rules (grules). These mechanisms are somewhat complex, however, and we suspect that only the most advanced users will be able to use them effectively.

Rewriting Theorems can be marked as rewrite rules, in which case Z/EVES can use them automatically during **rewrite** and **reduce** proof steps. Rewrite rules have conclusions that are equalities ($x = y$) or equivalences ($P \Leftrightarrow Q$). When rewriting a goal formula, instances of the left side of the equivalence are replaced by the corresponding instance of the right side. For example, the rule

> **Theorem** rule inUnit:
> $x \in \{y\} \Leftrightarrow x = y$

applies to the predicate $1 \in \{2\}$ and results in the equivalent predicate $1 = 2$.

There are about 500 rewriting rules in the current Z/EVES version of the Mathematical Toolkit. These rewriting rules, together with the simplifier, allow most obvious deductions to be made fully automatically with no user guidance. For example, the formula $\{1\} \subseteq \mathbb{N}$ is transformed to *true* by rewriting.

Some rewriting rules are added automatically when paragraphs are introduced. For example, if the declaration of a schema S is added, rewriting rules such as $\theta S \in S \Leftrightarrow S$ and $\theta S = \theta S' \Leftrightarrow x = x' \wedge \cdots$ are automatically generated.

Rewrite rules can also be applied as directed by the user. This can be useful if an automatic proof command goes awry, or if a proof needs to be directed by the user.

Reduction The **reduce** command applies rewrite rules and simplification, and furthermore will expand any defined abbreviations or schemas. In its pure form this command is seldom appropriate, as it expands the goal to use only primitive terms. However, its effect can be controlled with modifiers. For example, by using a modifier to temporarily "disable" one or more definitions, the **reduce** command can be made to stop before it fully reduces a formula.

User Control The Z/EVES user need not rely on the reduction commands (**simplify**, **rewrite**, and **reduce**) to complete a proof, but has several ways to control the prover or direct the proof. Command modifiers can be used to change the effect of the reduction commands. The reduction commands can be applied to a subpredicate or subexpression rather than the entire goal. Definitions and theorems can be declared *disabled* so that they will not be used automatically (but can still be used manually). Large goals can sometimes be split into independent cases which can then be worked on separately.

Our view is that a proof can be a collaboration between the user and the system, with the user providing guidance on the line of proof (by breaking the proof into a series of lemmas, and in individual proofs by directing the proof) and the system looking after bookkeeping details, eliminating trivial goals and subgoals, and guarding against logical errors.

Here is a simple example of a collaborative proof of the fact

$$\forall a, b : \mathbb{Z} \mid a \in \mathbb{N} \bullet a \mathbin{..} b \subseteq \mathbb{N}.$$

The proof begins with a `try` command that establishes the goal:

```
=> try \forall a,b: \num | a \in \nat @
        a \upto b \subseteq \nat ;
```

Z/EVES automatically applies the rule of generalization, removing the universal quantification and giving the goal

$$a \in \mathbb{Z} \wedge b \in \mathbb{Z} \wedge a \in \mathbb{N} \Rightarrow a \mathbin{..} b \subseteq \mathbb{N}.$$

The `rewrite` command results in the new goal

$$a \in \mathbb{Z} \wedge b \in \mathbb{Z} \wedge a \geq 0 \Rightarrow a \mathbin{..} b \in \mathbb{P}\mathbb{N},$$

where some small changes have been made, but the proof is far from complete. Z/EVES needs some guidance. The key to this proof is to use the definition of subset and to consider elements of the two sets involved. (Actually, at this stage "subset" has been eliminated and "in powerset of" appears instead.[2] This does not matter; the proof plan still works.) The Z/EVES reduction commands do not use this definition (as it was explicitly marked "disabled" when it was introduced). The user can, however, direct Z/EVES to apply this axiom:

```
=> apply inPower;
```

resulting in the new goal

$$a \in \mathbb{Z} \wedge b \in \mathbb{Z} \wedge a \geq 0 \Rightarrow (\forall e : a \mathbin{..} b \bullet e \in \mathbb{N}).$$

Prenexing removes the quantifier (and converts $e : a \mathbin{..} b$ into $e \in a \mathbin{..} b$), resulting on the goal

$$a \in \mathbb{Z} \wedge b \in \mathbb{Z} \wedge a \geq 0 \wedge e \in a \mathbin{..} b \Rightarrow e \in \mathbb{N}.$$

A `rewrite` command completes the proof, by rewriting $e \in a \mathbin{..} b$ to $a \leq e \leq b$, rewriting $e \in \mathbb{N}$ to $e \geq 0$, and applying the decision procedure for arithmetic.

---

[2] The rules of the Toolkit use "$S \in \mathbb{P} \, T$" as the preferred way of stating that S is a subset of T. The more natural looking "$S \subseteq T$" is in fact a ternary relation between S, T, and the (usually implicit) generic actual parameter for $\_ \subseteq \_$. This ternary relation is less easy to use in proofs.

4 Analysis of Specifications using Z/EVES

In the introduction, we mentioned five ways that Z/EVES can be used to analyse Z specifications. The first, syntax and type checking, is straightforward and will not be described further. In the following sections, we will give examples of the remaining types of analysis.

4.1 Domain Checking

For each paragraph in a specification, Z/EVES derives a *domain condition*. This domain condition is a predicate asserting that all the expressions appearing in the paragraph are meaningful. There are two ways an expression can fail to have a meaning: first, by applying a partial function to an element not in its domain (e.g., $1 \operatorname{div} 0$ or $\#\mathbb{N}$), and second, by using an improper definite description (e.g., $\mu x : \mathbb{N} \mid x \neq x$ for which there are too few possible xs, or $\mu x : \mathbb{Z} \bullet x > 0$, for which there are too many).

If a paragraph has a nontrivial domain condition, Z/EVES generates a "domain checking" conjecture for it. If this conjecture can be proved, then the paragraph has a well-defined meaning in the following sense: the proposed ISO standard for Z only partly specifies the semantics of expressions, leaving open the meanings of expressions failing the domain checking conditions. When the domain checking condition holds, then the standard defines a unique meaning for the paragraph.[3]

Consider the following specification, which describes a simple personnel database that records a set of employees, and for each employee records their boss and their salary. There are two constraints: each employee has a salary, and an employee's salary is less than their boss' salary.

$[PERSON]$

```
__ Personnel _____
  employees : ℙ PERSON
  boss_of : PERSON ↛ PERSON
  salary : PERSON ↛ ℕ
 _____
  dom salary = employees

  ∀ e : employees • salary(e) < salary(boss_of e)
_____
```

---

[3] We have sketched an argument for this assertion, but do not have a fully detailed proof. As the proposed standard is still in flux, we are waiting for it to stabilize before completing the detailed proof.

A non-trivial domain checking condition is generated for Schema *Personnel*:

$employees \in \mathbb{P} \ PERSON$
$\wedge \ boss\_of \in PERSON \nrightarrow PERSON$
$\wedge \ salary \in PERSON \nrightarrow \mathbb{N}$
$\wedge \ \text{dom} \ salary = employees$
$\wedge \ e \in employees$
$\Rightarrow e \in \text{dom} \ salary$
$\quad \wedge \ e \in \text{dom} \ boss\_of$
$\quad \wedge \ boss\_of \ e \in \text{dom} \ salary)$

The three conjuncts of the conclusion of this formula correspond to the three function applications in the last predicate of the schema.

We can simplify the condition somewhat by using a `rewrite` or `reduce` command; here we will rewrite, which produces the formula

$employees \in \mathbb{P} \ PERSON$
$\wedge \ boss\_of \in PERSON \nrightarrow PERSON$
$\wedge \ salary \in PERSON \nrightarrow \mathbb{N}$
$\wedge \ \text{dom} \ salary = employees$
$\wedge \ e \in employees$
$\Rightarrow e \in \text{dom} \ boss\_of$
$\quad \wedge \ boss\_of \ e \in \text{dom} \ salary$

Z/EVES has not made much progress, and for a good reason: this goal is not a theorem. The condition shows that we have forgotten some conditions in the schema. Both $e \in \text{dom} \ boss\_of$ and $boss\_of \ e \in \text{dom} \ salary$ concern the well-definedness of the final condition in the schema, $\forall e : employees \bullet salary(e) < salary(boss\_of \ e)$. Firstly, not every employee has a boss (in particular, the president would not). So, in the quantification, e should be specified to range only over those employees with bosses. Secondly, $boss\_of \ e$ might not be in the domain of *salary* (which we know is the set of employees). Indeed, we have forgotten to specify that the range of *boss_of* includes only employees.

As this example shows, domain conditions often fail when a specifier has been sloppy. The domain checking analysis therefore provides a useful error screen that goes beyond type checking to uncover semantic errors in a specification.

We can correct the errors in the *Personnel* schema and try again:

Personnel

$employees : \mathbb{P} \ PERSON$
$boss\_of : PERSON \nrightarrow PERSON$
$salary : PERSON \nrightarrow \mathbb{N}$

$\text{dom} \ salary = employees$
$\text{dom} \ boss\_of \subseteq employees$
$\text{ran} \ boss\_of \subseteq employees$

$\forall e : \text{dom} \ boss\_of \bullet salary(e) < salary(boss\_of \ e)$

After rewriting, the domain condition is

$$
\begin{aligned}
& employees \in \mathbb{P}\, PERSON \\
& \land\ boss\_of \in PERSON \nrightarrow PERSON \\
& \land\ salary \in PERSON \nrightarrow \mathbb{N} \\
& \land\ \mathrm{dom}\ salary = employees \\
& \land\ \mathrm{dom}\ boss\_of \in \mathbb{P}\, employees \\
& \land\ \mathrm{ran}\ boss\_of \in \mathbb{P}\, employees \\
& \land\ e \in \mathrm{dom}\ boss\_of \\
& \Rightarrow boss\_of\ e \in \mathrm{dom}\ salary
\end{aligned}
$$

Z/EVES was not able to complete the proof automatically and some user direction is required. The conclusion of this goal is a simple consequence of the fact that a function application has a value in the range of the function. This fact appears in the Z/EVES Toolkit as[4]

Theorem applyInRanPfun $[X, Y]$:
$\forall f : X \nrightarrow Y \bullet \forall a : \mathrm{dom}\, f \bullet f(a) \in \mathrm{ran}\, f \land f(a) \in Y$

The **use** command can be applied to add an instance of this theorem as a hypothesis to the goal:

```
=> use applyInRanPfun @
        f == boss\_of, X == PERSON, Y == PERSON, a == e;
```

which adds the hypothesis

$$
\begin{aligned}
& boss\_of \in PERSON \nrightarrow PERSON \land e \in \mathrm{dom}\ boss\_of \\
& \Rightarrow boss\_of(e) \in \mathrm{ran}\ boss\_of \land boss\_of(e) \in PERSON
\end{aligned}
$$

Now rewriting gives the result *true*, and we have shown that there are no domain errors in the schema.

Here is a second example, derived from the Library example in Potter, Sinclair, and Till's book [7]:

$[Copy, Book, Reader]$

$\mid\ maxloans : \mathbb{N}$

A simple definition of the library state schema takes the following form:

---

[4] We hope to write this fact in such a way that Z/EVES will be able to use it automatically, in some future version of the system. The main obstacle is the generic parameters of function ran, which must be guessed. Some changes to our translation may be needed to overcome this difficulty.

```
┌─ Library ─────────────────────────────────────────────
│ stock : Copy ⇸ Book
│ issued : Copy ⇸ Reader
│ shelved : ℙ Copy
│ readers : ℙ Reader
├───────────────────────────────────────────────────────
│ shelved = dom stock \ dom issued
│
│ ran issued ⊆ readers
│
│ dom issued ⊆ dom stock
│
│ ∀ r : readers • #(issued ▷ {r}) ≤ maxloans
└───────────────────────────────────────────────────────
```

The conclusion of the domain condition for this schema can be transformed by rewriting to $issued \triangleright \{r\} \in \mathbb{F}(Copy \times Reader)$—that is, $issued \triangleright \{r\}$ must be finite (so that its size can be determined). This condition is not implied by the schema. This error is easily repaired by declaring $issued : Copy \nrightarrow\!\!\!\!\rightarrow Reader$, so that only finitely many books can be on loan. With this change, a single **rewrite** command completes the proof of the domain condition.

The Z/EVES approach to domain errors is conservative; different Z users adopt different conventions. One common convention is that any atomic formula containing an expression with a domain error is simply false. This works fine for the library example above, but can lead to surprises in other specifications—certainly in the employee database example it gives the wrong result. The draft Z Standard leaves the meaning of expressions containing domain errors unspecified; our conservative approach therefore seems to be the safest treatment.

We have applied Z/EVES to a number of industrial specifications or published examples, and in most cases find unsatisfied domain checking conditions. These are not always errors, as noted above, but a surprising number of real errors are found.

4.2   Schema Expansion

The schema calculus, and schema inclusion, allow for very compact descriptions of systems. Sometimes, though, these descriptions are rather too compact. Z/EVES can be used to expand schemas and simplify the result, showing the full meaning of a schema. For example, continuing the library example, we might define the class of loan transactions (including loans and returns)

```
┌─ LoanTransaction ─────────────────────────────────────
│ ΔLibrary
├───────────────────────────────────────────────────────
│ readers' = readers
│ stock' = stock
└───────────────────────────────────────────────────────
```

and an operation for making a loan:

```
┌─ Issue ────────────────────────────────────────────────
│ LoanTransaction
│ c? : Copy
│ r? : Reader
├─────────────────────────────────────────────────────────
│ c? ∈ shelved
│ r? ∈ readers
│
│ issued' = issued ⊕ {c? ↦ r?}
└─────────────────────────────────────────────────────────
```

We can use Z/EVES to expand this schema by starting with it as a goal (thus, strictly speaking, we are simplifying the predicate associated with the schema). By using the **reduce** command, or a series of **invoke** commands, we can arrive at the formula

$$
\begin{aligned}
&stock \in Copy \twoheadrightarrow Book \\
\wedge\ &issued \in Copy \twoheadrightarrow Reader \\
\wedge\ &shelved \in \mathbb{P}\ Copy \\
\wedge\ &readers \in \mathbb{P}\ Reader \\
\wedge\ &shelved = \operatorname{dom} stock \setminus \operatorname{dom} issued \\
\wedge\ &\operatorname{ran} issued \subseteq readers \\
\wedge\ &\operatorname{dom} issued \subseteq \operatorname{dom} stock \\
\wedge\ &(\forall r : readers \bullet \#(issued \rhd \{r\}) \leq maxloans) \\
\wedge\ &stock' \in Copy \twoheadrightarrow Book \\
\wedge\ &issued' \in Copy \twoheadrightarrow Reader \\
\wedge\ &shelved' \in \mathbb{P}\ Copy \\
\wedge\ &readers' \in \mathbb{P}\ Reader \\
\wedge\ &shelved' = \operatorname{dom} stock' \setminus \operatorname{dom} issued' \\
\wedge\ &\operatorname{ran} issued' \subseteq readers' \\
\wedge\ &\operatorname{dom} issued' \subseteq \operatorname{dom} stock' \\
\wedge\ &(\forall r\_0 : readers' \bullet \#(issued' \rhd \{r\_0\}) \leq maxloans) \\
\wedge\ &readers' = readers \\
\wedge\ &stock' = stock \\
\wedge\ &c? \in Copy \\
\wedge\ &r? \in Reader \\
\wedge\ &c? \in shelved \\
\wedge\ &r? \in readers \\
\wedge\ &issued' = issued \oplus \{c? \mapsto r?\}
\end{aligned}
$$

Schema expansion is also effective when schemas have been defined using schema calculus.

4.3 Precondition Calculation

Z/EVES can be used to calculate the precondition of a schema. Given a state schema S and an operation declared by the schema

$$
\begin{array}{|l}
\underline{OP}\\
\quad S\\
\quad S'\\
\quad in? : X\\
\quad out! : Y\\
\hline
\quad condition
\end{array}
$$

the precondition is defined to be the formula $\exists\,S';\ out! : Y \bullet Op$, which guarantees that there is a possible output and final state. It is most convenient to work with the formula $S \wedge in? \in X \Rightarrow \exists\,S';\ out! : Y \bullet Op$, which includes the reasonable assumption that the initial state and input are suitable. This formula gives the "missing" part of the precondition (and, if the operation is total, is equivalent to *true*). Z/EVES can be used to simplify this predicate.

For example, a City Hall marriage registry might contain a record of all married couples:

$[Man, Woman]$

The relation *wife_of* records this information, it is a partial (not all men or women are married) injection (bigamy is illegal) from men to women.

$$
\begin{array}{|l}
\underline{Registry}\\
\quad wife\_of : Man \rightarrowtail Woman
\end{array}
$$

The *Wed* operation adds new couple to the registry:

$$
\begin{array}{|l}
\underline{Wed}\\
\quad \Delta Registry\\
\quad bride? : Woman\\
\quad groom? : Man\\
\hline
\quad wife\_of' = wife\_of \cup \{groom? \mapsto bride?\}
\end{array}
$$

Obviously, this is not a total operation. We can explore the precondition by working on the formula

$$Registry \wedge bride? \in Woman \wedge groom? \in Man \Rightarrow \text{pre } Wed.$$

Through reduction and rewriting,[5] we ultimately arrive at the formula

$wife\_of \in Man \rightarrowtail Woman \wedge bride? \in Woman \wedge groom? \in Man$
$\Rightarrow (\text{if } groom? \in \text{dom } wife\_of$
$\quad \text{then } wife\_of(groom?) = bride?$
$\qquad \wedge\ (bride? \in \text{ran } wife\_of \Rightarrow wife\_of^{\sim}(bride?) = groom?)$
$\quad \text{else } bride? \in \text{ran } wife\_of \Rightarrow wife\_of^{\sim}(bride?) = groom?)$

---

[5] Actually, a single **reduce** command should produce this formula. We expect it to do so in the next release of the system, as the situation that prevents the reduction from succeeding is now understood.

As can be seen, if either the bride or groom were already married, the two must already be a couple.

These precondition calculations work well when the operation schema uses explicit equalities to define the values of the components of the result state and outputs, as the one-point rule can then be applied to eliminate the existential quantifier used in the precondition. In other cases, more user direction is needed, and suitable result values need to be supplied by hand.

The satisfiability of initialization schemas can be shown in a similar fashion. For example, given the schema

---
InitialRegistry

$Registry'$

$wife\_of' = \emptyset$

---

we can work on the formula $\exists\, Registry' \bullet InitialRegistry$ to try to show that an initial state is possible. After reducing, Z/EVES produces the formula *true*— *Registry* and *InitialRegistry* were expanded, the one-point rule was applied, and some rewrite rules of the Toolkit were applied.

4.4 Theorem Proving

Specifications can be validated by *challenge theorems*. These are facts that should be consequences of the specification, but are not stated explicitly. Z/EVES allows for the statement of such theorems, and can be used to try to prove them.

Here is a simple example for the registry. We have decided to make the *Wed* operation applicable only for couples who are not already married:

Wed

$\Delta Registry$
$bride? : Woman$
$groom? : Man$

$bride? \notin \operatorname{ran} wife\_of$
$groom? \notin \operatorname{dom} wife\_of$
$wife\_of' = wife\_of \cup \{groom? \mapsto bride?\}$

Obviously, after a wedding, the bride should be recorded as the wife of the groom. We can check that by proving $\forall\, Wed \bullet wife\_of'(groom?) = bride?$. A sequence of three proof commands can be used to complete the proof:[6] **reduce** (expands the definition of *Wed*), **equality substitute** (uses the equation defining *wife_of'*), and **rewrite** (applies rewrite rules and simplification).

---

[6] This sequence, and some closely related sequences using prenexing, are commonly needed in Z/EVES, and a new proof command that does all the steps automatically will likely be added in the next version of the system.

5 Future Work

The development of Z/EVES is not complete, but the current version has been released so that we can see how effective the system is when applied in industrial and academic settings. Our experience is still limited, although we have been able to process a small number of realistic specifications. We have been surprised at the number of domain errors we identify in these specifications.

We should be able to use Z/EVES to prove refinement steps, as the proof obligations are expressible in Z. We have not yet experimented in this area, but plan to do so.

As we experiment with different specifications and different types of analysis, we are discovering some ways in which EVES is not perfectly adapted to the Z style. We are trying to improve the applicability of the system by adjusting the translation and some internal EVES settings.

The continuing development of Z/EVES will improve the toolkit and improve the power, usability, documentation, and interface of the system.

Z/EVES is freely available (but a license must be signed). Details can be found at URL http://www.ora.on.ca/z-eves/, or by sending electronic mail to eves@ora.on.ca.

6 Acknowledgements

The Z interface for EVES was developed with the sponsorship of the United States Department of Defense, by Irwin Meisels and the author. EVES was developed with the sponsorship of the Canadian Department of National Defence by Dan Craigen, Sentot Kromodimoeljo, Irwin Meisels, Bill Pase, and the author.

References

1. J. P. Bowen and M. J. C. Gordon. Z and HOL. In Bowen and Hall (eds.) *Z Users Workshop*, Springer Verlag Workshops in Computing, 1994.
2. D. Craigen, S. Kromodimoeljo, I. Meisels, W. Pase and M. Saaltink. EVES: An Overview. In *Proceedings of VDM '91 (Formal Software Development Methods)*, Noordwijkerhout, The Netherlands (October 1991), Lecture Notes in Computer Science 551, Springer Verlag, Berlin, 1991.
3. Kolyang, T. Santen, and B. Wolff. A Structure Preserving Encoding of Z in Isabelle/HOL. In J. von Wright, J. Grundy, and J. Harrison (eds.), *Theorem Proving in Higher Order Logics — 9th International Conference*, Lecture Notes in Computer Science 1125, Springer Verlag, 1996.
4. Sentot Kromodimoeljo, Bill Pase, Mark Saaltink, Dan Craigen and Irwin Meisels. The EVES System. In *Functional Programming, Concurrency, Simulation and Automated Reasoning*, Lecture Notes in Computer Science 693, Springer-Verlag, Berlin, 1993.
5. Irwin Meisels and Mark Saaltink. *The Z/EVES Reference Manual*. ORA Canada Technical Report TR-96-5493-03b, 1996.

6. Greg Nelson and Derek C. Oppen. Simplification by Cooperating Decision Procedures. *ACM Transactions on Programming Languages and Systems*, Vol. 1, No. 2, October 1979, 245–257.

7. Ben Potter, Jane Sinclair, and David Till. *An Introduction to Formal Specification and Z*. Prentice Hall, 1991.

8. J. M. Spivey. *The Z Notation: A Reference Manual, Second Edition*. Prentice Hall, 1992.

9. J. M. Spivey. *The fuzz Manual, Second Edition*. J. M. Spivey Computing Science Consultancy, May 1993.

Applications I

Taking Z Seriously

Anthony Hall

Praxis Critical Systems, UK

Extended Abstract

Z is in trouble. It is true that there is a flourishing annual meeting, a welter of books, a standardisation effort and lots of academic teaching and research interest. It is also true that there is pathetically little use of Z in industry. Even the famous success stories and Queens Awards are not followed by enthusiastic take up in the organisations responsible.

There is no chance of changing this situation until we take Z more seriously. We need to have more respect for what Z offers and we need to think seriously about all the things it does not offer. We need to have serious reasons for using Z and not something else. We need to recognize that if it is to be of serious use, it must be integrated into a complete development lifecycle and we need to provide tools for serious engineers. We need to teach it as a serious part of software engineering, not an interesting academic option.

If systems developers are to use Z, they will want to know what benefits it brings them. Saying, for example, that formal methods allow proofs of correctness is not a serious answer. Saying that formal methods save money is. But in order to understand why Z saves money, and whether it saves more or less money than, say, VDM, we have to understand just what it is that Z brings to development that other methods do not. I suggest that the most important characteristic of Z, which singles it out from every other formal method, is that it is completely independent of any idea of computation. In my opinion the key difference between Z and VDM, for example, is not two versus three valued logic or abstruse proof rules: it is, as Tony Hoare pointed out many years ago, that Z offers you the operator combinators AND and NOT.

This seemingly small difference means that you can use logic (otherwise known as ordinary language) to define your requirements: "I want the system to update the screens AND secure the data"; "I want the system NOT to disclose information to unauthorised users". And this means that Z is most useful for writing system requirements and system specifications. On the other hand, it makes it quite unsuitable for design. It may also lead to difficulty in modularising proofs.

If, then, we are to take Z seriously we should use it where these properties are important to us. We should use Z where the thing that is going to cost us money is getting the requirements or the system specification wrong–which is, fortunately for Z, one of the most common problems developers face. It is not only the potential users of Z who need to take it seriously. Those who would standardise it, and researchers who want to extend it, must do so too. Standardisation should recognize that the primary purpose of Z is to make comprehensible, unambiguous specifications, and all their efforts should be devoted to that end; not, for example, to simplifying its proof rules.

Sadly, one of the most popular ways in which Z has been extended (and I, alas, am one of the guilty men) is to bow to the current fashion for object orientation. I now believe this is a profound mistake: object orientation is essentially a design and implementation technique, and to make an object-oriented Z is to fail to take seriously the very fact that Z relieves us of such considerations. In contrast, object-oriented methods like Fusion, which have taken on board the real lessons of formal methods while avoiding the problems introduced by unfamiliar mathematical notation, are closer to the spirit of Z than some object-oriented versions of Z itself.

I do not wish to imply that Z is perfect and does not need to evolve. On the contrary, even in the area where it is strongest, system specification, Z needs to have some serious work done. The "established strategy" for using Z assumes that the specificand is an abstract data type. However, real systems are much more complicated than that. For example, in a well-known book on Z there is an example of a telephone exchange which contains one operation, *connect*, which is initiated by the system, not by the user. Now the meaning of this operation specification–for example the significance of its precondition, the rule we should use to refine it–is quite different from the meaning of a user-invoked operation. So we need to have more strategies, and possibly a richer language of operations, to distinguish different kinds of operations of the system.

When we step back from the system specification to the original requirements, there is even more work to do. We must, as Jackson has pointed out, distinguish statements about the system from statements about the environment. Furthermore we must distinguish statements of fact from statements of need, for they play completely different roles in the development. Again ,we need to enhance either the language of Z or our ways of using it to capture these distinctions.

Going the other way, towards design, we have to recognize that Z is simply not the right language to use. Design is essentially programming: saying how the system is going to work. Furthermore all real systems are implemented using concurrency. So we need to go from our Z specification, through some computational model of the specificand (such as an action system) to a refinement in, for example, a concurrent version of the refinement calculus. Of course we may find ourselves re-introducing Z or Z- like specifications of modules in the design, but there is more to a design than specifying its components.

All this means that Z is certainly not going to be our only development notation, and Z therefore needs to be integrated with other methods. Here too, I am afraid that there have been some approaches which have not taken seriously the real benefits of Z. For example using Z for low-level specifications within a deeply nested structured model is unlikely to provide a clear and unambiguous specification of the behaviour of the system as a whole. However, some integration is most certainly needed, for example with scenarios for requirements expression, with user interface definitions for system specification and with design notations for development.

One of the best ways of not taking Z seriously is to fail to integrate it into the development, by setting up a separate "formal specification" team; I fear such teams will inevitably be on the sidelines of the "real" development work.

Integration of any sort implies that the tools used for Z must interwork with the tools used elsewhere in the development. Another excellent way of not being taken seriously is to tell a company whose documentation is all in a PC word processor that for just one part of their development they must all learn LaTeX.

I believe that taking Z seriously needs to start with education and training. It is important that Z should be taught, if it is taught at all, as an integral part of a software engineering education, not just an interesting advanced option for the more mathematically curious. It means, above all, teaching not just how to write Z specifications but when and why to write them.

In summary, if Z is to flourish we must take it seriously and that means understanding its strengths, not trying to use it where it is weak, but developing it and using it in ways that give real benefit.

A Formal OO Method Inspired by Fusion and Object-Z

Klaus Achatz and Wolfram Schulte

Department of Computer Science, University of Ulm, 89069 Ulm, Germany
E-mail: {achatz,wolfram}@informatik.uni-ulm.de, Tel: +49 731 502-4166, Fax: -4162

Abstract. We present a new formal OO method, called $\mathcal{F}o\chi$, which is a synergetic combination of the semi-formal Fusion method and the formal specification language Object-Z. To manage complexity and to foster separation of concerns, $\mathcal{F}o\chi$ distinguishes between analysis and design. In each phase structure and behaviour specifications are developed step-by-step. The specifications may be graphical or textual. We give proof obligations to guarantee that the developed models are formally consistent and complete, and that the resulting system conforms to the original specification. By walking through a simple example – a graph editor – we illustrate the application of $\mathcal{F}o\chi$.

1 The Need for a Formal OO Method

Semi-formal OOA/D methods, such as Booch's object-oriented design [1] or Rumbaugh's OMT [20], are widely accepted in practice. Generally, they have a planned procedure; that is, step-by-step the software developer can approach a specific goal. This fact, combined with the support for creating visual appealing, abstract models of required structure and behaviour, makes them so attractive. However, due to their semi-formal nature, the notations of these methods lack a firm semantic basis. Thus, it is difficult to reason about the contents of the produced documents: their formal consistency, completeness, and conformance.

For high-integrity systems this lack of imprecision is unacceptable. Such systems need a way to verify the development. Formal methods add the mathematical rigour to the development of these systems. However, their take-up in the real world is still relatively small for several reasons. First, users of formal methods must have a sound mathematical background. Second, formal methods often lack structuring tools for program development in the large. Finally, formal methods often lack a detailed process.

The combination of semi-formal OOA/D methods with formal methods has been suggested as a good solution to overcome the aforementioned problems. Lano [15] and Ruiz-Delgado and associates [19] discuss the benefits of this combination at length.

The work done by various groups in this new area can be classified as either incorporating object-oriented features (i.e., objects, classes, and inheritance) into a formal method, or as integrating a formal language into a semi-formal OO method. Lano and Haughton [16] give a collection of case studies of the former approach, noteworthy examples of the latter approach are [5,8,12,14]. However, both approaches are not satisfying: Incorporating OO concepts in a formal method does not add the necessary process to develop OO software. Integrating a formal notation into an semi-formal OO method is not adequate because often neither the formal language fits the development paradigm nor the method is adapted to the new potentials for automation or verification, which formal specifications offer. We need a new synergetic combination of both methods with equal rights.

The purpose of this paper is to introduce the main features of our new formal object-oriented method $\mathcal{F}$o$\mathcal{X}$. We show which models to build, which steps to take, and which notation to use. Fusion supplies the inspiration of the process and the graphical notation, as formal notation we use Object-Z.

The principle contribution of $\mathcal{F}$o$\mathcal{X}$ is its process to develop good formal specifications. In particular:

- To foster separation of concerns, $\mathcal{F}$o$\mathcal{X}$ distinguishes analysis and design. It presents criteria when to stop the analysis, and what to do in the design.
- To master complexity, $\mathcal{F}$o$\mathcal{X}$ models state and behaviour in complementary views.
- To show the system's static and dynamic architecture, $\mathcal{F}$o$\mathcal{X}$ supports visual appealing graphical models as well as textual specifications in Object-Z.
- To verify the developed system, $\mathcal{F}$o$\mathcal{X}$ defines Object-Z proof obligations for all deliverables.

The rest of this section sketches the proposed method and gives an overview of Object-Z. We assume the reader to be familiar with a semi-formal OO method and Z.

1.1 $\mathcal{F}$o$\mathcal{X}$: Overview of the Method

$\mathcal{F}$o$\mathcal{X}$ is inspired by Fusion, a second generation method advocated by Hewlett Packard [7,17]. $\mathcal{F}$o$\mathcal{X}$ is a step-by-step process, which guides a development team from an initial requirements document to the verified design of an OO software system. The method distinguishes analysis, which produces a declarative specification of what the system does, and design, which produces an abstract OO model of how the system realizes the required behaviour. The reuse phase, which considers what classes should be specialized or generalized, and the implementation, which encodes the design in a programming language, is currently not part of the method. Figure 1 gives an overview of the $\mathcal{F}$o$\mathcal{X}$ process.

Analysis. Starting from a requirements document, the analysis process produces two models, which specify how the system interacts with its environment. An expressive entity-relationship diagram, called analysis class diagram, captures the *state* of the problem domain and the boundary of the system. The translation into Object-Z defines the semantics of this model.

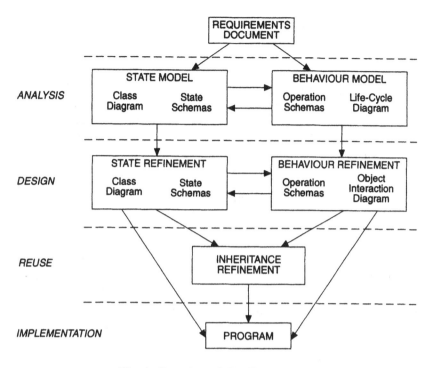

Fig. 1. Overview of the $\mathcal{F}$o$\mathcal{X}$ process

Object-Z operation schemas working on the system's state describe the *behaviour* of the system operations. A variant of statecharts defines the allowable sequences of system operations, that is the system's life-cycle. After the construction of the analysis models, the developer must validate them against the requirements and must check them for formal completeness and consistency.

Design. The design process uses the analysis models and produces two further models describing how the state and the behaviour is refined. A set of refined domain, controller, interface, and toolkit classes, expressed graphically in the design class diagram, represents the *state*.

The interaction of individual objects, specified using operation schemas supplemented by object interaction diagrams, realizes the required *behaviour*. After construction of the design models, the user must check whether the design conforms to the analysis.

$\mathcal{F}$o$\mathcal{X}$ adopts the principal process as well as several of its notations from Fusion. However, it differs from Fusion in the following respects. First, we extend Fusion's notations and give them a transformational semantics. Second, due to the use of Object-Z, we can prove internal completeness, consistency, and conformance of $\mathcal{F}$o$\mathcal{X}$'s models. Third, we change the design process and its notations: Whereas Fusion's process is driven by the development of object-interactions,

$\mathcal{F}o\mathcal{X}$ considers state *and* behaviour refinement as two interacting steps with equal rights.

The basic notations and process steps of $\mathcal{F}o\mathcal{X}$ can be illustrated most effectively by walking through a simple example. We will consider a small drawing application, which allows the user to draw lines and boxes, and to move them. This example is necessarily simple. The point is to illustrate each step and the modelling notation of the $\mathcal{F}o\mathcal{X}$ analysis and design phase. Section 2 and 3 presents the development of this example in depth.

1.2 Object-Z: Extensions from Z

$\mathcal{F}o\mathcal{X}$ uses as its formal notation Object-Z [10,11]. The major extension from Z [23] to Object-Z is the class schema, which encapsulates a single state schema and all operations that may affect its variables. The class schema is not only a syntactic extension, but also defines a type whose instances are object-references.

The specification of an object-oriented system consists of a number of named classes, which might stand in multiple inheritance relation.

In this paper, the basic structure of a class is as follows:

ClassName[generic parameters] ——————————————————
inherited class designators
type and constant definitions

  primary variables
  Δ
  derived variables

  state invariant

initial state schema
operation schemas

The local type and constant definitions have the same syntax as in Z. The nameless state schema, in the following called *STATE*, defines the attributes of the class and the class invariant. Object-Z supports derived variables, which are separated by a Δ from primary variables. The initial state schema is named *INIT* and has only a predicate part.

Both constants and variables can be object-references. They are declared either as a reference to a member of a particular class (e.g., $o : C$), or to a member of a collection of classes belonging to a particular inheritance hierarchy (e.g., $o :\downarrow C$). Object-Z uses the dot notation to reference constants, variables, and operations: $o.x$ denotes the constant or variable x of the object referenced by o; $o.Op$ applies the operation Op to the object o.

The operations are defined either as operation schemas or operation expressions, the latter using Object-Z schema operators. An operation schema extends the notion of a Z schema by adding a Δ-list: a list of variables that the operation may change; all other variables remain unchanged.

When a class is inherited, its local types, constants, and variables are available in the inheriting class. Any types and constants, having the same name, must have compatible definitions. The inherited class' state schema and initialization are conjoined implicitly in the inheriting class. Likewise, the operations are available implicitly, except when 'redefined' in the inheriting class. In this case, the inherited operation is conjoined with that of the inheriting class.

Logic. Object-Z's logic is based on $\mathcal{W}$, a Gentzen style sequent calculus proposed for Z [24]. Axioms and theorems are expressed using sequents. Further theorems are derived by the application of inference rules. Side-conditions on the inference rules are given using meta-functions, which return information from the specification text.

In opposite to Z, Object-Z consists not only of a global environment but also of several local environments, one for each class. To reason about properties within these local environments, Smith [22] extends the notion of the logic so that sequents can be interpreted within a particular class context as follows (A is a class):

$$A :: d \mid \Psi \vdash \Phi$$

A sequent is valid if any one of the predicates in the consequent Φ is true in all the environments enriched by the declaration d and the class A, and satisfying all the predicates in Ψ.

Because Object-Z's logic is based on the one of Z, it is apparent to adapt Z's proof obligations to Object-Z. The following sections show this adaptation.

2 Analysis

$\mathcal{F}\!o\kappa$, like most OOA/D methods, assumes that a natural language document (possibly supplemented by diagrams, forms, use cases, etc.) captures the requirements for the system to be built. Figure 2 shows the requirements document for the drawing application.

> A graph is composed of rectangular *boxes* of different sizes situated in different positions on a screen, and of *lines*. A line is *attached* to two boxes but cannot be attached to the same box twice. A set of boxes is connected if there is a path between them. Lines can optionally have a colour or be dashed – but not both.
> The user should be able to create new boxes, connect two boxes by a line, and to move a box with all its attached lines and connected boxes.

Fig. 2. A requirements document

The purpose of $\mathcal{F}o\chi$'s analysis is to develop a specification of what the system does in terms of how the system interacts with its environment. The analysis process consists of two steps.

Step 1: Develop the System State Model. The purpose of the state model, also called structural or object model, is to capture the problem domain relevant classes, their attributes, and their relationships.

Technique. You can use the Fusion or OMT technique to identify appropriate classes and associations between them. You add attributes as necessary, and improve the model by aggregation and inheritance.

Notation. In $\mathcal{F}o\chi$, you use Fusion's system object model notation, extended by invariants, a predicate on the initial state, and an explicit system class, which substitutes Fusion's system boundary. This notation is translated into Object-Z classes.

Verification. After construction of the analysis class model, you check that each class C is consistent, that is the class invariant is not contradictory. In formal terms:

$$C :: \vdash \exists \, STATE \bullet STATE \qquad (1)$$

Provided that the class C has defined an explicit initial schema $INIT$, you also have the proof obligation that an initial state exists:

$$C :: \vdash \exists \, STATE \bullet INIT \qquad (2)$$

Although we use an object-oriented approach to analysis, we must not forget that it is the functionality that users are interested in.

Step 2: Develop the System Behaviour Model. A system cooperates with active agents in its environment. Agents invoke atomic system operations, which manipulate the system's state and cause output to be generated.

Technique. You must identify a list of operations that the application can carry out. You can easily obtain this list by decomposing use cases into component operations, cf. [6], or by building scenarios of usage [7].

Next, you describe atomic system operations in Object-Z operation schemas and their allowable succession in a life-cycle diagram.

Notation. You specify the intended behaviour of atomic system operations in the system class using Object-Z schema expressions. Their description should follow the style of writing good specifications as advocated by Wordsworth [25]. You define the allowable sequencing of atomic system operations by a variant of statecharts.

Verification. After constructing the analysis behaviour model, you check that every system operation is consistent (i.e., implementable). In formal terms: There must be at least one poststate for any operation *Op* with input $x?$ and output $y!$ whose implicit precondition is satisfied:

$$System :: \vdash \exists STATE; \ x? : X \bullet \text{pre } Op \qquad (3)$$

where the implicit precondition pre *Op* is defined as the schema $\exists STATE'; \ y! : Y \bullet Op$.

After developing the life-cycle, you check that every system operation is sufficient complete; that is that the precondition of every system operation is fulfilled when the user requests the operation. Currently, we investigate the corresponding proof obligations.

Of course, verification does not help if the developed models do not capture the documented requirements. This discussion, however, is outside the scope of this paper.

The rest of this section proceeds by looking at the analysis of our example application.

2.1 Example: Developing the System State Model

Analysis normally starts by constructing lists of entities of different kinds occurring in the requirements document. A possible list may be *box, line, coloured, dashed, connected, attached*, and *user*. The next step is to go through the list, classifying each entry as a class, an attribute, a relationship, a generalization, or an aggregation. An analysis class diagram, which shows the concepts necessary for understanding the problem, using an extended entity-relationship diagram, captures these decisions.

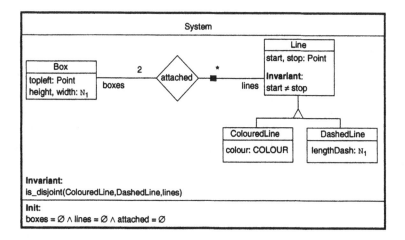

Fig. 3. Analysis class diagram for the drawing application

Figure 3 shows the analysis class diagram for the example problem. *Box* and *Line* are classes, which are related to each other by means of the association *attached*. The model specifies that each line must be attached to 2 boxes, whereas a box can be attached to multiple lines. *Box* has attributes determining its position and size, and *Line* has attributes fixing its start and stop point. *ColouredLine* and *DashedLine* are specializations of *Line*; each subclass extends its superclass by an additional attribute.

The diagram depicted in Fig. 3 differs from the usual object-model notation in three respects. First, we explicitly introduce an aggregation class *System*, which accommodates all classes and associations that contribute to the system state. *System* is the only class visible from the outside, and therefore constitutes the interface to the environment. Second, we annotate the graphical notation with invariant expressions (e.g., $start \neq stop$), constraining attributes and associations. Finally, we define the initial state of class instances. In our example, the *System*'s initial state requires that all system attributes are empty sets. The resulting models are more expressive, which supports the formalization process. (We adapted the extended notation of invariants from the Syntropy notation, developed by Cook and Daniels [8]).

The given sets and datatypes, used in the analysis class model, are *COLOUR* and *Point*, the latter being an abstract datatype with the usual numerical operations $+, -$, and so on:

$$[COLOUR, Point]$$

The extended entity-relationship diagram contains all information needed for an automatic translation into Object-Z classes. This results in:

__System__

$boxes : \mathbb{P}\, Box$ [instances of *Box*]
$lines : \mathbb{P}(\downarrow Line)$ [instances of $\downarrow Line$]
$attached : Box \leftrightarrow \downarrow Line$ [relationship between *Box* and $\downarrow Line$]

$is\_valid(attached, boxes, lines)$ [validity of association *attached*]
$\text{ran } attached = lines$ [totality constraint]
$\forall\, l : \text{ran } attached \bullet \#attached^{-1}(\!|\, \{l\}\, |\!) = 2$ [cardinality constraint]
$is\_disjoint(ColouredLine, DashedLine, lines)$ [disjoint subclasses]

$INIT \mathrel{\widehat{=}} [boxes = \varnothing \wedge lines = \varnothing \wedge attached = \varnothing]$
---

where *is_valid* asserts referential integrity. The predicate *is_disjoint* is an abbreviation for $\{l : ColouredLine \mid l \in lines\} \cap \{l : DashedLine \mid l \in lines\} = \varnothing$, which guarantees that the subtypes *ColouredLine* and *DashedLine* do not overlap. *Box* and *Line* are translated into:

```
┌─ Box ──────────────────────┐     ┌─ Line ──────────────────────┐
│ topleft : Point   [attributes]│   │ start, stop : Point          │
│ width, height : N₁           │    ├──────────────────────────────┤
└──────────────────────────────┘   │ start ≠ stop      [invariant] │
                                    └──────────────────────────────┘
```

The formalization of the specializations of *Line* reads:

```
┌─ ColouredLine ──────────────┐    ┌─ DashedLine ─────────────────┐
│ Line          [superclass]   │   │ Line          [superclass]   │
├──────────────────────────────┤   ├──────────────────────────────┤
│ colour : COLOUR              │    │ lengthDash : N₁              │
└──────────────────────────────┘   └──────────────────────────────┘
```

After constructing the state model, we show that the model is consistent (i.e., (1) and (2) hold). Let us show the consistency of the class *System*. A possible state of *System* is, for instance, defined by the binding $\langle$ *boxes* $\leadsto \varnothing$, *lines* $\leadsto \varnothing$, *attached* $\leadsto \varnothing \rangle$, which fulfills (1). Because this binding is the result of the *INIT* schema, the initialization proof obligation (2) of *System* is fulfilled, too.

2.2 Example: Development of the System Behaviour Model

The behaviour of a system determines the possible patterns of interaction between the system and its environment, and the effect those interactions have on the system's state. First, we consider the specification of single system operations. Then, we describe allowable sequences of those operations.

Operation schemas. The operation schemas define the behaviour of the system by specifying declaratively how each system operation affects the system state.

In our case study, the most interesting system operation is to move a subgraph. The operation *MoveSubGraph*, which is part of the system class, has two inputs: *box?* denotes a box of the subgraph to be moved, and *dv?* denotes the vector, by which the subgraph is moved:

```
┌─ in System ─────────────────────────────────────────────────┐
│ ┌─ MoveSubGraphInterface ─────────────────────────────────┐ │
│ │ box? : Box;  dv? : Point                      [input]   │ │
│ └─────────────────────────────────────────────────────────┘ │
└──────────────────────────────────────────────────────────────┘
```

MoveSubGraph makes use of a derived association *connected*, which is introduced as part of *System*'s state:

```
┌─ in System ──────────────────────────────────────────────────┐
│ ┌───────────────────────────────────────────────────────────┐ │
│ │ ...                                        [primary variables] │ │
│ │ Δ                                                           │ │
│ │ connected : Box ↔ Box                      [derived relation] │ │
│ │ ──────────────────────────────────────────────────────────  │ │
│ │ connected = attached ⨾ attached⁻¹                          │ │
│ └───────────────────────────────────────────────────────────┘ │
└───────────────────────────────────────────────────────────────┘
```

The association *connected* relates two boxes that are connected by a line.

The intended effect of *MoveSubGraph* is that *box?* with all its attached lines and connected boxes are moved by the vector *dv?*:

```
┌─ in System ──────────────────────────────────────────────────────┐
│ ┌─ MoveSubGraph ──────────────────────────────────────────────┐  │
│ │ MoveSubGraphInterface                                        │  │
│ │ ───────────────────────────────────────────────────────────  │  │
│ │ box? ∈ boxes                                   [precondition] │  │
│ │ let cBs == connected*(| {box?} |) ∪ {box?} in  [postcondition]│  │
│ │     let aLs == ⋃{b : cBs • attached(| {b} |)} in              │  │
│ │         ∀ b : cBs • b.SetTopleft(b.GetTopleft + dv?)          │  │
│ │         ∀ l : aLs • l.SetStartStop(l.GetStartStop + dv?)      │  │
│ └─────────────────────────────────────────────────────────────┘  │
└───────────────────────────────────────────────────────────────────┘
```

We describe the case where the input *box?* is part of *System*'s state. First, the transitive closure of connected boxes is computed, next, the associated lines are calculated, and finally, the attributes of the determined lines and boxes are adapted.

The expression $b.SetTopleft(b.GetTopleft + dv?)$ is an abbreviation of the schema expression $b.GetTopleft \bullet [topleft? : Point \mid topleft? = topleft! + dv?] \bullet b.SetTopleft$. Likewise, $l.SetStartStop(l.GetStartStop + dv?)$ is short for the schema expression $l.GetStartStop \bullet [start?, stop? : Point \mid start? = start! + dv? \land stop? = stop! + dv?] \bullet l.SetStartStop$. The operation names starting with *Get* and *Set* denote observer and modifier operations of the appropriate classes. They are defined as follows:

```
┌─ in Box ──────────────────────────────────────────────────────┐
│ GetTopleft ≙ [topleft! : Point | topleft! = topleft]           │
│ SetTopleft ≙ [Δ(topleft); topleft? : Point | topleft' = topleft?] │
└───────────────────────────────────────────────────────────────┘
```

```
┌─ in Line ─────────────────────────────────────────────────────┐
│ GetStartStop ≙ [start!, stop! : Point | start! = start ∧ stop! = stop] │
│ SetStartStop ≙ [Δ(start, stop); start?, stop? : Point |        │
│                      start' = start? ∧ stop' = stop?]          │
└───────────────────────────────────────────────────────────────┘
```

MoveSubGraph defines the normal case of the operation. Whether we have to handle mistakes in the input, that is whether we have to make the operation robust, is a question of sufficient completeness and is sketched at the end of this section.

In a similar way, we specify the other system operations *CreateBox* and *ConnectBoxes*.

After the construction of the operation schemas, we check each operation whether it is consistent (i.e., whether (3) holds). For the operation *MoveSubGraph*, we note that the implicit precondition contains the invariant of the system and the explicit precondition $box? \in boxes$, both do not contradict. Furthermore, note that *MoveSubGraph* does not change the attributes of the system, but changes the positions of boxes and lines. We make the following observations about the effect of *MoveSubGraph*. First, boxes do not have any associated invariant. Second, the invariant of lines is not affected, because the attributes *start* and *stop* are moved by the same vector. Finally, the system's attributes remain the same. From these facts we can conclude that the state after execution will satisfy the relationship specified in the body of the schema.

Life-cycle diagram. Life-cycles define the allowable sequences of system operations and events. They are used to determine the necessary preconditions that operations must have to be robust.

Life-cycles are expressed by a variant of statecharts, where following Selic and associates [21], transitions are based on the concept of request and provision of operations rather than the concept of events, as proposed originally by Harel [13].

We use formulae derived from life-cycle diagrams to prove sufficient completeness of the system operations; that is, the precondition of every system operation is fulfilled, when the user requests the operation.

This approach differs from the usual one proposed for Z (cf. [9]), where every system operation is made total using the schema calculus to define the various error schemas, which say what happens if the preconditions of system operations are not satisfied.

To shorten the presentation, we do not go further into details in this paper.

3 Design

The purpose of design is to introduce OO software structures, which satisfy the abstract definitions produced from analysis.

The transition from analysis to design is based on two important prerequisites on the system state and the system behaviour, respectively. First, every object of an analysis class C_A preserves its identity and becomes an object of a corresponding design class C_D with a possibly different state structure. In formal terms, for all analysis classes C_A and their corresponding design classes C_D, there exists a bijection $\Phi : C_A \rightarrowtail\!\!\!\rightarrow C_D$ (as a notational convention, the subscripts A and D distinguish, whether an entity belongs to analysis or design). Second, the interface of the system specified in the analysis is preserved in the design.

This means, the set of system operations, which the environment can invoke, and the outputs that the environment can receive, do not change. The design process comprises two steps.

Step 1: Refining the System State. The purpose of the design state model is to show how the system state is distributed among individual problem domain objects and new design objects. The latter may be introduced to address system requirements, such as performance, modularity, or adaptation to interfaces.

Technique. When refining the system state, you have to make five design decisions. First, you decide, which objects should provide an interface to the environment, perhaps by introducing new interface classes. Second, you allocate the responsibility for holding the state expressed by relationships to individual objects. Third, you determine how the types of the mathematical toolkit should be implemented. Fourth, you decide whether derived attributes should be recomputed over and over again, or whether their state should be stored in objects. Finally, you refine the initial state of the analysis classes.

Notation. The design class diagram shows, which design classes exist, and which static access pathes between them must be supported. A design class diagram is thus similar to an analysis class diagram, however, relationships are replaced by references labeled with role names. Design decision schemas in Object-Z make the relation between the analysis and design classes precise. These schemas (often called forward simulation [25]) explain how an analysis class is to be found from the corresponding design class by inheriting both classes and establishing invariants between them.

Verification. There are three aspects to check the state refinement. First, the state of every design class must be consistent, which is equivalent to equation (1) given in the analysis (cf. Sect. 2).

Second, the design state must conform to the analysis state. State conformance means that according to the design decision schema C_{Abs} the invariant of every design class C_D implies the invariant of its corresponding analysis class C_A. State conformance is expressed in formal terms by:

$$\frac{C_A :: \vdash STATE = AS \qquad C_D :: \vdash STATE = DS}{C_{Abs} :: \vdash STATE = \uparrow STATE \wedge Abs} \tag{4}$$

$$DS \vdash \exists AS \bullet AS \wedge Abs$$

The premise in the preceding proof obligation names the state of an analysis class C_A and its corresponding design class C_D. $\uparrow STATE$ and Abs are the inherited state and the invariant, respectively, of C_{Abs}.

Third, the initial state of every design class C_D must be correct. It is correct, when according to the design decision schema C_{Abs}, it represents an initial state of the corresponding analysis class C_A. This requirement is formally expressed by the theorem:

$$\frac{C_A :: \vdash STATE = AS \wedge INIT = AI \qquad C_D :: \vdash INIT = DI}{C_{Abs} :: \vdash STATE = {\uparrow}STATE \wedge Abs} \tag{5}$$

$$\overline{DI \vdash \exists AS \bullet AI \wedge Abs}$$

AI and DI denote the initial schema of C_A and C_D, respectively.

Step 2: Refining the System Behaviour. The purpose of the design behaviour model is to work out, how the interaction of individual objects realizes the system behaviour, specified by system operations.

Technique. For each system operation, you must identify the objects involved in the computation and how they cooperate, for example, as controller or collaborator, to realize the functionality specified in the schema for the operation. The goal is to distribute the functionality across the objects of the system.

Notation. Like Fusion, $\mathcal{F}ox$ requires the construction of object interaction diagrams to show how objects communicate. Nodes represent design objects, and arrows represent message passing. Each object interaction graph comes with Object-Z operation schemas, which are defined within the appropriate design classes to give meaning to the design operations introduced.

Verification. For every system operation, you have to guarantee the conformance of the analysis operation with its behaviour refinement. To proof this, you have to show that if the analysis operation Aop terminates, then the corresponding design operation Dop must also terminate:

$$\frac{C_A :: \vdash Op = Aop, \; C_D :: \vdash Op = Dop, \; C_{Abs} :: \vdash STATE = {\uparrow}STATE \wedge Abs}{\text{pre } Aop \wedge Abs \vdash \text{pre } Dop} \tag{6}$$

Furthermore, if both the analysis operation Aop and the corresponding design operation Dop are guaranteed to terminate, then every possible state after the design operation must be related by Abs' to a possible state after the analysis operation:

$$\frac{C_A :: \vdash STATE = AS \wedge Op = Aop \qquad C_D :: \vdash Op = Dop}{C_{Abs} :: \vdash STATE = {\uparrow}STATE \wedge Abs} \tag{7}$$

$$\overline{\text{pre } Aop \wedge Abs \wedge Dop \vdash \exists AS' \bullet Abs' \wedge Aop}$$

The rest of this section continues with refining the system state and behaviour of our drawing application.

3.1 Example: Refining the State

First, we distribute the state of the system among objects. The design class diagram captures the result of the distribution. It shows the entities necessary for object interaction. The model uses a similar form as the analysis class diagram. Figure 4 shows a design class diagram for the drawing application, where reused toolkit classes are drawn as grey boxes.

One task of the analysis class $System_A$ was to provide an interface to the environment. In the design, we introduce a new interface class $System_D$ to fulfill this task.

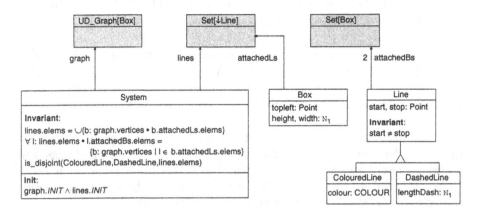

Fig. 4. Design class diagram for the drawing application

Next, we have to distribute the state, expressed by relationships, to individual objects. In our drawing example, we distribute the state that is expressed by the relationship $attached_A : Box_A \leftrightarrow\downarrow Line_A$ among the objects of Box_D and $Line_D$. Both classes obtain an object reference $attachedLs_D : Set[\downarrow Line_D]$ and $attachedBs_D : Set[Box_D]$, respectively. A cardinality constraint on $attachedBs$ preserves the requirement "a line is attached to two boxes" (cf. Fig. 2).

In $System_D$, a reference $lines_D$ to a reusable toolkit class $Set[\downarrow Line_D]$ replaces the attribute $lines_A : \mathbb{P} \downarrow Line_A$ in $System_A$.

Then, we have to decide about derived attributes, such as $connected_A$. In the drawing example, we want to store it instead of recomputing it repeatedly. Therefore, we introduce the object $graph_D$ – an instance of the reusable toolkit class UD_Graph of undirected graphs.

The first predicate of the invariant of class $System_D$ states that all $Line_D$ objects are stored as attached lines of particular Box_D objects. The second predicate guarantees that all boxes that have attached lines are stored as attached boxes of these lines. The third predicate formalizes the disjointness of $ColouredLine$ and $DashedLine$. The initialization of $System_D$ comprises the initialization of the respective containers $graph_D$ and $lines_D$.

From Fig. 4, we obtain the following specifications:

```
┌─ System ─────────────────────────────────────────────────
│ ┌──────────────────────────────────────────────────────
│ │ graph : UD_Graph[Box]
│ │ lines : Set[↓Line]
│ ├──────────────────────────────────────────────────────
│ │ lines.elems = ⋃{b : graph.vertices • b.attachedLs.elems}
│ │ ∀ l : lines.elems • l.attachedBs.elems =
│ │     {b : graph.vertices | l ∈ b.attachedLs.elems}
│ │ is_disjoint(ColouredLine, DashedLine, lines.elems)
│ └──────────────────────────────────────────────────────
│ INIT ≘ [graph.INIT ∧ lines.INIT]
└──────────────────────────────────────────────────────────
```

```
┌─ Box ──────────────────────┐   ┌─ Line ──────────────────────┐
│ ┌────────────────────────  │   │ ┌────────────────────────── │
│ │ topleft : Point          │   │ │ start, stop : Point        │
│ │ width, height : ℕ₁        │   │ │ attachedBs : Set[Box]      │
│ │ attachedLs : Set[↓Line]  │   │ ├────────────────────────── │
│ └────────────────────────  │   │ │ start ≠ stop               │
└────────────────────────────┘   │ │ #attachedBs.elems = 2      │
                                  │ └────────────────────────── │
                                  └──────────────────────────────┘
```

We assume that the toolkit classes *Set* and *UD_Graph* are given, and that they have the following state schemas and initial states:

```
┌─ Set[X] ─────────────────────────────
│ ┌──────────────────────────────────
│ │ elems : ℙ X
│ ├──────────────────────────────────
│ INIT ≘ [elems = ∅]
└──────────────────────────────────────
```

```
┌─ UD_Graph[X] ────────────────────────────────────────
│ ┌─────────────────────────────────────────────────
│ │ vertices : ℙ X
│ │ edges : X ↔ X
│ │ Δ
│ │ connected : X → ℙ X
│ ├─────────────────────────────────────────────────
│ │ is_reflexive(edges) ∧ is_valid(edges, vertices, vertices)
│ │ ∀ x : vertices • connected x = edges*⦇ {x} ⦈ ∪ {x}
│ └─────────────────────────────────────────────────
│ INIT ≘ [vertices = ∅ ∧ edges = ∅]
└───────────────────────────────────────────────────────
```

As we have already seen, the transition from analysis to design requires numerous design decisions, which must be documented in design decision schemas. The following Object-Z specification shows the decision schema for the refinement of class $System_A$:

$$
\begin{array}{|l}
\hline
\rule{0pt}{1em}System_{Abs} \\
\quad System_A \\
\quad System_D \\
\hline
\quad boxes_A = graph_D.vertices \\
\quad lines_A = lines_D.elems \\
\quad attached_A = \{b : graph_D.vertices;\ l : b.attachedLs.elems\ | \\
\qquad\qquad\qquad b.attachedLs.elems \neq \varnothing \bullet b \mapsto l\} \\
\hline
\end{array}
$$

The specialized class $System_{Abs}$ has no state itself but only the inherited state of $System_A$ and $System_D$. The invariant formalizes the preceding design decisions.

Note that the design decision schema $System_{Abs}$ denotes a functional data refinement; that is, the relation between design state and analysis state is a total function. Therefore, we can use a simpler set of conditions as proof obligations (see the subsequent verification step).

The remaining design classes $Line_D$ and Box_D are only extensions of their corresponding analysis classes. Thus, the appropriate design decision diagrams refer to projections from the design state to the analysis state.

After the state refinement, we must first check the refined classes for consistency. This task is similar to the corresponding verification in the analysis (cf. Sect. 2).

Second, we must check the conformance of every design class C_D with its corresponding analysis class C_A. This is expressed by proof obligation (4). In the drawing example, where all design decision diagrams denote total functions, we can avoid the existential quantifier in the conclusion of (4). Thus, the conclusion simplifies to $DS \wedge Abs \vdash AS$. The proof is straightforward.

Finally, we must verify that the initial state of every design class is correct. This requires the verification of (5) for every domain class. Again, we can use a simpler version of (5) by avoiding the existential quantifier in the conclusion, which then reads: $CI \wedge Abs \vdash AI$. For example, to proof the simplified version of (5) for the domain class $System_D$, we only have to show that $graph_D.vertices = \varnothing$ and $lines_D.elems = \varnothing$ with Abs implies the emptiness of the sets and relations $boxes_A$, $lines_A$, and $attached_A$, which obviously holds.

3.2 Example: Refining the Behaviour

In the second step of the design process, we work out how the interaction of individual objects realizes the system behaviour.

To show how the objects of the design communicate with each other during the computation of $MoveSubGraph$, we construct the object interaction diagram as depicted in Fig. 5.

The diagram determines the following message passing sequence:

(1) The interface object *system* checks, whether the actual parameter *box?* denotes a valid object by sending the query *IsVertice* to the server *graph*.

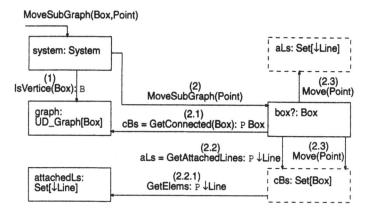

MoveSubGraph(Box,Point)

Fig. 5. Object interaction diagram for *MoveSubGraph*

(2) In case of success, *system* forwards the message *MoveSubGraph* to *box?*.

   (2.1) *box?* finds out, which boxes are connected to itself by sending a message *GetConnected* to the server object *graph*.

   (2.2) Then, all the attached lines of all connected boxes are determined by sending the message *GetAttachedLine* to all connected boxes.

      (2.2.1) Those boxes must send a message *GetElems* to the container object *attachedLs* to get its attached lines.

   (2.3) Finally, the identified lines and boxes are moved by means of the message *Move* to all connected boxes and attached lines, respectively.

Next, we define precise specifications of the operations involved in the object interaction diagram.

This results in a design operation schema for *MoveSubGraph* in the interface class *System*. Remember that all system operations of the design must have the same interface as their corresponding analysis operations:

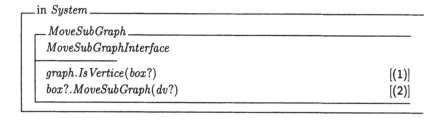

MoveSubGraph in *System* invokes a method of the same name in *Box* to move a box with all its attached lines and connected boxes by *dv?*:

```
┌─ in Box ─────────────────────────────────────────────────────────
│  ┌─ MoveSubGraph ───────────────────────────────────────────────
│  │  dv? : Point
│  ├────────────────────────────────────────────────────────────
│  │  let cBs == graph.GetConnected(self) in                    [(2.1)]
│  │     let aLs == ⋃{b : cBs • b.GetAttachedLines} in          [(2.2)]
│  │        ∀ b : cBs • b.Move(dv?) ∧ ∀ l : aLs • l.Move(dv?)   [(2.3)]
```

The method *GetAttachedLines* in *Box* returns all the attached lines of the receiver:

```
┌─ in Box ─────────────────────────────────────────────────────────
│  ┌─ GetAttachedLines ───────────────────────────────────────────
│  │  out! : ℙ(↓Line)
│  ├────────────────────────────────────────────────────────────
│  │  out! = attachedLs.GetElems                                [(2.2.1)]
```

The remaining methods *Move* in *Box* and *Line* move the target object to its new position:

```
┌─ in Box ──────────────────────        ┌─ in Line ──────────────────────
│  ┌─ Move ──────────────────          │  ┌─ Move ──────────────────────
│  │  Δ(topleft)                        │  │  Δ(start, stop)
│  │  dv? : Point                       │  │  dv? : Point
│  ├────────────────────────           │  ├──────────────────────────
│  │  topleft' = topleft + dv?          │  │  start' = start + dv?
│                                        │  │  stop' = stop + dv?
```

The preceding schemas use operations defined in the given toolkit classes *Set* and *UD_Graph*. We assume that they are specified as follows:

```
┌─ in Set[X] ──────────────────────────────────────────────────────
│  GetElems ≙ [elems! : ℙ X | elems! = elems]
```

```
┌─ in UD_Graph[X] ─────────────────────────────────────────────────
│  GetConnected ≙ [v? : X; out! : ℙ X | out! = connected(v?)]
│  IsVertice ≙ [v? : X; out! : 𝔹 | out! = v? ∈ vertices]
```

After the behaviour refinement, we must proof the conformance of each analysis system operation *Aop* with the refined operation *Dop*. This requires a number of consistency constraints to be shown. For example, to verify that $MoveSubGraph_D$

in $System_D$ conforms to $MoveSubGraph_A$ in $System_A$, we must show both (6) and (7). First, if $MoveSubGraph_A$ terminates, then $MoveSubGraph_D$ must also terminate. This is easy to show, because if $box? \in boxes_A$ then due to $boxes_A = graph_D.vertices$ in Abs it also holds that $box? \in graph_D.vertices$. This concludes the proof of (6).

Second, every possible state after $MoveSubGraph_D$ must be related to a possible state after $MoveSubGraph_A$. Since Abs denotes a total function, we can simplify (7) by avoiding the existential quantifier in the conclusion. The simpler conclusion reads: pre $Aop \land Abs \land Dop \land Abs' \vdash Aop$. The proof consists mainly of fold–unfold steps. We omit the details of the proof.

4 Conclusion and Future Work

We have presented the formal object-oriented method $\mathcal{F}ox$, which describes how to develop good formal analysis and design specifications. The underlying development method, inspired by Fusion, guides the developer through the process; its results are described using Object-Z. This makes rigorous verification possible. For every deliverable, we have given proof obligations.

Summarizing, $\mathcal{F}ox$'s analysis and design proceeds step-by-step. Each step is concluded by a formal check for consistency, completeness, or conformance.

The developed graph editor is a rather small example. We have worked out two more elaborate ones, namely an elevator controller, and an information system to distribute medical students over hospitals. These examples show that formal development is possible but rather costly.

The work presented in this paper cannot be complete. The synergy of a formal object-oriented notation — currently we adopt the notation to the future standard of the Unified Method [2] — and an object-oriented development process provides a large potential for further research: We must explore the completeness check for life-cycles in depth. As we observed, the transition from analysis to design (and its verification) is a restricted form of state and behaviour refinement. We must study this fact in more detail to provide further aid, for instance, in the form of formal design patterns for the transition and its verification. Finally, the process needs tool-support for validation and verification of the produced models.

References

1. G. Booch. *Object-Oriented Analysis and Design with Applications.* Benjamin/Cummings, 2nd edition, 1993.
2. G. Booch and J. Rumbaugh. Unified method for object-oriented development, version 0.9. Technical report, Rational Software Corporation, 1996.
3. J. P. Bowen and J. A. Hall, editors. *Z User Workshop, Cambridge 1994*, Workshops in Computing. Springer-Verlag, 1994.
4. J. P. Bowen and M. G. Hinchey, editors. *ZUM'95: The Z Formal Specification Notation, 9th International Conference of Z Users*, volume 967 of *Lecture Notes in Computer Science*, 1995.

5. J-M. Bruel, B. Chintapally, R. B. France, and G. Raghvan. FuZE–draft of the user's guide. Technical report, Department of Computer Science & Engineering, Florida Atlantic University, 1996.

6. D. Coleman. Fusion with use cases – extending Fusion for requirements modelling. Internal report, HP Labs, October 1995.

7. D. Coleman, P. Arnold, S. Bodoff, Ch. Dollin, H. Gilchrist, F. Hayes, and P. Jeremes. *Object-Oriented Development, The Fusion Method.* Prentice-Hall, 1994.

8. S. Cook and J. Daniels. *Designing Object Systems: Object-Oriented Modelling with Syntropy.* Prentice-Hall, 1994.

9. A. Diller. *An Introduction to Formal Methods.* Wiley, 2nd. edition, 1994.

10. R. Duke, P. King, G. A. Rose, and G. Smith. The Object-Z specification language: Version 1. Technical Report 91-1, Department of Computer Science, University of Queensland, St. Lucia 4072, Australia, April 1991.

11. R. Duke, G. Rose, and G. Smith. Object-Z: A specification language advocated for the description of standards. Technical Report 94-45, Department of Computer Science, Software Verification Centre, December 1994.

12. J. A. R. Hammond. Producing Z specifications from object-oriented analysis. In Bowen and Hall [3], pages 316–336.

13. D. Harel. Statecharts: A visual formalism for complex systems. *Science of Computer Programming*, 8:231–274, 1987.

14. F. Hayes and D. Coleman. Coherent models for object-oriented analysis. In *OOPSLA'91*, pages 171–183. ACM, ACM Press, 1991.

15. K. Lano. *Formal Object-Oriented Development.* Springer-Verlag, 1995.

16. K. Lano and H. Haughton, editors. *Object-Oriented Specification Case Studies.* Prentice-Hall, 1994.

17. R. Malan, R. Letsinger, and D. Coleman. *Object-Oriented Development at Work – Fusion in the Real World.* Prentice-Hall, 1995.

18. J. E. Nicholls, editor. *Z User Workshop, York 1991*, Workshops in Computing. Springer-Verlag, 1992.

19. A. Ruiz-Delgado, D. Pitt, and C. Smythe. A review of object-oriented approaches in formal methods. *The Computer Journal*, 38(10):777–784, 1996.

20. J. Rumbaugh, M. Blaha, W. Premerlani, F. Eddy, and W. Lorensen. *Object-Oriented Modeling and Design.* Prentice-Hall, 1991.

21. B. Selic, G. Gullekson, and P. T. Ward. *Real-time Object-Oriented Modeling.* Wiley, 1994.

22. G. Smith. A logic for Object-Z. Technical Report 94-48, Department of Computer Science, University of Queensland, St. Lucia 4072, Australia, December 1994.

23. J. M. Spivey. *The Z Notation, A Reference Manual.* Prentice-Hall, 2nd edition, 1992.

24. J. C. P. Woodcock and S. M. Brien. W: A logic for Z. In Nicholls [18], pages 77–96.

25. J. B. Wordsworth. *Software Development with Z: A Practical Approach to Formal Methods in Software Engineering.* Addison-Wesley, 1993.

$\mathcal{W}$ Reconstructed

Jon Hall[1] and Andrew Martin[2]

[1] High Integrity Systems Engineering Group
Department of Computer Science
The University of York
Heslington, York Y01 5DD
U.K.
[2] Software Verification Research Centre
Department of Computer Science
The University of Queensland
Brisbane, Queensland 4072
Australia

Abstract. An early version of the Z Standard included the deductive system $\mathcal{W}$ for reasoning about Z specifications. Later versions contain a different deductive system. In this paper we sketch a proof that $\mathcal{W}$ is *relatively sound* with respect to this new deductive system. We do this by demonstrating a semantic basis for a correspondence between the two systems, then showing that each of the inference rules of $\mathcal{W}$ can be simulated as derived rules in the new system. These new rules are presented as tactics over the the inference rules of the new deductive system.

1 Introduction

An important part of the Z Standardization activity has been the definition of a logical deductive system for Z. Whilst some have sought to provide support for reasoning about Z specifications by embedding the language in an existing well-understood framework (HOL, Eves, PVS, Isabelle, for example; [BG94,Jon92,Saa92,KSW96,ES94]), other research has attempted to provide support for reasoning *within* Z, making use of Z's type system and supporting directly the rather unusual use of variables (bound and free) which arise in Z schemas.

$\mathcal{W}$ [WB92] was the first published deductive system to accomplish this. It has formed the basis of a number of projects, including the first version of the Jigsa$\mathcal{W}$ tool [Mar93], the reasoning support in Cadi$\mathbb{Z}$ [TM91,TH95], a logic for Object-Z [Smi95], and the Z-in-Isabelle work [KB95]. A similar logical system is implemented in *Balzac/Zola* [Har91], and some unpublished work demonstrates that it is sound relative to $\mathcal{W}$. Draft 1.0 of the Z Standard incorporated a deductive system based on $\mathcal{W}$. Subsequent work [Bri95] has led to the inclusion in the most recent version [Nic95] (the Standard) of a slightly different logic.

This paper attempts to demonstrate that those with a considerable investment in $\mathcal{W}$ need not be left 'high and dry' by the change in the Standard. We exhibit a correspondence between the theorems of the two logical systems, by showing that the new one in the Standard is sufficiently broad as to be able to synthesize most of the inference rules of W.[1] We thus begin the demonstration that W is *relatively sound*[2] with respect to the new deductive system, i.e., that it is sound *if* the new deductive system is sound. A secondary aim is to bring an understanding of these two inference systems, and their relationship to the Standard's semantics for Z, to a wider audience.

The next section of this paper offers an overview of the two inference systems, and an account of the simple language of *tactics* which will be used to describe derived rules. Section 3 discusses the means by which we will relate the inferences of one system in the other, and Section 4 uses the methods described there to explain, in some detail, the embedding of W's predicate calculus in the new system. The following sections undertake a similar analysis for expressions and schemas, and the paper concludes with a discussion.

This paper will frequently refer to the new deductive system (as published in [Nic95, Annex F]) in contrast to W. We do not wish to be responsible for naming this logic, but will avoid cumbersome descriptions by referring to it from time to time as ν.

2 The Two Deductive Systems

2.1 W

W is a logic in the style of Gentzen (many rules; few axioms).

The judgements of the logic are *typed, multiple-conclusion* sequents.

Declarations | Predicates ⊢ Predicates

The intended interpretation of such a sequent is as a conjecture stating that when the current environment is enriched by the †-separated list of declarations, by assuming all of the predicates on the left of the 'turnstile' ⊢ (the antecedents), it is possible to prove one of those on the right (the consequents). Where either of these collections of predicates is empty, it is denoted by white space, so the simplest judgement of the logic is (⊢) (which is unprovable). Other simple judgements are (⊢ true) and (false ⊢), both of which are provable.

Inference rules are written as

$$\frac{\text{premisses}}{\text{conclusion}} \text{ rulename(proviso)} \quad \text{or} \quad \frac{\text{premisses}}{\text{conclusion}} \text{ rulename(proviso)}$$

the second form being used when the rule can be applied in both upwards and downwards directions. This has also been indicated using a single rule, and the annotation ↑↓. The proviso (possibly empty) is a decidable condition on

---

[1] This paper addresses most of the inference rules of W but is not exhaustive in its coverage.

[2] That is to say, *faithful*

the alphabets and the free variables of the declarations and predicates. Special meta-functions are provided to compute these alphabets (the alphabet of the declaration d is written αd; so $\alpha(x : X; \; y : Y) = \{x, y\}$) and free variables (the free variables of predicate p are $\phi_p p$; we have for example $\phi_p(\forall x : N \bullet x > y) = \{y\}$).

Rule-lifting permits a simple presentation of most of W's inference rules:

Theorem 1 (Rule-Lifting). *If*

$$\frac{E \dagger D' \mid \Phi' \vdash \Psi'}{E \dagger D \mid \Phi \vdash \Psi}$$

is sound, then so is

$$\frac{F \dagger E \dagger D' \mid p, \Phi' \vdash q, \Psi'}{F \dagger E \dagger D \mid p, \Phi \vdash q, \Psi}$$

provided that $(\alpha D \cup \alpha D' \cup \alpha E) \cap (\phi p \cup \phi q) = \varnothing$.

As a result, many rules can be presented in a minimalist form. The cut rule is a good example:

$$\frac{\vdash P \qquad\qquad P \vdash}{\vdash} \; cut \; .$$

The abstraction to this form will not be exploited here: when ν is used later to synthesize the rules of W, the full form of the W rules will be used. Thus, for instance, *cut* will be the rule

$$\frac{D \mid \Phi \vdash P, \Psi \qquad\qquad D \mid P, \Phi \vdash \Psi}{D \mid \Phi \vdash \Psi} \; cut \; .$$

Another meta-rule justifies the use of semantic equivalences as inference rules. This is put to good use in W, in its distinctive treatment of *substitution*. Substitution is usually a metalogical notion but it may be accomplished *within* Z through use of bindings. As an example, if the name x is to be replaced by expression e in predicate p, this may be written as the Z term $\langle x := e \rangle \odot p$. Tables of equivalences [BN+92, Annex F] permit the simplification of such expressions, so that for example we may derive the semantic equivalence ($\equiv$)

$$\langle x := e \rangle \odot (\forall y : N \bullet x < y) \equiv (\forall y : N \bullet e < y) \; .$$

As usual in the Gentzen presentation, comma and conjunction (on the left of the turnstile) are semantically equivalent. This semantic equivalence can be seen either by reference to the semantic definition for a sequent given in [BN+92], or by the bidirectional nature of the $\wedge\vdash$ rule. Similar remarks apply to disjunction on the right of the turnstile, and the $\vdash\vee$ rule. One of the ramifications of this equivalence is that any conjecture in W can be stated without the appearance of these commas. Indeed, the Standard's syntax for including conjectures in a specification document is simply

$\vdash$ Predicate .

2.2 The new deductive system

The new deductive system, ν, is similar to $\mathcal{W}$ in many respects. The notation for substitution and the free variable function is changed slightly, to remove possible ambiguities between predicates and expressions. ($\odot$ is written for substitution into predicates; $_\odot$ for expressions; the free variables of expression e are ϕe, and those of predicate p are Φp.) In essence though, they remain the same.

The observation that declarations are necessarily *ordered*, and that a later declaration may use variables declared by an earlier one, led to the introduction of a dagger as a separator in the Standard's account of $\mathcal{W}$.[3] A further observation was that it is sometimes useful to consider generic judgements (for example, in stating laws about the generic toolkit operators). A generic parameter to the judgement is in effect a locally-added given set. In fact, any specification paragraph can be a useful antecedent to a conjecture, and so the new deductive system is based around judgements which have as antecedents 'local paragraphs' (separated by $\ddagger$, so that the presentation is horizontal, rather than vertical) to be added to the current specification for the purpose of proving the consequent, which is a single predicate.

$$\text{Par} \ddagger \cdots \ddagger \text{Par} \vdash \text{Predicate} \ .$$

The entire preceding specification part is assumed as antecedent to a conjecture. For example, in the following specification:

$$
\begin{array}{|l}
\hline
\_S \underline{}\\
\quad name_1?, name_2? : NAME \\
\hline
\quad name_1? \neq name_2? \\
\hline
\end{array}
$$

$$S \vdash \#\{name_1?, name_2?\} = 2$$

the conjecture is actually

$$S := [name_1?, name_2? : NAME \mid name_1? \neq name_2?] \ddagger$$
$$S$$
$$\vdash$$
$$\#\{name_1?, name_2?\} = 2$$

Moreover, the toolkit definition of the $\#$ operator is also implicitly present in the antecedent (it being an implicit part of the specification[4], and can be used when necessary.)

---

[3] A semicolon is typically used to separate declarations in Z. It does not introduce *order*, however, so that $X : \mathbb{P}N$; $x : X$ is not typically acceptable. In proof, such declarations may be desirable, and so the dagger symbol is introduced, so that in $X : \mathbb{P}N \dagger x : X$, the declaration of x occurs within the scope of variable X, and so the declaration is well-formed.

[4] The notion of a **section** makes this formal, but we overlook that detail here.

2.3 Tactics to relate the two

We borrow here the notion of a *tactic* from the proof tool community. There, a tactic is a program which accomplishes a proof; here we will use a restricted notion of tactics to describe derived inference rules. In this way, we make the scope of the derived rules quite explicit. The language of tactics used here will be a subset of *Angel* [MGW96].

The simplest tactics will be the primitive inference rules. Additionally, it will sometimes be useful to refer to the tactic which leaves its goal unchanged: **skip**. These simple tactics may be composed sequentially and in parallel.

The application of one tactic after another corresponds to their sequential composition, so the proof tree on the left below is constructed by the tactic on the right (notice that our rules and tactics are understood to operate in a backwards (goal-directed) sense).

$$\cfrac{\cfrac{\cfrac{Goal_4}{Goal_3}\ t_3}{Goal_2}\ t_2}{Goal_1}\ t_1 \qquad\qquad t_1\ ;\ t_2\ ;\ t_3$$

When a tactic (rule) application causes the tree to *bifurcate*, a parallel composition of tactics is needed; a different tactic may be applied to each of the resulting branches:

$$\cfrac{\cfrac{Goal_4}{Goal_3}\ t_2 \qquad \cfrac{\cfrac{Goal_5}{Goal_2}\ t_3}{}\ t_1}{Goal_1} \qquad\qquad t_1\ ;\ (t_2\ \|\ t_3)$$

We will also use *pattern-matching* in defining tactics. When rules and tactics are parameterised, it may be necessary to supply a parameter which is dependent upon the current goal. The π operator does this:

$$\pi\, x_1, \ldots x_n \bullet pat(x_1, \ldots x_n) \longrightarrow tac(x_1, \ldots x_n)\ .$$

The terms $x_1, \ldots, x_n$ are bound to expressions in the current goal according to the pattern *pat*, and supplied as parameters to tactic *tac*.

As a concrete example of the action of these tactics, consider the implication introduction rule of the new deductive system:

$$\cfrac{\Gamma \ddagger P \vdash Q}{\Gamma \vdash P \Rightarrow Q}\ impI\ .$$

We may derive the reverse (downward) version of this law by using the rules of implication elimination, assumption and thinning:

$$\cfrac{\cfrac{}{\Gamma \ddagger P \vdash P}\ AssumPred \qquad \cfrac{\cfrac{\Gamma \vdash P \Rightarrow Q}{\Gamma \ddagger P \vdash P \Rightarrow Q}\ Thinr}{}\ }{\Gamma \ddagger P \vdash Q}\ impE$$

This may be written as a tactic in which the introduced predicate is given explicitly:

$$impI^{\downarrow} := \pi\, p \bullet (\Gamma \ddagger p \vdash q) \longrightarrow (impE\, p\,;\,(AssumPred \parallel Thinr))$$

The p in the π term matches the appropriate predicate in the antecedent (the other matches are mere place-holders). This predicate is supplied as a parameter to the rule of implication elimination, which is followed by *AssumPred* applied to the left-hand branch of the tree, and *Thinr* applied to the right-hand.

The sequential composition binds more strongly than the parallel, so $(t_1;t_2 \parallel t_3)$ should be read as $\big((t_1;t_2) \parallel t_3\big)$.

3 Semantic Basis for the Correspondence

The role of a logic in the presentation of Z is to present semantic relationships between predicates in a convenient format for producing proofs, i.e., as sequents and inference rules. To be able to compare ν and $\mathcal{W}$, we must relate their different presentations of of sequents and inference rules. We will do this here and the following sections.

A direct approach to relating the presentations is to show that each application of a $\mathcal{W}$ inference rule within the proof can also be seen as a deduction in ν. Unfortunately, and as will be clear to the reader, the syntaxes of sequents in the two presentations differ so that such a direct approach is not possible.

However, using the semantic equivalence (mentioned above) of comma and conjunction on the left of the turnstile and comma and disjunction on the right of the turnstile, we may show that to any $\mathcal{W}$ sequent there is a precisely equivalent ν sequent. In the next section, we extend this equivalence to proofs.

We begin by defining the ν equivalent of a $\mathcal{W}$ sequent:

Definition 2 ($\nu/\mathcal{W}$ Equivalence). Let $S = D_1 \ddagger \ldots \ddagger D_l \mid P_1, \ldots, P_m \vdash Q_1, \ldots, Q_n$ be a $\mathcal{W}$ sequent. Define

$$\nu(S) = \begin{cases} D_1 \ddagger \ldots \ddagger D_l \ddagger P_1 \ddagger \ldots \ddagger P_m \vdash \mathit{false} & n = 0 \\ D_1 \ddagger \ldots \ddagger D_l \ddagger P_1 \ddagger \ldots \ddagger P_m \vdash Q_1 \vee \cdots \vee Q_n & n \geq 1 \end{cases}$$

Clearly, $\nu(S)$ is a ν sequent. Moreover:

Proposition 3 ($\nu/\mathcal{W}$ Equivalence). *For S as above, S and $\nu(S)$ have the same semantics.*

We show the equivalence in the case when $n \geq 1$ only. Then case when $n = 0$ is similar (and simpler).

Proof. The reversibility of the following deductions is the basis of the equivalence. The first is a deduction in $\mathcal{W}$, the second in ν[5]:

$$\frac{\dfrac{\dfrac{D_1 \dagger \ldots \dagger D_l \mid P_1, \ldots, P_m \vdash Q_1, \ldots, Q_n}{D_1 \dagger \ldots \dagger D_l \mid P_1, \ldots, P_m \vdash Q_1 \vee \ldots \vee Q_n} \vdash \vee^*}{D_1 \dagger \ldots \dagger D_l \mid P_1 \wedge \ldots \wedge P_m \vdash Q_1 \vee \ldots \vee Q_n} \wedge \vdash^*}{\vdash \forall D_1 \bullet \ldots \bullet \forall D_l \mid P_1 \wedge \ldots \wedge P_m \bullet Q_1 \vee \ldots \vee Q_n} \vdash \forall^*$$

Also:

$$\frac{\dfrac{\dfrac{D_1 \ddagger \ldots \ddagger D_l \ddagger P_1 \ddagger \ldots \ddagger P_m \vdash Q_1 \vee \ldots \vee Q_n}{D_1 \ddagger \ldots \ddagger D_l \ddagger P_1 \wedge \ldots \wedge P_m \vdash Q_1 \vee \ldots \vee Q_n} \; PredConj^*}{D_1 \ddagger \ldots \ddagger D_l \mid P_1 \wedge \ldots \wedge P_m \vdash Q_1 \vee \ldots \vee Q_n} \; SchPred}{\vdash \forall D_1 \bullet \ldots \bullet \forall D_l \mid P_1 \wedge \ldots \wedge P_m \bullet Q_1 \vee \ldots \vee Q_n} \; AllI^*$$

Given these reversible deductions, the result follows from the obvious semantic equivalence of $\vdash \forall D_1 \bullet \ldots \bullet \forall D_l \mid P_1 \wedge \ldots \wedge P_m \bullet Q_1 \vee \ldots \vee Q_n$ as a $\mathcal{W}$ sequent and as a ν sequent. □

3.1 Example

Using Definition 2 and Proposition 3 we see that the $\mathcal{W}$ conjecture

$$S : \mathbb{P}\mathbb{Z} \dagger x : \mathbb{Z} \mid x \in S \vdash x \in S, \neg x \in S$$

and the ν conjecture

$$S : \mathbb{P}\mathbb{Z} \ddagger x : \mathbb{Z} \ddagger x \in S \vdash x \in S \vee \neg x \in S$$

are equivalent.

Since we have shown that sequents in $\mathcal{W}$ may be embedded in ν, our task of demonstrating relative soundness is reduced to showing that each inference rule in $\mathcal{W}$ may be expressed as a combination of rules (i.e., as a tactic) of ν. As proofs consist of applications of inference rules, a simple inductive proof (which we do not give here) then provides that all $\mathcal{W}$ proofs may be expressed in ν.

3.2 ν is Classical

Given the presentation of the Standard, it is not immediately clear as to whether ν is a classical or an intuitionistic logic. The characteristic difference between the two is that in a classical logic we have the law of the excluded middle: $\vdash p \vee \neg p$ (for any predicate p); in an intuitionistic logic we do not. $\mathcal{W}$ is very

---

[5] The notation N^*, where N is an inference rule, denotes repeated applications of N.

clearly classical so that for the correspondence we wish to show, a necessary condition is that ν should be classical. We show that ν is classical in this section.

The rule

$$\frac{\Gamma \ddagger \neg p \vdash q}{\Gamma \vdash p \vee q} \; not-or$$

is clearly sufficient (with the rule of assumption) to prove the law of the excluded middle, and so to determine that ν is indeed classical.

A tactic to derive $impI^{\downarrow}$ was given above; we may derive a tactic for $notE^{\downarrow}$ as follows:

$$notE^{\downarrow} := \pi\, p \bullet (\Gamma \ddagger \neg p \vdash \mathsf{false}) \longrightarrow$$
$$impE\, p \, ; (Thinr \parallel notDef^{\downarrow}\, ; AssumPred)$$

$$\cfrac{\cfrac{\Gamma \vdash p}{\Gamma \ddagger \neg p \vdash p}\; Thinr \quad \cfrac{\cfrac{}{\Gamma \ddagger \neg p \vdash \neg p}\; AssumPred}{\Gamma \ddagger \neg p \vdash p \Rightarrow \mathsf{false}}\; notDef}{\Gamma \ddagger \neg p \vdash \mathsf{false}}\; impE$$

Using these tactics, we may prove a version of de Morgan's law:

$$\frac{\Gamma \vdash \neg\,(\neg\,p \wedge \neg\,q)}{\Gamma \vdash p \vee q}\; deMorgan$$

The proof trees are omitted here for reasons of brevity, but tactics to construct them are as follows:

$$deMorgan^{\uparrow} := \pi\, p, q \bullet (\Gamma \vdash p \vee q) \longrightarrow$$
$$impE(\neg\,(\neg\,p \wedge \neg\,q));$$
$$(\mathbf{skip} \parallel impI\, ; notE\, ; Shift\, ; notE^{\downarrow}\, ; AndI;$$
$$(notDef\, ; impI\, ; Shift\, ; notE^{\downarrow}\, ; orIr\, ; AssumPred$$
$$\parallel notDef\, ; impI\, ; Shift\, ; notE^{\downarrow}\, ; orIl\, ; AssumPred))$$

$$deMorgan^{\downarrow} := \pi\, p, q \bullet (\Gamma \vdash \neg\,(\neg\,p \wedge \neg\,q)) \longrightarrow$$
$$notDef\, ; impI\, ; PredConj\, ; OrE\, p\, q;$$
$$(Thinr\, ; Thinr$$
$$\parallel impI^{\downarrow}\, ; notDef\, ; Thinr\, ; AssumPred$$
$$\parallel impI^{\downarrow}\, ; notDef\, ; AssumPred)$$

Finally, the law we require:

$$not-or := deMorgan\, ; notDef\, ; impI\, ; PredConj\, ; notE$$

$$\cfrac{\cfrac{\cfrac{\cfrac{\cfrac{\Gamma \ddagger \neg p \vdash q}{\Gamma \ddagger \neg p \ddagger \neg q \vdash \text{false}} \; notE}{\Gamma \ddagger \neg p \wedge \neg q \vdash \text{false}} \; PredConj}{\Gamma \vdash (\neg p \wedge \neg q) \Rightarrow \text{false}} \; impl}{\Gamma \vdash \neg (\neg p \wedge \neg q)} \; notDef}{\Gamma \vdash p \vee q} \; deMorgan$$

Definition 2 defines the ν equivalent of the comma-separated list of consequent predicates of a $\mathcal{W}$ sequent to be a disjunction of those predicates. The *not–or* rule we have just derived allows the splitting of a disjunction on the right of the turnstile in a ν sequent. The ability to split disjunctions in ν sequents will be particularly useful in giving ν derivations of $\mathcal{W}$ inference rules as, with it, we will be able to isolate the 'predicate of interest' (that manipulated by the rule) in a $\mathcal{W}$ inference rule.

3.3 Presentational Conventions

As there are no global declarations in the predicate list of the ν equivalent of a $\mathcal{W}$ sequent, the P_i in such that list may be arbitrarily reordered (using *Shift*). Therefore, we may assume that the antecedent predicate of interest occupies position m, i.e., is P_m. We will also assume that the consequent predicate of interest occupies position Q_1. If Γ is the antecedent list $D_1 \ddagger \ldots \ddagger D_n \mid P_1, \ldots, P_m$, then by $\overline{\Gamma}$ we denote its ν equivalent, i.e., $D_1 \ddagger \ldots \ddagger D_n \ddagger P_1 \ddagger \ldots \ddagger P_m$; if Φ is the consequent list $Q_1, \ldots, Q_n$, then by $\overline{\Phi}$ we denote its ν equivalent, i.e., $Q_1 \vee \cdots \vee Q_n$ when $n \geq 1$ and false otherwise.

In the following deductions, we will often assume that Q_1 is the only disjunct to appear on the right. The derived rule *not–or* above justifies this assumption in allowing us to move $Q_2 \vee \cdots \vee Q_n$ to the left of the turnstile, and back again.

The task, then, is to show that for each rule r in $\mathcal{W}$, which relates the sequents $S, S_1, \ldots, S_n$:

$$\frac{S_1 \quad \ldots \quad S_n}{S} \; r$$

it is possible to derive the deduction

$$\frac{\nu(S_1) \quad \ldots \quad \nu(S_n)}{\nu(S)} .$$

We will do this by sketching a suitable proof tree in ν, and also by giving the formal definition of the derived rule as a tactic over the rules of ν. When $\mathcal{W}$'s rule r is encoded by tactic t over ν in this sense, we will write $r \approx t$.

4 Embedding $\mathcal{W}$ in the new system

In this section, we demonstrate how to implement most of the deductive rules of $\mathcal{W}$ using the new deductive system. The order of presentation here follows that of the Z Standard v1.0 [BN+92, Page 193, Sections 7.5-7.9]. For each translated rule of $\mathcal{W}$, we exhibit a proof tree which simulates its action in ν. We also give a tactic, according to the conventions described previously, to demonstrate that these translations can be mechanically encoded.

Other than for the Leibniz rule, we have complete coverage of the Structural, Expression and Predicate Calculus rules. For the Schema we outline a possible approach but, due to the regularity of the presentation, omit many details. We also note here that the Definition rules are implicitly given in ν as a whole specification may appear on the left of the turnstile.

Due to incompleteness in the presentations of the two logics, the translation of a few rules do not appear. Where possible we have completed the presentation of the logics, and given translations of these complete rules.

4.1 An Example

To illustrate the approach of the following section we present the following simple example of proofs in the two systems. The sequent $S : \mathbb{P}\mathbb{Z} \dagger x : \mathbb{Z} \vdash x \in S, \neg\, x \in S$ has an almost trivial proof in $\mathcal{W}$:

$$\frac{\dfrac{}{S : \mathbb{P}\mathbb{Z} \dagger x : \mathbb{Z} \mid x \in S \vdash x \in S} \; Assumption}{S : \mathbb{P}\mathbb{Z} \dagger x : \mathbb{Z} \vdash x \in S, \neg\, x \in S} \vdash \neg$$

The ν form of the sequent is $S : \mathbb{P}\mathbb{Z} \ddagger x : \mathbb{Z} \vdash x \in S \vee \neg\, x \in S$. Using the *not–or* rule of the previous section, the proof in ν is of the same length, but somewhat different in that, rather than move the negative disjunct ($\neg\, x \in S$) on to the left of the turnstile as in the $\mathcal{W}$ proof, we move the positive disjunct ($x \in S$):

$$\frac{\dfrac{}{S : \mathbb{P}\mathbb{Z} \ddagger x : \mathbb{Z} \ddagger \neg\, x \in S \vdash \neg\, x \in S} \; AssumPred}{S : \mathbb{P}\mathbb{Z} \ddagger x : \mathbb{Z} \vdash x \in S \vee \neg\, x \in S} \; not\text{--}or$$

After the presentation of the transformation of the $\mathcal{W}$ rules into ν, we will be able to translate the above $\mathcal{W}$ proof into its ν equivalent for comparison.

4.2 Embedding the General (structural) rules

Thin The precise rule of *thin* given in the account of $\mathcal{W}$ is odd in that it requires the simultaneous thinning of declarations, antecedents and consequents. Its analogue in the new deductive system is given below, but we expect other forms to be more useful.

thin $\approx$ *falseE* ; *Thinr* ; *Thinr*

To see the effect of this tactic, a proof tree is needed in which some of the artifacts of the encoding are also made visible.

$$\frac{\dfrac{\dfrac{\dfrac{\dfrac{\dfrac{\dfrac{\overline{\Gamma} \vdash \overline{\Phi}}{\overline{\Gamma} \ddagger \neg \overline{\Phi} \vdash false} \; not\text{-}or}{\overline{\Gamma} \ddagger \neg \overline{\Phi} \ddagger d \vdash false} \; Thinr}{\overline{\Gamma} \ddagger d \ddagger \neg \overline{\Phi} \vdash false} \; Shift(\alpha d \cap \phi \overline{\Phi} = \varnothing)}{\overline{\Gamma} \ddagger d \ddagger \neg \overline{\Phi} \ddagger p \vdash false} \; Thinr}{\overline{\Gamma} \ddagger d \ddagger p \ddagger \neg \overline{\Phi} \vdash false} \; Shift}{\overline{\Gamma} \ddagger d \ddagger p \ddagger \neg \overline{\Phi} \vdash q} \; falseE}{\overline{\Gamma} \ddagger d \ddagger p \vdash q \vee \overline{\Phi}} \; not\text{-}or$$

The proviso to the *Shift* rule, i.e., $(\alpha d \cap \phi \overline{\Phi} = \varnothing)$, is as predicted by an application of Rule-Lifting to *thin*.

Assumption

$$assumption \approx AssumPred$$

$$\frac{}{\overline{\Gamma} \ddagger \neg \overline{\Phi} \ddagger p \vdash p} \; AssumPred$$

Cut has been described previously. Its tactic is

$$cut\ p \approx impE\ (p \vee \overline{\Phi});$$
$$(\textbf{skip} \parallel impI\ ; orE\ p\ \overline{\Phi};$$
$$(AssumPred \parallel Shift\ ; Thinr \parallel AssumPred))\ .$$

$$\dfrac{\dfrac{}{\overline{\Gamma} \ddagger p \vee \overline{\Phi} \vdash p \vee \overline{\Phi}} \; AssumPred \quad \dfrac{\dfrac{\dfrac{\overline{\Gamma} \ddagger p \vdash \overline{\Phi}}{\overline{\Gamma} \ddagger p \ddagger p \vee \overline{\Phi} \vdash \overline{\Phi}} \; Thinr}{\overline{\Gamma} \ddagger p \vee \overline{\Phi} \ddagger p \vdash \overline{\Phi}} \; Shift \quad \dfrac{}{\overline{\Gamma} \ddagger p \vee \overline{\Phi} \ddagger \overline{\Phi} \vdash \overline{\Phi}} \; AssumPred}{\dfrac{\overline{\Gamma} \ddagger p \vee \overline{\Phi} \vdash \overline{\Phi}}{\dfrac{\overline{\Gamma} \vdash p \vee \overline{\Phi}}{} } } \; orE}{\dfrac{\overline{\Gamma} \vdash p \vee \overline{\Phi} \quad \dfrac{\dfrac{\overline{\Gamma} \ddagger p \vee \overline{\Phi} \vdash \overline{\Phi}}{\overline{\Gamma} \vdash (p \vee \overline{\Phi}) \Rightarrow \overline{\Phi}} \; impI}{}}{\overline{\Gamma} \vdash \overline{\Phi}} \; impE}$$

Equality We have, trivially,

$$reflection \approx Refl\ .$$

The tree presentation is:

$$\frac{\overline{\frac{}{\varGamma \ddagger \neg \overline{\varPhi} \vdash e = e}} \; Refl}{\overline{\varGamma} \vdash e = e \vee \overline{\varPhi}} \; not\text{-}or$$

We note the use of *not–or* to isolate the consequent disjunct $e = e$. This step (and its inverse, which was not necessary here) will be omitted in the sequel.

Truth Disconcertingly, proving $\vdash$ true requires a little effort:

$$truth \approx impE \; (\text{true} \Rightarrow \text{true}) \; ; (\text{trueDef} \parallel \text{trueDef}) \;.$$

$$\frac{\overline{\frac{}{\overline{\varGamma} \ddagger \neg \overline{\varPhi} \vdash \text{true} \Rightarrow \text{true}}} \; trueDef \quad \overline{\frac{}{\overline{\varGamma} \ddagger \neg \overline{\varPhi} \vdash (\text{true} \Rightarrow \text{true}) \Rightarrow \text{true}}} \; trueDef}{\overline{\varGamma} \ddagger \neg \overline{\varPhi} \vdash \text{true}} \; impE$$

Falsehood That false $\vdash$ is a theorem is slightly easier to show:

$$contradiction \approx falseE \; ; AssumPred \;.$$

$$\frac{\overline{\frac{}{\overline{\varGamma} \ddagger \text{false} \vdash \text{false}}} \; AssumPred}{\overline{\varGamma} \ddagger \text{false} \vdash \overline{\varPhi}} \; falseE$$

Negation The encoding of the 'cross-over' rules for negation relies on the choice to make $\vdash$ false correspond with the judgement with an empty consequent.

$$(\vdash \neg) \approx notDef \; ; impI$$

$$\frac{\overline{\frac{\overline{\frac{\overline{\varGamma} \ddagger \neg \overline{\varPhi} \ddagger p \vdash \text{false}}{\overline{\varGamma} \ddagger \neg \overline{\varPhi} \vdash p \Rightarrow \text{false}}} \; impI}{\overline{\varGamma} \ddagger \neg \overline{\varPhi} \vdash \neg p}} \; notDef}{}$$

The reversibility of the deduction is justified the bidirectional nature of the *impI* rule which was show in Section 2.3.

The demonstration of the bidirectional nature of the other *Negation* rule is more work in the upwards direction than in the downwards direction:

$$(\neg \vdash^\uparrow) \approx notE$$

$$\frac{\overline{\varGamma} \ddagger \neg \overline{\varPhi} \ddagger \neg p \\ \vdash \text{false}}{\overline{\varGamma} \ddagger \neg \overline{\varPhi} \vdash p} \; notE$$

$$(\neg \vdash^{\downarrow}) \approx \pi\,p \bullet (\Gamma \ddagger \neg p \vdash \text{false}) \longrightarrow impE\ p\,;\,(Thinr \parallel notDef^{\downarrow}\,;\,AssumPred)$$

$$
\cfrac{
  \cfrac{\overline{\Gamma} \ddagger \neg \overline{\Phi} \vdash p}{\overline{\Gamma} \ddagger \neg \overline{\Phi} \ddagger \neg p \vdash p}\ \textit{Thinr}
  \qquad
  \cfrac{
    \cfrac{\overline{\Gamma} \ddagger \neg \overline{\Phi} \ddagger \neg p \vdash \neg p}{\overline{\Gamma} \ddagger \neg \overline{\Phi} \ddagger \neg p \vdash p \Rightarrow \text{false}}\ \textit{notDef}
  }{}\ \textit{AssumPred}
}{\overline{\Gamma} \ddagger \neg \overline{\Phi} \ddagger \neg p \vdash \text{false}}\ impE
$$

Conjunction

Here the encoding is particularly straightforward.

$$(\wedge \vdash) \approx PredConj$$

$$
\cfrac{\overline{\Gamma} \ddagger \neg \overline{\Phi} \ddagger p \ddagger q \vdash \text{false}}{\overline{\Gamma} \ddagger \neg \overline{\Phi} \ddagger p \wedge q \vdash \text{false}}\ \textit{PredConj}
$$

$$(\vdash \wedge) \approx AndI$$

$$
\cfrac{\overline{\Gamma} \ddagger \neg \overline{\Phi} \vdash p \qquad \overline{\Gamma} \ddagger \neg \overline{\Phi} \vdash q}{\overline{\Gamma} \ddagger \neg \overline{\Phi} \vdash p \wedge q}\ \textit{AndI}
$$

Disjunction

Disjunction in the antecedent simplifies to two goals, each gaining one of the disjuncts as a new antecedent.

$$(\vee \vdash) \approx \pi\,p,q \bullet (\Gamma \ddagger p \vee q \vdash \text{false}) \longrightarrow$$
$$orE\ p\ q\,;\,(AssumPred \parallel Shift\,;\,Thinr \parallel Shift\,;\,Thinr)$$

$$
\cfrac{
  \cfrac{\overline{\Gamma} \ddagger \neg \overline{\Phi} \ddagger p \vee q \vdash p \vee q}{}\ \textit{AssumPred}
  \quad
  \cfrac{
    \cfrac{
      \cfrac{\overline{\Gamma} \ddagger \neg \overline{\Phi} \ddagger p \vdash \text{false}}{\overline{\Gamma} \ddagger \neg \overline{\Phi} \ddagger p \ddagger p \vee q \vdash \text{false}}\ \textit{Thinr}
    }{\overline{\Gamma} \ddagger \neg \overline{\Phi} \ddagger p \vee q \ddagger p \vdash \text{false}}\ \textit{Shift}
    \quad
    \cfrac{
      \cfrac{\overline{\Gamma} \ddagger \neg \overline{\Phi} \ddagger q \vdash \text{false}}{\overline{\Gamma} \ddagger \neg \overline{\Phi} \ddagger q \ddagger p \vee q \vdash \text{false}}\ \textit{Thinr}
    }{\overline{\Gamma} \ddagger \neg \overline{\Phi} \ddagger p \vee q \ddagger q \vdash \text{false}}\ \textit{Shift}
  }{}
}{\overline{\Gamma} \ddagger \neg \overline{\Phi} \ddagger p \vee q \vdash \text{false}}\ \textit{OrE}
$$

Disjunction in the consequent is, surprisingly, equivalent to **skip**, because of the encoding we have chosen.

$$(\vdash \vee) \approx \textbf{skip}$$

Notice that the corresponding rules of ν (*OrIl* and *OrIr*) are used in the definition of *not–or* (Section 3.2)—that is to say, they are not superfluous, despite not corresponding in the obvious way to $\vdash \vee$.

Implication on the right is entirely straightforward:

$$(\vdash \Rightarrow) \approx impI \ .$$

$$\frac{\overline{\varGamma} \ddagger \neg \overline{\varPhi} \ddagger p \vdash q}{\overline{\varGamma} \ddagger \neg \overline{\varPhi} \vdash p \Rightarrow q} \ impII$$

Implication on the left requires a little more work.

$$(\Rightarrow \vdash) \approx \pi \, p, q \bullet (p \Rightarrow q \vdash) \longrightarrow$$
$$impE \ p \ ; (\mathbf{skip} \parallel impI \ ; impE \ q;$$
$$(impI \ ; AssumPred \parallel Thinr \ ; Thinr \ ; impI))$$

$$\cfrac{\cfrac{\cfrac{\overline{\varGamma} \ddagger \neg \overline{\varPhi} \ddagger p \Rightarrow q \vdash p \Rightarrow q}{\cfrac{}{\begin{array}{c}\overline{\varGamma} \ddagger \neg \overline{\varPhi} \ddagger p \Rightarrow q \\ \vdash p \Rightarrow q\end{array}}\ AssumPred} {\begin{array}{c}\overline{\varGamma} \ddagger \neg \overline{\varPhi} \ddagger p \Rightarrow q \ddagger p \\ \vdash q\end{array}}\ impI \quad \cfrac{\cfrac{\cfrac{\cfrac{\overline{\varGamma} \ddagger \neg \overline{\varPhi} \ddagger q \vdash false}{\overline{\varGamma} \ddagger \neg \overline{\varPhi} \vdash q \Rightarrow false}\ impI}{\begin{array}{c}\overline{\varGamma} \ddagger \neg \overline{\varPhi} \ddagger p \Rightarrow q \\ \vdash q \Rightarrow false\end{array}}\ Thinr}{\begin{array}{c}\overline{\varGamma} \ddagger \neg \overline{\varPhi} \ddagger p \Rightarrow q \ddagger p \\ \vdash q \Rightarrow false\end{array}}\ Thinr}{\begin{array}{c}\overline{\varGamma} \ddagger \neg \overline{\varPhi} \ddagger p \Rightarrow q \ddagger p \\ \vdash false\end{array}}\ impE}{\begin{array}{c}\overline{\varGamma} \ddagger \neg \overline{\varPhi} \ddagger p \Rightarrow q \\ \vdash p \Rightarrow false\end{array}}\ impI \quad \overline{\varGamma} \ddagger \neg \overline{\varPhi} \ddagger p \Rightarrow q \vdash p}{\begin{array}{c}\overline{\varGamma} \ddagger \neg \overline{\varPhi} \ddagger p \Rightarrow q \\ \vdash false\end{array}}\ impE$$

Equivalence The $\mathcal{W}$ rule for iff is very similar to the new definition of iff; a following *AndI* is needed in order to separate the resulting conjuncts.

$$(\vdash \Leftrightarrow) \approx iffDef \ ; AndI$$

$$\cfrac{\cfrac{\overline{\varGamma} \ddagger \neg \overline{\varPhi} \vdash p \Rightarrow q \quad \overline{\varGamma} \ddagger \neg \overline{\varPhi} \vdash q \Rightarrow p}{\overline{\varGamma} \ddagger \neg \overline{\varPhi} \vdash p \Rightarrow q \wedge q \Rightarrow p}\ AndI}{\overline{\varGamma} \ddagger \neg \overline{\varPhi} \vdash p \Leftrightarrow q}\ iffDef$$

More work is needed to accomplish the same result in the antecedent.

$$(\Leftrightarrow \vdash) \approx \pi \, p, q \bullet (p \Leftrightarrow q \vdash) \longrightarrow$$
$$impE \ (p \Rightarrow q \wedge q \Rightarrow p);$$
$$(iffDef^{\downarrow} \ ; AssumPred \parallel Thinr \ ; impI \ ; PredConj)$$

$$
\cfrac{
\cfrac{
\cfrac{
\cfrac{\overline{\Gamma} \ddagger \neg \overline{\Phi} \ddagger p \Rightarrow q \ddagger q \Rightarrow p}{\vdash \mathsf{false}}
}{
\cfrac{\overline{\Gamma} \ddagger \neg \overline{\Phi} \ddagger p \Rightarrow q \wedge q \Rightarrow p}{\vdash \mathsf{false}} \; PredConj
}}{ }}{ } \; impI
$$

$$
\cfrac{
\cfrac{\overline{\Gamma} \ddagger \neg \overline{\Phi} \ddagger p \Leftrightarrow q \vdash p \Leftrightarrow q}{
\cfrac{\overline{\Gamma} \ddagger \neg \overline{\Phi} \ddagger p \Leftrightarrow q}{\vdash p \Rightarrow q \wedge q \Rightarrow p}
} \; AssumPred \quad
\cfrac{
\cfrac{\overline{\Gamma} \ddagger \neg \overline{\Phi}}{\vdash (p \Rightarrow q \wedge q \Rightarrow p) \Rightarrow \mathsf{false}}
}{
\cfrac{\overline{\Gamma} \ddagger \neg \overline{\Phi} \ddagger p \Leftrightarrow q}{\vdash (p \Rightarrow q \wedge q \Rightarrow p) \Rightarrow \mathsf{false}} \; Thinr
}
}{\overline{\Gamma} \ddagger \neg \overline{\Phi} \ddagger p \Leftrightarrow q \vdash \mathsf{false}} \; impE
$$

iffDef ... *impE* (as labelled in the tree)

Universal Quantification

$(\forall\vdash) \approx impE(b \odot P);$
$\qquad (AllE \,; (AssumPred \parallel (Shift \,; AssumPred))$
$\qquad \parallel impI)$

The effect of the tactic is illustrated by the following proof tree:

$$
\cfrac{
\cfrac{\overline{\Gamma} \ddagger \neg \overline{\Phi} \ddagger b \in [St] \ddagger \forall St \bullet P}{\vdash \forall St \bullet P} \; AssumPred \quad
\cfrac{
\cfrac{\overline{\Gamma} \ddagger \neg \overline{\Phi} \ddagger \forall St \bullet P \ddagger b \in [St]}{\vdash b \in [St]} \; AssumPred
}{\cfrac{\overline{\Gamma} \ddagger \neg \overline{\Phi} \ddagger b \in [St] \ddagger \forall St \bullet P}{\vdash b \in [St]} \; Shift}
}{
\cfrac{\overline{\Gamma} \ddagger \neg \overline{\Phi} \ddagger b \in [St] \ddagger \forall St \bullet P}{\vdash b \odot P} \; AllE \qquad
\cfrac{
\cfrac{\overline{\Gamma} \ddagger \neg \overline{\Phi} \ddagger b \in [St] \ddagger \forall St \bullet P \ddagger b \odot P}{\vdash \mathsf{false}}
}{
\cfrac{\overline{\Gamma} \ddagger \neg \overline{\Phi} \ddagger \Sigma \ddagger b \in [St] \ddagger \forall St \bullet P}{\vdash (b \odot P) \Rightarrow \mathsf{false}} \; impI
} \; impE
}{\overline{\Gamma} \ddagger \neg \overline{\Phi} \ddagger b \in [St] \ddagger \forall St \bullet P \vdash \mathsf{false}}
$$

The rule $(\vdash\forall)$ is present (with another name) in the new deductive system.

$(\vdash\forall) \approx AllI$

$$
\cfrac{\overline{\Gamma} \ddagger \neg \overline{\Phi} \ddagger S \vdash P}{\overline{\Gamma} \ddagger \neg \overline{\Phi} \vdash \forall S \bullet P} \; AllI
$$

Existential Quantification The effect of $\exists\vdash$ is largely reproduced by *ExistsE* and *AssumPred*, but a little more work is needed to re-create the *W* rule exactly— dropping the original quantified predicate.

$(\exists\vdash) \approx \pi \, S, p \bullet (\exists S \bullet p \vdash \mathsf{false}) \longrightarrow$
$\qquad ExistsE \, S \, p \,; (AssumPred \parallel Shift2 \,; Thinr)$

$$
\cfrac{
\cfrac{\overline{\Gamma} \ddagger \neg \overline{\Phi} \ddagger \exists S \bullet P \vdash \exists S \bullet P}{} \; AssumPred \qquad
\cfrac{
\cfrac{
\cfrac{\overline{\Gamma} \ddagger \neg \overline{\Phi} \ddagger S \ddagger P \vdash \mathsf{false}}{\overline{\Gamma} \ddagger \neg \overline{\Phi} \ddagger S \ddagger P \ddagger \exists S \bullet P \vdash \mathsf{false}} \; Thinr
}{\overline{\Gamma} \ddagger \neg \overline{\Phi} \ddagger \exists S \bullet P \ddagger S \ddagger P \vdash \mathsf{false}} \; Shift
}{} \; ExistsE
}{\overline{\Gamma} \ddagger \neg \overline{\Phi} \ddagger \exists S \bullet P \vdash \mathsf{false}}
$$

It is safe to *Shift* the quantified predicate past the declaration (schema) paragraph S (i.e. the side-conditions to *Shift* are satisfied) because the predicate cannot have any free variables captured by the declaration, due to its form.

Conversely, the most straightforward implementation of $\vdash \exists$ removes the quantifier, but the rule is stated in [BN+92] as retaining it.

$$(\vdash\exists) \approx \pi\, b, S, p \bullet (b \in S \vdash \exists S \bullet p) \longrightarrow$$
$$notE\,;\,impE(\exists S \bullet p);$$
$$(ExistsI\,;\,(not{-}or\parallel Shift\,;\,AssumPred)\parallel notDef\,;\,AssumPred)\ .$$

$$
\cfrac{
  \cfrac{
    \cfrac{\overline{\Gamma}\ddagger\neg\overline{\Phi}\ddagger b\in S\vdash\exists St\bullet P\vee b\odot P}
          {\overline{\Gamma}\ddagger\neg\overline{\Phi}\ddagger b\in S\ddagger\neg\exists St\bullet P\vdash b\odot P}\ {not{-}or}
    \qquad
    \cfrac{
      \cfrac{\overline{\Gamma}\ddagger\neg\overline{\Phi}\ddagger\neg\exists St\bullet P\ddagger b\in S\vdash b\in S}{}\ {AssumPred}
    }
    {\overline{\Gamma}\ddagger\neg\overline{\Phi}\ddagger b\in S\ddagger\neg\exists St\bullet P\vdash b\in S}\ {Shift}
  }
  {
    \cfrac{
      \cfrac{\overline{\Gamma}\ddagger\neg\overline{\Phi}\ddagger b\in S\ddagger\neg\exists St\bullet P\vdash\exists S\bullet P}
            {\begin{array}{c}\overline{\Gamma}\ddagger\neg\overline{\Phi}\ddagger b\in S\ddagger\neg\exists S\bullet P\\ \vdash\exists S\bullet P\\ \vdots\end{array}}\ {ExistsI}
      \qquad
      \cfrac{
        \cfrac{\overline{\Gamma}\ddagger\neg\overline{\Phi}\ddagger b\in S\ddagger\neg\exists St\bullet P\vdash\neg\exists St\bullet P}{}\ {AssumPred}
        \quad
        \cfrac{\begin{array}{c}\overline{\Gamma}\ddagger\neg\overline{\Phi}\ddagger b\in S\ddagger\neg\exists St\bullet P\\ \vdash\exists S\bullet P\Rightarrow false\end{array}}{}\ {notDef}
      }
      {\overline{\Gamma}\ddagger\neg\overline{\Phi}\ddagger b\in S\ddagger\neg\exists St\bullet P\vdash\exists S\bullet P}\ {impE}
    }
    {\begin{array}{c}\overline{\Gamma}\ddagger\neg\overline{\Phi}\ddagger b\in S\ddagger\neg\exists St\bullet P\\ \vdash false\end{array}}
  }
}
{\overline{\Gamma}\ddagger\neg\overline{\Phi}\ddagger b\in S\vdash\exists St\bullet P}\ {notE}
$$

4.3 Expressions

Many of ν's rules are direct analogues of those in $\mathcal{W}$. Thus we have, for example[6],

$extmem \approx Extmem$

$compre \approx Setcomp$

$tuple \approx TupleSel$

$cartmem \approx Prodmem$

$binding \approx Bindsel$

$bindingsel \approx BindEqu$.

Powerset The rule *powerset* was stated incorrectly in [BN+92]. Its correct statement includes the proviso that y should not appear free in u[7]

$$\cfrac{y:t\vdash y\in u}{y:t\vdash t\in\mathbb{P}\,u}\ powerset\ (y\notin\phi u)$$

---

[6] We note that *bindingsel* was erroneously named *tuplesel* in [BN+92].

[7] Which prevents, for instance, the incorrect deduction

$$\cfrac{y:\{3,4\}\vdash y\in\{y\}}{y:\{3,4\}\vdash\{3,4\}\in\mathbb{P}\{y\}}\ powerset\ .$$

This allows us to synthesize the following

$$powerset \approx Thinr \; ; Powerset \; ; AllI$$

$$\cfrac{\cfrac{\cfrac{\overline{\Gamma} \ddagger \neg \overline{\Phi} \ddagger y : t \vdash y \in u}{\overline{\Gamma} \ddagger \neg \overline{\Phi} \vdash \forall y : t \bullet y \in u} \; AllI}{\overline{\Gamma} \ddagger \neg \overline{\Phi} \vdash t \in \mathbb{P}u} \; Powerset(y \notin \phi u)}{\overline{\Gamma} \ddagger \neg \overline{\Phi} \ddagger y : t \vdash t \in \mathbb{P}u} \; Thinr \; (y \notin \phi u)$$

A similar construction can be used to synthesize *extension*.

Tuple Selection has a relatively simple derivation:

$$tuplesel \approx Tupleequ \; ; Refl$$

$$\cfrac{\cfrac{}{\overline{\Gamma} \ddagger \neg \overline{\Phi} \vdash u_i = u_i} \; Refl}{\overline{\Gamma} \ddagger \neg \overline{\Phi} \vdash (u_1, \ldots, u_i, \ldots, u_n).i = u_i} \; Tupleequ$$

Theta Removing the erroneous $\Gamma \vdash S$ premiss (the schema does not need to have a satisfying substitution for θS to exist), and adding the proviso that constrains the alphabet of e to be that of S, we have a simple synthesis. In the case when the alphabet of S has only a single member we have that:

$$theta \approx ThetaEqu; \; impE \; (n_1 = u_1) \; ; (\textbf{skip} \parallel impI \; ; BindingEqu)$$

$$\cfrac{\cfrac{\overline{\Gamma} \ddagger \neg \overline{\Phi} \vdash n_1 = u_1 \quad \cfrac{\cfrac{\cfrac{}{\overline{\Gamma} \ddagger \neg \overline{\Phi} \ddagger n_1 = u_1 \vdash \langle\!| \; n_1 := u_1 \; |\!\rangle.n_1 = u_1} \; BindingEqu}{\overline{\Gamma} \ddagger \neg \overline{\Phi} \vdash n_1 = u_1 \Rightarrow \langle\!| \; n_1 := u_1 \; |\!\rangle.n_1 = u_1} \; impI}{\overline{\Gamma} \ddagger \neg \overline{\Phi} \vdash \langle\!| \; n_1 := u_1 \; |\!\rangle.n_1 = u_1} \; impE}{\overline{\Gamma} \ddagger \neg \overline{\Phi} \vdash \langle\!| \; n_1 := u_1 \; |\!\rangle = \theta S} \; ThetaEqu \; (\alpha S = \{n_1\})$$

Other signatures are simple extensions of this, as is the reverse case of the rule. The extension to the case when S is decorated requires an extension to the *ThetaEqu* rule which we do not give here.

Schema Expression was presented without the proviso that $\alpha b = \alpha S$. Including this we have the direct correspondence

$$schemaexp \approx SchBindMem$$

$$\cfrac{\overline{\Gamma} \ddagger \neg \overline{\Phi} \vdash b \odot S}{\overline{\Gamma} \ddagger \neg \overline{\Phi} \vdash b \in S} \; SchBindMem$$

4.4 Schemas

The new deductive system has rules (derived rules, in most cases) for all of the schema calculus operators. These cover both the use of schemas as predicates and as expressions. The rules for schema expressions are the more general; they can be used to derive those for schema predicates—which are identical to those in $\mathcal{W}$.

For example, writing $[S \wedge T]$ (to indicate that the conjunction is a schema conjunction and not merely a logical one), $\mathcal{W}$ has

$$\frac{\vdash [S] \qquad \vdash [T]}{\vdash [S \wedge T]} \ (\vdash[\wedge]) \ .$$

The following proof over ν accomplishes the same:

$$\frac{\dfrac{\Gamma \vdash S}{\Gamma \vdash b \odot S} \ (1) \qquad \dfrac{\Gamma \vdash T}{\Gamma \vdash b \odot T} \ (1)}{\dfrac{\Gamma \vdash b \odot S \wedge b \odot T}{\Gamma \vdash [S \wedge T]}} \ BindSch(\alpha(S \wedge T) = \{x_1, \dots, x_n\})$$

where $b = \langle\!| x_1 == x_1, \dots, x_n == x_n |\!\rangle$. The step marked (1) is simply the effect of substituting the binding b (the identity) into the predicate to which it is attached.[8]

Similar derivations may be constructed for all of the schema calculus operators, and the rules in 'Declarations' (Section F.9 of [BN+92]).

4.5 Example Revisited

Earlier, we proved the conjecture $S : \mathbb{P}\,\mathbb{Z} \dagger x : \mathbb{Z} \vdash x \in S, \neg\, x \in S$, in both $\mathcal{W}$ and ν. Using the above translation, we have the following translation of the $\mathcal{W}$ proof, which we present for comparison:

$$\frac{\dfrac{\dfrac{\dfrac{\dfrac{\rule{6cm}{0.4pt}}{S : \mathbb{P}\,\mathbb{Z} \ddagger x : \mathbb{Z} \ddagger x \in S \vdash x \in S} \ AssumPred}{S : \mathbb{P}\,\mathbb{Z} \ddagger x : \mathbb{Z} \ddagger x \in S \ddagger \neg\, x \in S \vdash false} \ notE}{S : \mathbb{P}\,\mathbb{Z} \ddagger x : \mathbb{Z} \ddagger \neg\, x \in S \ddagger x \in S \vdash false} \ Shift}{S : \mathbb{P}\,\mathbb{Z} \ddagger x : \mathbb{Z} \ddagger \neg\, x \in S \vdash \neg\, x \in S} \ notE}{S : \mathbb{P}\,\mathbb{Z} \ddagger x : \mathbb{Z} \vdash x \in S \vee \neg\, x \in S} \ not{-}or$$

---

[8] This is not an exact copy of the effect of the proof rule in $\mathcal{W}$, since there the premisses are schemas (expressions) not predicates. However, repeated application of these rules (to remove all schema calculus operators) will yield the same end result.

5 Discussion

We have shown that most of W is relatively sound with respect to the new deductive system. Moreover, we do not expect the completion of this task to be overly difficult. At present, we do not have a proof that the new system is itself sound, but Brien's thesis [Bri95] goes a long way towards this. Future work clearly includes producing a good soundness proof for the new system.

We have also demonstrated that the new system is classical.

The work in this paper shows that our translation of W into ν produces proofs which are rather contrived. There are two main reasons for this:

- we do not have the comma separator, which forces us in general to precede a deduction with an 'isolate interesting consequent predicate' step, and follow it by the inverse step;
- most of ν's inference rules manipulate the consequent predicate, meaning that proofs which involve the manipulation of antecedent predicates are more difficult (or at least longer).

Neither of these convolutions derives from our translation, but they reflect the difference in styles of working appropriate to each logical system. The issue of which style is generally most appropriate (or whether another style is more natural) is still an open question.

The deductive systems differ significantly in their treatment of the context in which the conjecture appears. In W, special rules appeal to the context to import definitions from the specification into the antecedent. In ν this is accomplished by having those definitions be implicitly part of the antecedent. A formal correspondence of these two styles should be straightforward to demonstrate, but does not appear to add anything of interest, so is omitted here.

Acknowledgments

The authors would like to thank Samuel Valentine, Ian Toyn and the anonymous referees for their careful reading and comments on this paper. Jon Hall works under the auspices of the British Aerospace funded Dependable Computing Systems Centre in the University of York, England. Andrew Martin's work is supported by core funds at the Software Verification Research Centre, a Special Research Centre of the Australian Research Council.

References

[BG94]   Jonathan P. Bowen and Mike J. C. Gordon. Z and HOL. In J. P. Bowen and J. A. Hall, editors, *Z User Workshop, Cambridge 1994*, Workshops in Computing, pages 141–167. Springer-Verlag, 1994.

[BH95]   Jonathan P. Bowen and Michael G. Hinchey, editors. *ZUM'95: The Z Formal Specification Notation*, volume 967 of *LNCS*. Springer-Verlag, 1995.

[BN+92] S. M. Brien, J. E. Nicholls, et al. Z base standard. ZIP Project Technical Report ZIP/PRG/92/121, SRC Document: 132, Version 1.0, Oxford University Computing Laboratory, Wolfson Building, Parks Road, Oxford, OX1 3QD, UK, November 1992.

[Bri95] Stephen M. Brien. *A Model and Logic for Generically Typed Set Theory (Z)*. D.Phil. thesis, University of Oxford, 1995. New version expected 1996.

[ES94] Marcin Engel and Jens Ulrik Skakkeebæk. Applying PVS to Z. ProCoS II Technical Report IT/DTU ME 3/1, Department of Computer Science, Technical University of Denmark, December 1994.

[Har91] W. T. Harwood. Proof rules for Balzac. Technical Report WTH/P7/001, Imperial Software Technology, Cambridge, UK, 1991.

[Jon92] R. B. Jones. ICL ProofPower. *BCS FACS FACTS*, Series III, 1(1):10–13, Winter 1992.

[KB95] Ina Kraan and Peter Baumann. Implementing Z in Isabelle. In Bowen and Hinchey [BH95], pages 355–373.

[KSW96] Kolyang, T. Santen, and B. Wolff. A structure preserving encoding of Z in Isabelle/HOL. In *1996 International Conference on Theorem Proving in Higher Order Logic*. Springer-Verlag, 1996.

[Mar93] Andrew Martin. Encoding W: A Logic for Z in 2OBJ. In J. C. P. Woodcock and P. G. Larsen, editors, *FME'93: Industrial-Strength Formal Methods*, volume 670 of *Lecture Notes in Computer Science*, pages 462–481. Springer-Verlag, 1993.

[MGW96] A. P. Martin, P. H. B. Gardiner, and J. C. P. Woodcock. A tactic calculus. *Formal Aspects of Computing*, 8(4):479–489. Springer-Verlag, 1996.

[Nic95] John Nicholls, editor. *Z Notation*. Z Standards Panel, ISO Panel JTC1/SC22/WG19 (Rapporteur Group for Z), 1995. Version 1.2, ISO Committee Draft.

[Saa92] M. Saaltink. Z and Eves. In J. E. Nicholls, editor, *Z User Workshop, York 1991*, Workshops in Computing, pages 223–242. Springer-Verlag, 1992.

[Smi95] Graeme Smith. Extending W for Object-Z. In Bowen and Hinchey [BH95], pages 276–295.

[TH95] I. Toyn and J.G. Hall. Proving Conjectures using CadiZ. CadiZ documentation (to appear as a York University Technical Report).

[TM91] I. Toyn and J.A. McDermid. CadiZ: An architecture for Z tools and its implementation. *Software—Practice and Experience*, 25(3):305–330, 1991.

[WB92] J. C. P. Woodcock and S. M. Brien. W: A Logic for Z. In *Proceedings 6th Z User Meeting*. Springer-Verlag, 1992.

Using the Rippling Heuristic in Set Membership Proofs

Ina Kraan
inak@ifi.unizh.ch

Department of Computer Science
University of Zürich
Switzerland

Abstract. We demonstrate how the rippling heuristic [Bundy *et al* 93], originally developed for inductive proofs, can be used to automate set membership proofs. Set membership proofs occur frequently as subgoals in Z proofs, and automating these goals would lift a significant burden of the proof off of users of proof tools. The approach is promising and is being integrated into the proof tool Z-in-Isabelle [Kraan & Baumann 95a].

1 Introduction

The availability of powerful tools is indispensable if formal methods are to gain wide-spread acceptance. We have been developing a proof tool for Z called Z-in-Isabelle, which is an embedding of Z (following [Brien & Nicholls 92]) in the logical framework Isabelle [Kraan & Baumann 95a]. In [Kraan & Baumann 95b], we point out several goals that must be achieved to have a usable tool. One of them is that the user must be relieved of the many tedious secondary proof obligations. A large number of these are set membership proofs, i.e., demonstrating that something is a member of some set, where the particular set may be known or not.

[Bundy *et al* 93] presents a heuristic called *rippling* to guide proofs. The fundamental idea underlying rippling is to measure the differences between the goal of a proof state and one or more of its hypotheses, and to constrain rewriting in such a way that the differences between them are reduced. Rippling was initially developed for use in inductive proofs, but has been applied successfully to a number of areas in non-inductive theorem proving as well [Walsh *et al* 92, Negrete-Yankelevich 96]. In this paper, we apply rippling to a new family of proofs, i.e., set membership proofs, using the techniques presented in [Basin & Walsh 96].

Set membership proofs are ubiquitous in Z proofs. Take, for instance, the definition of *dom* from the mathematical toolkit

$$
\begin{array}{l}
\underline{\quad[X, Y]\quad} \\
\mathrm{dom} : (X \leftrightarrow Y) \to \mathbb{P}\, X \\
\hline
\forall R : X \leftrightarrow Y \bullet \mathrm{dom}\, R = \{x : X;\ y : Y \mid (x \mapsto y) \in R \bullet x\}
\end{array}
$$

and the lemma

$$a \in A, b \in B \vdash \mathrm{dom}\,\{a \mapsto b\} = \{a\} \ . \tag{1}$$

To unfold the definition of dom, we would have to prove, among others,

$$a \in A, b \in B \vdash \{a \mapsto b\} \in A \leftrightarrow B \ . \tag{2}$$

This is precisely the type of subgoal that should be taken care of automatically. While many of these subgoals are trivial, some are not.

In this paper, we show how the rippling heuristic can automate such proofs. The following section provides a brief introduction to proof planning and rippling. Section 3 demonstrates how rippling can be made applicable to set membership proofs, and Section 4 presents a more complex example. Section 5 presents conclusions.

2   Proof Planning and the Rippling Heuristic

To use the built-in heuristics common in theorem provers more flexibly, [Bundy 88] suggests using a meta-logic to reason about and to plan proofs. Proof plans are constructed in the meta-logic by successively applying *methods* to a conjecture until a combination of methods has been found that forms a complete plan. A method is a partial specification of a tactic in the following sense: If a sequent matches the input pattern, and the pre-conditions are met, the tactic is applicable; if the tactic succeeds, the output conditions will be true of the resulting sequents. Explicit proof planning has been implemented in *CI4M* [Bundy *et al* 90]. The plans are executable in *Oyster* [Bundy *et al* 90] or *Mollusc* [Richards *et al* 94], both sequent-style interactive proof checkers.

The advantages of the meta-logic approach are twofold. First, search is less expensive, since methods capture the effects of the corresponding tactics, while avoiding the possibly considerable cost of executing them. More importantly, however, the meta-level representation of the proof can be augmented with additional information on the proof to restrict the search space. The information is passed from method to method, which gives a global rather than a local view of the proof.

Rippling is a heuristic that guides rewriting by annotating terms to mark the differences between the goal to be proven and the available hypotheses, and by using the annotation to restrict rewriting. Rippling was first developed for inductive proofs, where the induction conclusion differs from the induction hypothesis by constructors (or destructors) applied to the induction variables. When proving some proposition P of the natural numbers, for instance, the step case of an inductive proof entails showing that $P(s(n))$ follows from $P(n)$. The difference between $P(s(n))$ and $P(n)$ is the application of the constructor s. For a complete description of rippling in inductive proofs, see [Bundy *et al* 93]. Rippling has since been generalized to other types of proofs. In the subsequent presentation, we follow the general terminology of [Basin & Walsh 96], and we refer to that paper for a more formal presentation. Here, we will simply illustrate the principles of rippling using the example of the associativity of $+$

$$\forall x, y, z. \, (x + y) + z = x + (y + z) \, ,$$

which we will prove by structural induction on x. The annotated step case of the proof is

$$(x + y) + z \quad = \quad x + (y + z)$$
$$\vdash$$
$$(\boxed{\underline{s(x)}}^{\uparrow} + y) + z = \boxed{\underline{s(x)}}^{\uparrow} + (y + z) \, .$$

The constructors that are present in induction conclusion but not in the induction hypothesis are what is boxed, but not underlined. These are called *wave fronts*. The arrows on the boxes indicate in which direction they are to be moved—up or down the term tree. The underlined terms in the boxes are *wave holes*. They are to be preserved in rewriting. What remains when all wave fronts are removed from the conclusion is the *skeleton*. The skeleton is identical to the induction hypothesis.

To ripple the conclusion is to rewrite it in such a way that wave fronts are removed or shifted to a more useful position. Annotated rewrite rules are *wave rules*. Wave rules not only preserve skeletons, but also decrease some appropriate measure on annotated terms. A simple measure, proposed in [Basin & Walsh 96] and sufficient for the purposes of this paper, is based the width of wave fronts, the level at which they occur, and their direction. As rippling is strictly measure-decreasing, it is guaranteed to terminate.

The recursive definition of $+$

$$\forall x, y. \, s(x) + y = s(x + y) \tag{3}$$

gives rise, for example, to the wave rule

$$\boxed{\underline{s(M)}}^{\uparrow} + N :\Rightarrow \boxed{\underline{s(M + N)}}^{\uparrow} \tag{4}$$

where M and N are free variables. The effect of applying wave rule (4) is to move the wave front s on the left-hand side outwards past the $+N$ to a position higher up the term tree. Note that $:\Rightarrow$ indicates annotated rewriting, not implication. Rippling reasons backwards from the conclusion to the hypotheses. Thus, wave rules may be based on equality, equivalence, or implication from right to left.

It is important to distinguish clearly between the proof planning level and the theorem proving level. The planner methods operate on annotated terms, and the rippling method in particular rewrites annotated terms using wave rules such as (4). Each method application causes a tactic to be built. This tactic, when applied in the theorem prover, tries to prove that the (unannotated) output sequents of the method follow from the (unannotated) input sequent by applying inference rules and appealing to axioms such as (3) or to previously proven lemmas of the object-level logic. Note that success or termination of the tactic generated by the method cannot be guaranteed.

For the step case of the associativity of $+$, we need two axioms, the recursive definition of $+$ given above and the replacement axiom of equality

$$\forall x, y.\ s(x) + y = s(x + y)$$
$$\forall x, y.\ x = y \Rightarrow s(x) = s(y)\ ,$$

which give rise to the wave rules

$$\boxed{s(\underline{M})}^{\uparrow} + N :\Rightarrow \boxed{s(\underline{M + N})}^{\uparrow} \tag{5}$$

$$\boxed{s(\underline{M})}^{\uparrow} = \boxed{s(\underline{N})}^{\uparrow} :\Rightarrow M = N\ . \tag{6}$$

Applying wave rule (5) three times to

$$\left(\boxed{s(\underline{x})}^{\uparrow} + y\right) + z = \boxed{s(\underline{x})}^{\uparrow} + (y + z)$$

yields

$$\boxed{s(\underline{x + y})}^{\uparrow} + z = \boxed{s(\underline{x})}^{\uparrow} + (y + z)$$

$$\boxed{s(\underline{(x + y) + z})}^{\uparrow} = \boxed{s(\underline{x})}^{\uparrow} + (y + z)$$

$$\boxed{s(\underline{(x + y) + z})}^{\uparrow} = \boxed{s(\underline{x + (y + z)})}^{\uparrow}\ .$$

Applying wave rule (6) once finally yields

$$(x + y) + z = x + (y + z)\ ,$$

which is identical to the induction hypothesis.

One variation of rippling we make use of in the following is *colored rippling* [Yoshida *et al* 94]. In some proofs, the conclusion needs to be rippled towards more than one hypothesis, e.g., when doing induction over trees. The conclusion will contain multiple skeletons, i.e., one for each hypothesis, and each skeleton is marked with a color to avoid confusion.

In inductive proofs, the annotation of the conclusion is determined by the type of induction—the constructor (destructor) in the conclusion (hypothesis) is the wave front. In non-inductive proofs, a more general approach is needed to determine the differences between hypotheses and conclusions. We compare and annotate hypotheses and conclusions using *difference matching* [Basin & Walsh 92], or possibly *difference unification* [Basin & Walsh 93].

Difference matching can be defined as follows: Given a term s and a term t, return annotations A_s and substitutions σ such that A_s is an annotation of s, and the skeleton of s with annotation A_s under substitution σ matches t

$$\sigma(skeleton(s, A_s)) = t\ .$$

For example, difference matching $f(X)$, where X is a variable, against a will yield $\boxed{f(\underline{a})}$.

Difference unification can be defined as following: Given a term s and a term t, return annotations A_s and A_t and substitutions σ such that A_s is an annotation of s, A_t is an annotation of t, the skeletons of s with annotation A_s and t with annotation A_t are equal under substitution σ, and σ is a most general unifier

$$\sigma(skeleton(s, A_s)) = \sigma(skeleton(t, A_t)) .$$

For example, difference unifying $f(s(a), a)$ and $f(a, s(a))$ yields $f(\boxed{s(\underline{a})}, a)$ and $f(a, \boxed{s(\underline{a})})$, respectively.

Matches and unifiers are not necessarily unique. However, by requiring minimal annotations, the number of matches and unifiers can be considerably reduced. For more details, and algorithms, see [Basin & Walsh 92, Basin & Walsh 93].

The general approach of rippling with difference matching (or unification) works as follows: The conclusion or goal is difference matched (or unified) against the available hypotheses, and rippling is used to rewrite the goal (and the hypotheses).

Set membership proofs are particularly suitable candidates for rippling, since the membership of some element in a set can often be decomposed into the membership of constituents of the element in the constituents of the set. We demonstrate this in the following sections.

3 Rippling in Set Membership Proofs

Figure 1 presents a straightforward proof of (2) after unfolding the abbreviational definition of $\leftrightarrow$, i.e., a proof of

$$a \in A, b \in B \vdash \{a \mapsto b\} \in \mathbb{P}(A \times B) . \tag{7}$$

Some rule applications have been collapsed into single steps for sake of brevity. This proof is not in the spirit of rippling, since it does not preserve structure, but rather destroys and then recreates it.

We now show how difference matching and rippling would prove (7). If we difference match the conclusion against the hypotheses, we obtain the colored, annotated sequent

$$\underline{a \in A}_{c_1}, \underline{b \in B}_{c_2} \vdash \left[\left\{ \boxed{\underline{a}_{c_1} \mapsto \underline{b}_{c_2}}^{\uparrow} \right\}_{c_1,c_2} \right]^{\uparrow} \in \left[\mathbb{P}\boxed{(\underline{A}_{c_1} \times \underline{B}_{c_2})}^{\uparrow} \right]_{c_1,c_2}^{\uparrow} . \tag{8}$$

One problem with the hand proof is the introduction and subsequent elimination of the variable x. It is the rule for power sets on the right

$$\frac{x : t \vdash x \in u}{\vdash t \in \mathbb{P} u} \quad [\mathbb{P}_R]$$

$$\frac{\dfrac{\rule{4cm}{0.4pt}}{a \in A, b \in B \vdash a \in A} \; assum}{a \in A, b \in B \vdash a \mapsto b.1 \in A} \; sel1 \qquad \frac{\dfrac{\rule{4cm}{0.4pt}}{a \in A, b \in B \vdash b \in B} \; assum}{a \in A, b \in B \vdash a \mapsto b.2 \in B} \; sel2$$

$$\frac{\rule{10cm}{0.4pt}}{a \in A, b \in B \vdash a \mapsto b \in A \times B} \; \times_R$$

$$\frac{a \in A, b \in B \vdash a \mapsto b \in A \times B}{a \in A, b \in B, x = a \mapsto b \vdash x \in A \times B} \; Leibniz$$

$$\frac{a \in A, b \in B, x = a \mapsto b \vdash x \in A \times B}{a \in A, b \in B, x \in \{a \mapsto b\} \vdash x \in A \times B} \; ext_L$$

$$\frac{a \in A, b \in B, x \in \{a \mapsto b\} \vdash x \in A \times B}{a \in A, b \in B \vdash \{a \mapsto b\} \in \mathbb{P}(A \times B)} \; \mathbb{P}_R$$

Fig. 1. "Hand" proof of (7)

that introduces x in the proof. To avoid introducing such a variable, we must know the structure of t. We can then specialize the inference rule for that particular structure. If t is an extension, as in (7), we can collapse the first three steps of the proof into one derived rule of inference[1]:

$$\frac{\vdash x_1 \in u \wedge \ldots \wedge x_n \in u}{\vdash \{x_1, \ldots, x_n\} \in \mathbb{P}\, u} \quad [\, \mathbb{P}_R\, ext \,]$$

This inference rule gives rise to a wave rule[2]

$$\boxed{\left\{ \underline{x_1}_{C_1}, \ldots, \underline{x_n}_{C_n} \right\}}^{\uparrow} \in \boxed{\mathbb{P}\,\underline{u}_{C_1 \ldots C_n}}^{\uparrow} \;:\Rightarrow\; \boxed{x_1 \in \underline{u}_{C_1}}^{} \wedge \ldots \wedge \boxed{x_n \in \underline{u}_{C_n}}^{\uparrow}.$$

Applying the wave rule to (8) yields

$$\underline{a \in A}_{c_1}, \underline{b \in B}_{c_2} \vdash \boxed{\underline{a}_{c_1} \mapsto \underline{b}_{c_2}}^{\uparrow} \in \boxed{\underline{A}_{c_1} \times \underline{B}_{c_2}}^{\uparrow} \tag{9}$$

In the hand proof, the inference rule for Cartesian products on the right

$$\frac{\vdash t.1 \in u_1 \wedge \ldots \wedge t.n \in u_n}{\vdash t \in (u_1 \times \ldots \times u_n)} \quad [\, \times_R \,]$$

---

[1] One of the main advantages of implementing a proof tool in a logical framework is the possibility of deriving new rules of inference in the tool itself, either interactively or automatically.

[2] Wave rules are computed automatically from rules of inference, lemmas, axioms, etc., by difference unifying the left-hand side and the right-hand side of the unannotated rule, lemma or axiom, and directing the wave fronts in such a way that the measure of the right-hand side is the maximum possible that is still smaller than the measure of the left-hand side.

was applied. This inference rule gives rise to a wave rule directly, albeit it a complicated one

$$t \in \boxed{\underline{u_1}_{C_1} \times \ldots \times \underline{u_n}_{C_n}}^{\uparrow} \;\Rightarrow$$

$$\boxed{\boxed{\underline{t}_{C_1}.1}^{\downarrow} \in u_1 \;\wedge\; \ldots \;\wedge\; \boxed{\underline{t}_{C_n}.n}^{\downarrow} \in u_n}^{\uparrow} \quad . \tag{10}$$

If we apply it to (9), we get

$$\underline{a \in A}_{c_1}, \underline{b \in B}_{c_2} \vdash$$

$$\boxed{\boxed{\underline{a}_{c_1} \mapsto b}^{\uparrow}.1 \in A \;\wedge\; \boxed{a \mapsto \underline{b}_{c_2}}^{\uparrow}.2 \in B}^{\uparrow} \quad . \tag{11}$$

Fortunately, the two wave rules

$$\boxed{\boxed{x \mapsto y}^{\uparrow}.1}^{\downarrow} \;\Rightarrow\; x$$

$$\boxed{\boxed{x \mapsto y}^{\uparrow}.2}^{\downarrow} \;\Rightarrow\; y$$

based on the lemmas

$$x \mapsto y.1 = x$$
$$x \mapsto y.2 = y$$

simplify matters considerably, yielding

$$\underline{a \in A}_{c_1}, \underline{b \in B}_{c_2} \vdash \boxed{\underline{a \in A}_{c_1} \wedge \underline{b \in B}_{c_2}}^{\uparrow} \quad .$$

Since we have now reassembled the two skeletons in their entirety, we can appeal to the hypotheses, and we obtain

$$\vdash \mathit{true} \wedge \mathit{true} \;,$$

which is trivial. We have thus completed a proof plan for (7).

While the application of the inference rule for Cartesian products $\times_R$ does lead to a proof, the goal (11) seems overly complex. In fact, there is a much simpler proof using a derived inference rule for the application of $\mapsto$

$$\frac{\vdash a \in A \wedge b \in B}{\vdash a \mapsto b \in A \times B} \quad [\mapsto\_\mathrm{appl\_type}\,]$$

which yields the straightforward wave rule

$$\boxed{\underline{a}_{C_1} \mapsto \underline{b}_{C_2}}^{\uparrow} \in \boxed{\underline{A}_{C_1} \times \underline{B}_{C_2}}^{\uparrow} :\Rightarrow \boxed{\underline{a \in A}_{C_1} \wedge \underline{b \in B}_{C_2}}^{\uparrow} . \tag{12}$$

The inference rule $\mapsto$_appl_type is automatically derived in Z-in-Isabelle[3]. Figure 2 presents example wave rules for a number of definitions from the mathematical toolkit.

$$\boxed{\underline{a}_{C_1} \mapsto \underline{b}_{C_2}}^{\uparrow} \in \boxed{\underline{A}_{C_1} \times \underline{B}_{C_2}}^{\uparrow} :\Rightarrow \boxed{\underline{a \in A}_{C_1} \wedge \underline{b \in B}_{C_2}}^{\uparrow}$$

$$\boxed{\underline{A}_{C_1} \cap \underline{B}_{C_2}}^{\uparrow} \in \mathbb{P} X :\Rightarrow \boxed{\underline{A \in \mathbb{P} X}_{C_1} \wedge \underline{B \in \mathbb{P} X}_{C_2}}^{\uparrow}$$

$$\boxed{first(\underline{x}, y)}^{\uparrow} \in X :\Rightarrow \boxed{\underline{x \in X} \wedge y \in Y}^{\uparrow}$$

$$\boxed{\underline{m}_{C_1} \ldots \underline{n}_{C_2}}^{\uparrow} \in \boxed{\mathbb{P} \underline{\mathbb{Z}}_{C_1, C_2}}^{\uparrow} :\Rightarrow \boxed{\underline{m \in \mathbb{Z}}_{C_1} \wedge \underline{n \in \mathbb{Z}}_{C_2}}^{\uparrow}$$

$$\boxed{head(\underline{s})}^{\uparrow} \in X :\Rightarrow s \in \boxed{\mathrm{seq}_1 \underline{X}}^{\downarrow}$$

$$\boxed{rev(\underline{s})}^{\uparrow} \in \mathrm{seq}\, X :\Rightarrow s \in \mathrm{seq}\, X$$

Fig. 2. Wave rules based on automatically derived toolkit lemmas

Applying wave rule (12) to the conclusion of (9) yields

$$\underline{a \in A}_{c_1}, \underline{b \in B}_{c_2} \vdash \boxed{\underline{a \in A}_{c_1} \wedge \underline{b \in B}_{c_2}}^{\uparrow}$$

directly in one step. Figure 3 summarizes the simplified proof of (7).

The situation in (9) is interesting, since it shows that choice points do occur in rippling, i.e., there is an non-trivial choice between rewrites. One way of chosing among them would be to compare the measures of the resulting terms and select the smallest—this gives preference to (12) over (10).

4 A More Complex Example

Set membership subgoals are usually generated by appealing to some universally quantified hypothesis or axiom, e.g., a definition in the mathematical toolkit. Z-in-Isabelle provides a tactic to unfold relations defined in the mathematical

[3] For every definition in the mathematical toolkit, Z-in-Isabelle automatically derives a suite of inference rules concerning the (basic) type of the operator, the type of an application of the operator, unfolding its definition, etc.

Wave rules:

$$\boxed{\left\{\underline{x_1}_{C_1}, \cdots, \underline{x_n}_{C_n}\right\}}^\uparrow \in \boxed{\mathbb{P}\,\underline{U}_{C_1 \ldots C_n}}^\uparrow :\Rightarrow \boxed{x_1 \in \underline{U}_{C_1} \wedge \ldots \wedge x_n \in \underline{U}_{C_n}}^\uparrow$$

$$\boxed{\underline{a}_{C_1} \mapsto \underline{b}_{C_2}}^\uparrow \in \boxed{\underline{A}_{C_1} \times \underline{B}_{C_2}}^\uparrow :\Rightarrow \boxed{a \in \underline{A}_{C_1} \wedge b \in \underline{B}_{C_2}}^\uparrow$$

Rippling:

$$\underline{a \in A}_{c_1}, \underline{b \in B}_{c_2} \vdash \boxed{\left\{\boxed{\underline{a}_{c_1} \mapsto \underline{b}_{c_2}}^\uparrow\right\}_{c_1, c_2}}^\uparrow \in \boxed{\mathbb{P}\,\boxed{\underline{A}_{c_1} \times \underline{B}_{c_2}}^\uparrow}_{c_1, c_2}^\uparrow$$

$$\underline{a \in A}_{c_1}, \underline{b \in B}_{c_2} \vdash \boxed{\underline{a}_{c_1} \mapsto \underline{b}_{c_2}}^\uparrow \in \boxed{\underline{A}_{c_1} \times \underline{B}_{c_2}}^\uparrow$$

$$\underline{a \in A}_{c_1}, \underline{b \in B}_{c_2} \vdash \boxed{a \in \underline{A}_{c_1} \wedge b \in \underline{B}_{c_2}}^\uparrow$$

$$a \in A, b \in B \vdash true \wedge true$$

Fig. 3. Proof of (7) via rippling

toolkit. In the current implementation, it tries a few simple techniques to solve the resulting set membership goals. These techniques are not capable of solving, for instance, the set membership proof we present in the following.

Lemma (1) is a typical law for toolkit definitions. Another such law[4] is

$$S \in \mathbb{P}\,X, A \in \mathbb{P}\mathbb{P}\,X \vdash S \cap \left(\bigcup A\right) = \bigcup \{T : A \bullet S \cap T\} \ .$$

This law expresses that the intersection of a set S with the big union of a set A is equal to the big union of the set of intersections of S and the member sets of A. To prove this law, we certainly need to unfold the definition of $\bigcup$, which is

$$\underline{\underline{=[X]}}$$
$$\bigcup : \mathbb{P}(\mathbb{P}\,X) \to \mathbb{P}\,X$$
$$\forall A : \mathbb{P}(\mathbb{P}\,X) \bullet \bigcup A = \{x : X \mid (\exists S : A \bullet x \in S)\}$$

To be able to unfold, we need to establish

$$S \in \mathbb{P}\,X, A \in \mathbb{P}(\mathbb{P}\,X) \vdash \{T : A \bullet S \cap T\} \in \mathbb{P}(\mathbb{P}\,X) \ .$$

First, as for the previous example, we need to apply the inference rule

$$\frac{x : t \vdash x \in u}{\vdash t \in \mathbb{P}\,u} \quad [\mathbb{P}_R]$$

---

[4] [Spivey 92] is a good source for such laws.

in some form or another. We again specialize the rule, this time for set comprehensions:

$$\frac{t \in S, P(t) \vdash Q(t) \in U}{\vdash \{t : S \mid P(t) \bullet Q(t)\} \in \mathbb{P}\, U} \quad [\, \mathbb{P}_R \text{ compre'}\,]$$

So far, we have rewritten only formulae in sequents, not entire sequents. Thus, the fact that we have formulae on both sides of the sequent symbol in the hypothesis is unfortunate. However, the derived rule

$$\frac{\vdash t \in S \wedge P(t) \Rightarrow Q(t) \in u}{\vdash \{t : S \mid P(t) \bullet Q(t)\} \in \mathbb{P}\, u} \quad [\, \mathbb{P}_R \text{ compre }]$$

follows easily from $\mathbb{P}_R$ compre'. This rule yields a wave rule

$$\boxed{\left\lfloor \{t : S \mid P(t) \bullet \underline{Q(t)}\} \in \mathbb{P}\,\underline{U} \right\rfloor^{\uparrow}} :\Rightarrow \boxed{t \in S \wedge P(t) \Rightarrow \underline{Q(t) \in U}}^{\uparrow} . \qquad (13)$$

The only other wave rule we need is the one for $\cap$ in Figure 2

$$\boxed{\underline{S}_{C_1} \cap \underline{T}_{C_2}}^{\uparrow} \in \mathbb{P}\, X :\Rightarrow \boxed{\underline{S \in \mathbb{P}\, X}_{C_1} \wedge \underline{T \in \mathbb{P}\, X}_{C_2}}^{\uparrow} . \qquad (14)$$

As before, we match the goal against the hypotheses. Difference matching against the first hypothesis yields

$$\underline{S \in \mathbb{P}\, X}, A \in \mathbb{P}\,(\mathbb{P}\, X) \vdash \boxed{\left\{ T : A \bullet \boxed{\underline{S \cap T}}^{\uparrow} \right\}}^{\uparrow} \in \boxed{\mathbb{P}\left(\underline{\mathbb{P}\, X}\right)}^{\uparrow} . \qquad (15)$$

Difference matching against the second hypothesis yields

$$S \in \mathbb{P}\, X, \underline{A \in \mathbb{P}\,(\mathbb{P}\, X)} \vdash \boxed{\left\{ \boxed{T : \underline{A}}^{\uparrow} \bullet S \cap T \right\}}^{\uparrow} \in \mathbb{P}\,(\mathbb{P}\, X) . \qquad (16)$$

In the example in the previous section, the matches against the two hypotheses were compatible, i.e., combining both annotations on the conclusion resulted in a well-annotated term. In this case, the two annotations are not compatible, and we must therefore select one of the matches and discard the other. In this case, the choice is simple—while no wave rule applies to the second match, wave rule (13) applies to the first match. Rewriting (15) with (13) results in

$$\underline{S \in \mathbb{P}\, X}, A \in \mathbb{P}\,(\mathbb{P}\, X) \vdash \boxed{T \in A \Rightarrow \boxed{\underline{S \cap T}}^{\uparrow} \in \mathbb{P}\, X}^{\uparrow} .$$

We can now apply wave rule (14) using only the color C_1

$$\underline{S \in \mathbb{P}\, X}, A \in \mathbb{P}\,(\mathbb{P}\, X) \vdash \boxed{T \in A \Rightarrow \boxed{\underline{S \in \mathbb{P}\, X} \wedge T \in \mathbb{P}\, X}^{\uparrow}}^{\uparrow}$$

and exploit the hypothesis $S \in \mathbb{P}\,X$

$$S \in \mathbb{P}\,X, A \in \mathbb{P}(\mathbb{P}\,X) \vdash T \in A \Rightarrow \mathit{true} \wedge T \in \mathbb{P}\,X \ .$$

This is as far as rippling takes us. To complete the proof, we still need to simplify the conclusion

$$S \in \mathbb{P}\,X, A \in \mathbb{P}(\mathbb{P}\,X), T \in A \vdash T \in \mathbb{P}\,X \ ,$$

apply the inference rule for power sets on the left, and appeal to the hypotheses:

$$\cfrac{\cfrac{\quad}{T \in A, T \in \mathbb{P}\,X \vdash T \in A}\;\mathit{assum} \qquad \cfrac{\quad}{T \in A, T \in \mathbb{P}\,X \vdash T \in \mathbb{P}\,X}\;\mathit{assum}}{A \in \mathbb{P}(\mathbb{P}\,X), T \in A \vdash T \in \mathbb{P}\,X}\;\mathbb{P}_L$$

The rippling heuristic completed a large part of the proof of this lemma, and left only few, simple steps to be solved.

5 Conclusions

In the previous sections, we have shown how the rippling heuristic can be used to automate set membership proofs in Z. Such proofs occur frequently as subgoals in more general proofs in Z, such as the lemmas on mathematical toolkit definitions or the data refinement proofs in [Spivey 92].

Using rippling has a number of advantages: It involves very little search, and it always terminates. If it succeeds, at least one hypothesis can be exploited. The success of rippling depends crucially on the availability of suitable wave rules. We have shown that it is possible to derive inference rules and lemmas which give rise to wave rules from existing rules and lemmas.

The general procedure in the previous sections can be summarized as follows:

1. Difference match the conclusion against the hypotheses.
2. Combine the annotations if possible; select one (possibly more) otherwise.
3. Ripple the conclusion.
4. Exploit the hypotheses.
5. Simplify the remaining subgoal, if any.

This procedure could be improved in various ways. [Negrete-Yankelevich 96] suggests difference unifying rather than difference matching goals and hypotheses, and allowing rippling on hypotheses as well. This requires taking into account polarity. He also proposes *balancing* sequents prior to rippling, i.e., applying introduction rules to maximize the amount of shared structure between hypotheses and goal. For instance, when proving

$$\vdash a \in A \Rightarrow b \in B \Rightarrow a \mapsto b \in \mathbb{P}(A \times B) \ ,$$

balancing would apply $\Rightarrow_R$ twice before difference unifying.

Another possibility would be rippling entire sequents rather than formulae or terms. This would avoid the awkwardness of rules (12) and (13). Rule (12) would become

$$\vdash \boxed{\underline{a}_{C_1} \mapsto \underline{b}_{C_2}}^{\uparrow} \in \boxed{\underline{A}_{C_1} \times \underline{B}_{C_2}}^{\uparrow} :\Rightarrow \begin{cases} \vdash a \in A \\ \vdash b \in B \end{cases},$$

rule (13)

$$\vdash \boxed{\{t : S \mid P(t) \bullet \underline{Q}(t)\} \in \mathbb{P}\,\underline{U}}^{\uparrow} :\Rightarrow t \in S, P(t) \vdash Q(t) \in U .$$

These extensions should improve the performance of rippling on set membership proofs.

In the long run, we would like to apply rippling to other types of proofs, e.g., the lemmas

$$a \in A, b \in B \vdash \mathrm{dom}\,\{a \mapsto b\} = \{a\}$$

and

$$S \in \mathbb{P}\,X, A \in \mathbb{P}\mathbb{P}\,X \vdash S \cap \left(\bigcup A\right) = \bigcup\{T : A \bullet S \cap T\}$$

themselves.

Rippling has turned out to be a useful heuristic for guiding set membership proofs. We are therefore integrating it in our existing tool Z-in-Isabelle. As a preliminary implementation, we are adapting $C\!I\!AM$ to the syntax of Z. In order to execute the resulting proof plans, we can either adapt the tactics generated by the rippling methods to be executable directly in Z-in-Isabelle, or, alternatively, implement in Isabelle the tacticals used in the tactics generated by $C\!I\!AM$.

References

[Basin & Walsh 92] D. Basin and T. Walsh. Difference matching. In Deepak Kapur, editor, *11th Conference on Automated Deduction*, pages 295–309, Saratoga Springs, NY, USA, June 1992. Published as Springer Lecture Notes in Artificial Intelligence, No 607.

[Basin & Walsh 93] D. Basin and T. Walsh. Difference unification. In *Proceedings of the 13th IJCAI*. International Joint Conference on Artificial Intelligence, 1993. Also available as Technical Report MPI-I-92-247, Max-Planck-Institut für Informatik.

[Basin & Walsh 96] David Basin and Toby Walsh. A calculus for and termination of rippling. *Journal of Automated Reasoning*, 16(1/2):147–180, 1996.

[Brien & Nicholls 92] S. M. Brien and J. E. Nicholls. Z base standard. Technical Monograph PRG-107, Oxford University Computing Laboratory, Wolfson Building, Parks Road, Oxford, UK, November 1992. Accepted for ISO standardization, ISO/IEC JTC1/SC22.

[Bundy 88]   A. Bundy. The use of explicit plans to guide inductive proofs. In R. Lusk and R. Overbeek, editors, *9th Conference on Automated Deduction*, pages 111–120. Springer-Verlag, 1988. Longer version available from Edinburgh as DAI Research Paper No. 349.

[Bundy *et al* 90]   A. Bundy, F. van Harmelen, C. Horn, and A. Smaill. The Oyster-Clam system. In M.E. Stickel, editor, *10th International Conference on Automated Deduction*, pages 647–648. Springer-Verlag, 1990. Lecture Notes in Artificial Intelligence No. 449. Also available from Edinburgh as DAI Research Paper 507.

[Bundy *et al* 93]   A. Bundy, A. Stevens, F. van Harmelen, A. Ireland, and A. Smaill. Rippling: A heuristic for guiding inductive proofs. *Artificial Intelligence*, 62:185–253, 1993. Also available from Edinburgh as DAI Research Paper No. 567.

[Kraan & Baumann 95a]   I. Kraan and P. Baumann. Implementing Z in Isabelle. In J.P. Bowen and M.G. Hinchey, editors, *Proceedings of ZUM'95: The Z Formal Specification Notation*, Lecture Notes in Computer Science 967, pages 355–373. Springer-Verlag, 1995.

[Kraan & Baumann 95b]   I. Kraan and P. Baumann. Logical frameworks as a basis for verification tools: A case study. In *Proceedings of the Tenth Knowledge-Based Software Engineering Conference*, 1995.

[Negrete-Yankelevich 96]   S. Negrete-Yankelevich. *Proof Planning with Logic Presentations*. Unpublished PhD thesis, Department of Artificial Intelligence, University of Edinburgh, 1996.

[Richards *et al* 94]   B.L. Richards, I. Kraan, A. Smaill, and G.A. Wiggins. Mollusc: a general proof development shell for sequent-based logics. In A. Bundy, editor, *12th Conference on Automated Deduction*, pages 826–30. Springer-Verlag, 1994. Lecture Notes in Artificial Intelligence, vol 814; Also available from Edinburgh as DAI Research paper 723.

[Spivey 92]   J. M. Spivey. *The Z Notation: A Reference Manual*. Series in Computer Science. Prentice Hall International, 2nd edition, 1992.

[Walsh *et al* 92]   T. Walsh, A. Nunes, and A. Bundy. The use of proof plans to sum series. In D. Kapur, editor, *11th Conference on Automated Deduction*, pages 325–339. Springer Verlag, 1992. Lecture Notes in Computer Science No. 607. Also available from Edinburgh as DAI Research Paper 563.

[Yoshida *et al* 94]   Tetsuya Yoshida, Alan Bundy, Ian Green, Toby Walsh, and David Basin. Coloured rippling: An extension of a theorem proving heuristic. Technical Report TBA, Dept. of Artificial Intelligence, Edinburgh, 1994.

System Development

A Practical Method for Rigorously Controllable Hardware Design

E. Börger and S. Mazzanti

Università di Pisa, Dipartimento di Informatica, Corso Italia, 40, 56125 Pisa, Italy
(boerger,mazzanti@di.unipi.it)

Abstract. We describe a method for rigorously specifying and verifying the control of pipelined microprocessors which can be used by the hardware designer for a precise documentation and justification of the correctness of his design techniques. We proceed by successively refining a one-instruction-at-a-time-view of a RISC processor to a description of its pipelined implementation; the structure of the refinement hierarchy is determined by standard instruction pipelining principles (grouped following the kind of conflict they are designed to avoid: structural hazards, data hazards and control hazards).

We illustrate our approach through a formal specification with correctness proof of Hennessy and Patterson's RISC processor *DLX* but the method can be extended to complex commercial microprocessor design where traditional or purely automatic methods do not scale up. The specification method supports incremental design techniques; the modular proof method offers reusing proofs and supports the designer's intuitive reasoning, in particular "local" argumentations typical for upgrading and optimizing machines. Since our models come in the form of Abstract State Machines, they can be made executable by ASM interpreters and can thereby be used for prototypical simulations.

1   Introduction

It is well known that microprocessors are subject to subtle design errors. Conventional methods like simulation to debug processors before fabrication consume enormous resources in terms of manpower and of machines. In recent years various formal verification techniques have been proposed to overcome the well-known theoretical and practical limits of such conventional techniques and have been applied to the analysis of a certain number of (usually rather simple and unpipelined) microprocessors. Some typical examples standing for many others are [JBG86] [Bow87] [C88] [C89] [Hunt89] [LC91] [Her92] [Be93] [Win94] and [Ta95] which includes an excellent detailed survey.

   We develop a practical method which reduces the labor required to do formally supported design and verification of microprocessors by orders of magnitude. The method allows one to define a hierarchy of refinement steps each of which is focussed on a specific feature of the processor to be constructed and comes with a correctness proof expressing the intuitive reasoning of (i.e. the justification given by) the designer. The guiding principle of these successively refined specifications is to mimic as closely as possible the incremental features in hardware design. We add to this incremental approach a locality principle (see the notions of projection and of relevant locations below) which supports the local reasoning typical for practical hardware design. As a by-product one can break the proof of the properties of interest into elementary inductions and a few natural case distinctions corresponding to the different pipelining conflict types and the methods to solve them; in this way we prepare the ground for additional support by a mechanical verification using automated proof development systems such as PVS, HOL, IMPS or model checking systems.

We concentrate our attention in this paper on control, where notoriously most errors are found during the design of a processor. We do this for the challenging case of microprocessors with an instruction pipeline, exemplified through the standard pipelined RISC processor DLX developed by Hennessy and Patterson [HP90] in order to illustrate the essential features of RISC processors like the MIPS R3000 (see [Hen93]), Intel i860, Sun SPARC, Motorola M88000. Pipelining is a key implementation technique used to make fast CPUs. It provides a simultaneous execution of multiple instructions which exploits the independence between (parts of) instructions, as a result of which the execution speed for programs is improved. Since pipelining is not visible to the programmer, the more it is crucial to ensure that the semantics of instructions is preserved by the concurrency of operations which is inherent in this technique. We prove the correctness of Hennessy and Patterson's pipelined processor with respect to its sequential model (one-instruction-at-a-time view of the processor). The task therefore consists in starting from a mathematical model for the datapath and the sequential control of DLX, refining this model to the pipelined version of DLX and proving the correctness of the refinement process.

The overall structure of our design-driven refinement hierarchy is determined by the major instruction pipelining principles which can be grouped following the kind of conflict they are designed to avoid: structural, data and control hazards. For the crucial transition from the sequential (programmer-view) DLX model to the parallel execution model of its pipelined variant DLX^p we provide a (local projection) technique for extracting from certain segments of a concurrent DLX^p-computation—where at each step many operations concerning different (types of) instructions are performed in parallel—an equivalent sequential DLX-subcomputation of the one instruction (type) under analysis (see the notions of *relevant* and *result locations* and of *instruction cycles* below). For this first refinement step we concentrate on the current techniques to make the pipelined version of DLX free from structural hazards and abstract from the more sophisticated data or control hazards and stalls, i.e. for the proofs we assume the compiler to organize the sequence of instructions in such a way that they are sufficiently independent upon entering the pipe. In the further refinement steps we show that for the refined models DLX^{data}, DLX^{ctrl} and DLX^{pipe} of DLX^p the compiler assumption on data and control hazard freeness can piecemeal be dispensed with. (For a transparent and easily manageable proof it turned out to be advantageous to distinguish data hazards for not jump instructions—solved in DLX^{data}—and data hazards for jumps instructions—solved in two steps in DLX^{ctrl} and in DLX^{pipe}.[1]) Alltogether we therefore justify the following claim.[2]

Main Theorem (Correctness of DLX^{pipe} with respect to DLX).[3]
For each DLX program P, the result of the sequential execution of P on the machine DLX is the same as the result of the pipelined execution of P on the machine DLX^{pipe}.

Due to the systematic use of successive refinements, organized around the different pipelining problems and the methods for their solution[4], our approach can be applied for the design-driven verification as well as for the verification-driven design of RISC cores (including their rigorous documentation) at any level of abstraction. The modularity of the specification and analysis method provides the possibility to reuse correctness theorems along the refinement hierarchy. Such a decomposition of a complex goal into simpler subgoals corresponds to well established mathematical and engineering practice. Our method is still practicable when instead of DLX one has to deal with more complex microprocessors, more advanced pipeling techniques or more sophisticated memory systems.[5]

The divide-and-conquer approach to design-driven formal verification advocated here has proved to be practically viable for complex systems where traditional approaches failed; see the proofs in [BR95] [BD95] for the correct-

---

[1] We are grateful to Sofiene Tahar for pointing out to us that an architecture which is similar to our DLX^{pipe} has been implemented in [DeTa94].

[2] It has been suggested to view our theorem as saying that DLX^{pipe} is sequentially consistent with respect to DLX in the sense of Lamport [La79]. This is an oversimplifying interpretation. Lamport's definition is phrased in terms of certain *"execution results"* being *"the same"*. One of the major problems we solve in this paper is to define in a rigorous but transparent manner *a)* what the computer architect understands by *"the result of a DLX execution with pipeling"* and *b)* precisely at which moments during the pipelined execution of DLX the results of this execution have to be checked for *"being the same"* as the result of the sequential DLX computation.

[3] This theorem has been announced in [Bo95].

[4] This is the basic methodological difference between our refinement hierarchy and the interesting hierarchical structuring proposed in [W90] and followed also in [WC95] and [Taku95]. The abstractions in these papers reflect some typical compiler hierarchy levels, leading from the assembler level through the level of microprogrammed code to (code formalizing) the electronic block model which constitutes the gate level of the hardware system. We define our abstractions in order to isolate and reflect as closely as possible the different hazard types and the methods for their solution. It is of course possible to combine the two structuring methods where this is needed to break down the complexity of the overall problem into pieces which can be handled relying on assistance from machines.

[5] The work on this paper grew out from a reverse engineering project of a parallel architecture (see [BoDC95]) where we faced pipelining together with VLIW parallelism. In [BoDC95] we have used our abstraction and refinement technique to structure a real-life processor into simple and rigorously defined basic components.

ness of compiling Prolog programs to the Warren Abstract Machine or Occam programs to the Transputer and the work on the machine checked versions of the WAM correctness proof using KIV [A95] and Isabelle [P96]. During
the last years theorem provers have been used to verify also pipelined processors, but either the processors are simple or the verification is rather complex
[Cy93, BB93, Ro92, SGGH91, SB90]. For two recent projects to formally verify
DLX using HOL and PVS see [TaKu95, Cy95][6]. In the model checking verification of a subset of the pipelined *DLX* in [BD94], Dill's goal is an automatic
verification procedure where the human intervention is confined to the development of operational descriptions of the specification and the implementation.
Our primary concern in this paper is to support the actual design work by a
simple method which can be used by the computer architect to lay down his design steps and to reason about their effect in a rigorous, checkable and falsifiable
way. To this purpose we provide a rigorous simple behavioral modelling of both
the specification and the implementation and relate the two by a hierarchy of
transparent definitions and (proofs of) properties; we try to break the complexity of the processor by revealing the structure of the run time interaction of its
main parts and by linking in an understandable hierarchical way the sequential
and the pipelined execution models.

To break the complexity of real-life non-toy systems it is crucial not to be
bound by the straitjacket of an a priori given formal framework and to be able
to separate the specification and its justification from mechanical verification
concerns. One thing is to rigorously support the designer's reasoning and the
structuring of his work into intellectually manageable parts; another thing is the
detailed logical encoding which is unavoidable to make the specification understandable and checkable not for a human user, but for a machine. Both forms of
"understanding" and "proving" have their own logic, needs and merits. Combining the two will enable us to master the complexity of current computer systems.
Once the largely creative and hardly mechanizable decomposition effort has led
to a hierarchy of stepwise refined rigorous models, related by lemmas stating
the properties of interest, the justification of the desired overall behavior of the
system can be split into separate, possibly mechanizable, proofs of such lemmas.
Flexible and sufficiently expressive systems for machine assisted verification will
incorporate such hierarchical decomposition techniques. We advocate a brain-
AND–brawn approach (see [Bo95]) for both design-driven post-verification and
verification-driven design using on-the-fly-verification.

It will help if the reader is familiar with the semantics of *Abstract State
Machines* defined in [G95][7] although what follows can be understood correctly

---

[6] Cyrluk's specification and implementation can be viewed as a PVS formalization of
the semantics of (some of) the rules of our models DLX and DLX[P]. Cyrluk, Tahar
and Kumar do not define our notions of relevant and of result locations which allow
us to structure and to localize the proof obligations boiling them down to the bare
minimum. Using these notions we can recover the sequential states from successive
pipelined states by simple projections which directly support the way the designer
reasons about the relation between sequential and parallel pipelined execution.

[7] Previously Gurevich's ASMs have been called evolving algebras.

by reading our ASM rules as pseudo–code over abstract data types. We therefore abstain from repeating here the definitions of [G95].

2 Parallelizing the sequential DLX to DLX^p

The one-instruction-at-a-time machine DLX can be constructed by a straightforward formalization of the control graphs in [HP90]. We define DLX as Abstract State Machine in the appendix[8] without commenting further and refer to [BM96] for explanations about how the abstractions of this sequential model make our proof method uniform with respect to the size of the register file, the width of the datapath, the instruction set, the memory access (bandwidth), etc.[9] We explain in the rest of this section the few changes which suffice to refine DLX to a machine DLX^p where at each clock cycle simultaneously five basic steps are executed, one for each of five instructions.

The five basic execution steps appearing in DLX are instruction fetch *(IF)*, instruction decode including the fetching of operands *(ID)*, execution proper for ALU operations and (data or branch) address calculation using the ALU *(EX)*, memory access *(MEM)* and writing the computed result back into the final register-file destination *(WB)*. The order in which these basic execution steps follow each other for the execution of an instruction is described in DLX by a 0-ary function *mode*. Ideally one can pipeline DLX by letting the processor execute during each clock cycle simultaneously five basic steps, one for each of five instructions. This can be realized by eliminating from DLX the sequential control by *mode* and by replacing where necessary the *mode* guards by operation code guards corresponding to the pipe stage of the instruction in question. In the resulting new machine, at each moment for each of the five basic execution steps a rule is applied (clock synchronized architectural parallelism).

However one has to guarantee that the five pipe stages which are active on every clock cycle do not compete for resources, each functional architectural unit being available at each step only once. We describe briefly how the rules of DLX can be refined to DLX^p rules which resolve these structural conflicts.

Resolving structural conflicts. The simplicity of the DLX instructions set results in limited resource competition and in simple datapath/control refinements to avoid it. Four major groups of resources have to be doubled so that any

---

[8] When using instruction related functions like *opcode, fstop, scdop, iop* etc., we usually suppress their standard argument, namely the content of the instruction register IR. Standard terminology and notation are adopted without explanation from [HP90].

[9] In order to concentrate on the essential features of the pipelining parallelism, we start here not with the instruction set architecture as seen by the programmer (assembly language one-instruction-at-a-time view), but with its refinement where it becomes visible that each instruction is executed in stages (pipelining steps). For reasons of simplicity we skip the floating point instructions of DLX; although the treatement of hazards is more complex with the (multicycle) floating point operations, the concepts are the same as for the integer pipeline. We do care however not to abstract away crucial control features like the user–requested interrupt handling.

combination of operations can occur in pipe stages which are executed simultaneously in one clock cycle, namely the memory access (to fetch instructions), an addition mechanism (to increment the program counter PC), the memory data register (for overlapping load and store instructions), and latches for the instruction register IR, for PC and for the ALU output C (to hold values which are needed later in the pipeline)[10].

Instruction fetching and incrementing PC. A memory access conflict between instruction fetching and load/store instructions is avoided by increasing the memory bandwith, formalized by an additional memory access function mem_{instr} used only for fetching instructions and supposed to be a subfunction of the DLX function mem; in this way we abstract from any particular implementation feature related to using separate instruction and data caches which we intend to treat in a later refinement step. [11] Another resource conflict which would appear at each clock cycle concerns the ALU had we to use it for incrementing PC. The usual solution consists in providing a separate PC-incrementer, namely our abstract function $next$. Thus we have the new rule $FETCH$ below, belonging to the pipe stage set IF; the condition $jumps$, defined by[12]

$$opcode(IR1) \in JUMP \vee (opcode(IR1) \in BRANCH \wedge \overline{opcode(IR1)}(A) = true),$$

ensures that PC can be updated by the $FETCH$–rule only when no jump or branch rule has to update PC in the execution phase. The new rule $OPERAND$, belonging to the pipe stage set ID, is obtained from the DLX homonym by deleting the $mode$ guards and updates.

FETCH	$IR \leftarrow mem_{instr}\,(PC),$	OPERAND	$A \leftarrow fstop\,(IR)$
	if $\neg jumps$ then $PC \leftarrow next\,(PC)$		$B \leftarrow scdop\,(IR)$

Latches for longer living values. Some of the values which appear during the execution of an instruction at a certain pipe stage are needed at later pipe stages and have to be copied in order not to get overwritten by a subsequent instruction occurring in the pipeline. This is the case for (segments of) IR. For reasons of simplicity we abstract from instruction format and decoding details and provide three additional registers $IR1$, $IR2$, $IR3$ to keep copies of a fetched

---

[10] The concept of simultaneous execution of multiple ASM rules allows us to abstract from the distinction of pipe stages into a writing and a subsequent reading phase (see [TaKu95]). This justifies the simultaneous execution of for example the rules $OPERAND$ and MEM_ADDR or $Pass\_B\_to\_MDR$. It also means that we consider the simultaneous read and write access of the register file (by the rules ID and WB) as not constituting a resource conflict. The explicit introduction of phases would come up to a routine extension of our rules.

[11] The DLX processor does not support self-modifying code. That feature, which can be found in older usually non-pipelined architectures, would require a much more subtle treatment of control hazards than the one present in pipelined processors.

[12] $IR1$ cointains the value IR had in the previous clock cycle, see below. By $\overline{opcode(\ldots)}$ we denote the function encoded by $opcode(\ldots)$. Registers A, B store outputs from register file registers for use in later clock cycles.

instruction through the pipe stages *EX*, *MEM*, *WB*, i.e. with the following new preservation rules belonging to the rule sets *ID*, *EX*, *MEM* respectively:

$$\boxed{\text{Preserv IR}} \; IR1 \leftarrow IR, \quad \boxed{\text{Preserv } IRi} \; IR(i+1) \leftarrow IRi \quad \text{with } i = 1, 2.$$

Two 0-ary functions *PC1*, *C1* are needed to save the values of *PC*, *C* for one pipe stage. *PC1* provides at pipe stage *EX* of an instruction *I* a copy of the value of *PC* after the FETCH stage of *I* (serving in case *I* is a jump instruction the execution of which triggers a transfer or an update of that *PC*–value). *C1* provides at pipe stage *WB* a copy of the ALU output value *C* computed in the pipe stage *EX* of *I* (for instructions with *ALU/SET*–operations, for *JLINK* instructions and for *MOVS2I*).

$$\boxed{\text{Preserv PC}} \quad PC1 \leftarrow PC \qquad \boxed{\text{Preserv C}} \quad C1 \leftarrow C$$

For reasons to be explained in the next section the rule for copying the current value of *PC* into *PC1* will have this form only in the last two models DLX^{ctrl} and DLX^{pipe} and a slightly extended form in DLX^p and DLX^{data}.

Doubling MDR. In *DLX* the memory data register *MDR* is the only interface between the register-file and the memory and serves for both loading and storing. In the pipelined version for *DLX* a load instruction *I* which in the pipeline immediately precedes a store instruction *I'* would compete with *I'* for writing into *MDR* in its pipe stage *MEM* (when *I'* in its pipe stage *EX* wants to write *B* into *MDR*). This resource conflict is resolved by doubling *MDR* into two registers *LMDR* and *SMDR* and by refining as follows the *DLX*–rules *MEM_ADDR* and *Pass_B_to_MDR*, both belonging to the set *EX*:[13]

> **if** *opcode* (*IR1*) ∈ *LOAD* ∪ *STORE* **if** *opcode* (*IR1*) ∈ *STORE*
> **then** *MAR* ← *A* + *ival* (*IR1*)   **then** *SMDR* ← *B*

The *DLX*–rule *MEM_ACC* is divided in DLX^p into the following two refined rules, one for *LOAD* and one for *STORE*, both belonging to the set *MEM*:

> $\boxed{\text{STORE}}$ **if** *opcode* (*IR2*) ∈ *STORE* $\boxed{\text{LOAD}}$ **if** *opcode* (*IR2*) ∈ *LOAD*
> **then** *mem* (*MAR*) ← *SMDR*   **then** *LMDR* ← *mem* (*MAR*)

The new rule *Pass_B_to_MDR* requires a new direct link from the exit of *B* to the entry of *SMDR* in order to avoid the use of the ALU for this data transfer.

Speeding up the pipe stages. Since all pipe stages proceed simultaneously and the time which is needed for moving an instruction one step down the pipeline is a machine cycle, the length of the latter is determined by the time required for the slowest pipe stage. The two *DLX*–rules *ALU*, *ALU'* are combined into the following DLX^p–rule *ALU* (belonging to the set *EX*), thus eliminating the intermediate step to put the right second operand into *TEMP*.[14]

---

[13] The register *MAR* stores the address for the memory access. The function *ival* yields the immediate value encoded in an instruction.

[14] The function *iop* detects operation code for immediate operations.

if $opcode\,(IR1) \in ALU \cup SET$ **then if** $iop\,(opcode\,(IR1)) = true$
$\qquad\qquad\qquad\qquad$ **then** $C \leftarrow \overline{opcode\,(IR1)}\,(A,\,ival\,(IR1))$
$\qquad\qquad\qquad\qquad$ **else** $C \leftarrow \overline{opcode\,(IR1)}\,(A,\,B)$

The DLX–rule $SUBWORD$ (which selects and outputs to C the required portion of the word loaded from the memory) is incorporated into the following $WRITE\_BACK$–rule under the guard that the value to be written comes through a loading instruction; if this value has been computed by executing an ALU/SET, $JLINK$, $MOVS2I$ instruction, it comes from $C1$. The price for this refinement is linking the exit of $LMDR$ directly (without passing through $C1$) to the entry of the register-file and adding to the latter a selector for choosing among $C1$ and (the required portion of) $LMDR$. Transfering a subword of $LMDR$ into a destination register in the following rule can be realized without using the ALU by relying upon the usual shift functions of registers like $LMDR$.

$\boxed{\text{WRITE\_BACK}}$ **if** $opcode\,(IR3) \in ALU \cup SET \cup \{MOVS2I\} \cup JLINK$
$\qquad\qquad\qquad\quad$ **then** $dest\,(IR3) \leftarrow C1$

$\qquad\qquad\qquad\quad$ **if** $opcode\,(IR3) \in LOAD$
$\qquad\qquad\qquad\quad$ **then** $dest\,(IR3) \leftarrow \overline{opcode\,(IR3)}\,(LMDR)$

The remaining DLX^p–rules—namely $MOVESPECIAL$, $JUMP$, $BRANCH$—all belong to the pipe stage EX and are obtained from their DLX-homonyms by deleting the $mode$ guards and updates and by replacing the arguments IR, PC by $IR1$, $PC1$ respectively. This concludes the specification of the ASM model DLX^p which is spelled out in full in the appendix.

3 Justifying the correctness of the parallelization

For the proof of the correctness of DLX^p with respect to DLX we start by defining the notions of result location, of used location and of relevant location which will allow us to recover DLX–states from successive pipelined DLX^p–states by simple projections. We consider only computations which are reachable from appropriate initial states. We say that two computations C in DLX and C^p in DLX^p *correspond* to each other if their *initializations* coincide on the common signature except where explicitly stated otherwise. For DLX–initializations we assume $reg\,(IR) = undef$ and $mode = FETCH$, for DLX^p–initializations $reg\,(PC1) = reg\,(C1) = reg\,(IRi) = undef$ for i=1,2,3. We often use $f\,(undef) = undef$, for each function f. We say that a computation is initialized or starts with an instruction $instr$ if $mem\,(PC) = mem_{instr}\,(PC) = instr$.

Instruction Cycles. We can justify the correctness claim by a series of simple local arguments—one for each instruction (class)—be decomposing computations into segments each of which constitutes a subcomputation during which a given instruction is executed completely. In DLX computations, an *instruction cycle* for $instr$ is any subcomputation which starts with $mode = FETCH$ and $mem_{instr}\,(PC) = instr$ and leads to the next state with $mode = FETCH$;

in DLX^p computations, an *instruction cycle* for *instr* is any subcomputation which starts with *instr* and ends with the first following pipe stage of *instr* at the end of which the values of all the result locations of *instr*, as defined below, are computed. We call this pipe stage the end (pipe) stage of *instr*; whether it is *EX(instr)*, *MEM(instr)* or *WB(instr)* depends on *instr* and is defined in table 1.

We prove the correctness of DLX^p with respect to DLX instructionwise by showing that in every pair (C, C^p) of corresponding DLX/DLX^p–computations, *corresponding instruction cycles* compute the same result. The correspondence between instruction cycles in C and in C^p is defined by the order in which they occur: if $I_1, I_2,...$and $I'_1, I'_2,...$are the instruction cycles of C and C^p respectively (in the order in which they appear there), then I_i and I'_i correspond to each other. By I_0, I'_0 we indicate the initial state. We say that I_0, I'_0 formalise the "result" of "no computation step". In particular we will show below that I_i and I'_i are instruction cycles for the same instruction.

Result Locations. The simplicity of the DLX instruction set makes it easy to localize, uniformly for a few classes of instructions, where and when the result of an instruction belonging to a class becomes visible in a DLX/DLX^p–computation, namely in certain registers or memory locations. The pair $< reg,$ $PC >$ is defined to be a *result location* for each instruction *instr*. The other result locations for *instr* are determined by table 1.[15] We assume *dest (instr) = R31* in case *instr* $\in$ *JLINK*. *reg (fstop (instr)) + ival (instr)* is supposed to be a memory address if *instr* $\in$ *LOAD* $\cup$ *STORE*.

The result of *instr* is given by the values *f(a)* assigned to the result locations $<f, a>$ for *instr* through the execution of *instr*. In DLX–computations it can be read off from the final state of the inspected instruction cycle for *instr*. In DLX^p–computations the result of (an occurrence of) *instr* is smeared over the whole instruction cycle of *instr* and must be collected from different pipe stages, depending on the instruction type. Table 1 defines which result is collected after which pipe stage[16]. This completes the definition of the result of the execution of occurrences of *instr*. The result of *instr* is also called the result of the instruction

---

[15] In DLX every instruction has only one result proper and this result is written at the end of the instruction's execution.

[16] In this way we provide a simple explicit and local definition of the global and implicit data and time abstraction functions which are introduced in [WC95] to "collect different pieces of the pipelined state stream at different times and package them into a state record to appear in the non–pipelined state stream at a particular time". [W90] could make successful use of the orthogonality of data and temporal abstraction functions in his hierarchical approach to microprogrammed (non pipelined) microprocessor verification. When pipelining is present these two abstractions are not orthogonal any more. [WC95] define an new abstraction function in order "to preserve the illusion that instructions execute sequentially in the architectural model even though the pipelined implementation performs operations in parallel". By using the notion of result locations defined here, together with the notion of relevant locations defined below, we reduce the complexity of such an abstraction function and boil it down to the consideration of local features which are familiar from the design practice.

cycle of (the given occurrence of) *instr*. For notational convenience, the result of
a computation is defined as the sequence of the results of its instruction cycles.

Result Location	Updated by *instr* in	to be collected after the end of the pipe stage
< reg, dest (instr) >	$ALU \cup SET \cup LOAD \cup JLINK \cup \{MOVS2I\}$	WB(*instr*)
< reg, IAR>	$\{TRAP, MOVI2S\}$	EX(*instr*)
< mem, arg>	$STORE$	MEM(*instr*)
< reg, PC>	$JUMP \cup BRANCH$	EX(*instr*)
	$\notin JUMP \cup BRANCH$	IF(*instr*)

Table 1. Result locations and their collection time. *arg* is an abbreviation for the value
of *reg (fstop (instr))* + *ival (instr)* at the moment of fetching *instr* in *DLX*.

Used Locations and Hazards. Given an instruction cycle for I in a DLX^p
computation, denote by $I \overset{1,2,3}{<} I'$ that an instruction cycle for I' is starting
1,2 or 3 steps after the one for I. Hazards can arise if $I \overset{1,2,3}{<} I'$ and I' uses
a result of I. Table 2 defines what is "used" by an *instr* in a run, namely—
besides static information like the one encoded in *instr* and accessed using the
functions *opcode, nthop, dest, iop, ival*—the content of operand registers in the
register file, of *PC*, of memory locations and of the interrupt address register
IAR. The table also defines the critical pipe stage during which the machine
needs the correct value of that location. A simple analysis of the DLX^p–rules
(see the definition of *Irrelev 1,2* below) shows that conflicts can arise in two
ways, namely *a)* if I' uses, as one of its operands, the content of the destination
register of a preceding instruction I in the pipe, *b)* if I' enters in the pipe shortly
after a jump or branch instruction. For the analysis of these *data and control
hazards* we distinguish whether or not the data dependence concerns a jump or
branch instruction.

Definition. *I' is data dependent on I iff* $I \overset{1,2,3}{<} I'$ *and one of (i), (ii) holds.*
(i) *dest (I)* $\in$ *{ fstop (I'), scdop (I') }* *and* $I' \notin JUMP \cup BRANCH$,
(ii) *dest (I)* = *fstop (I')* *and* $I' \in JUMP \cup BRANCH$.

A DLX^p computation is *data hazard free* if it contains no occurrence of an
instruction which is in the pipe together with an occurrence of an instruction on
which it is data dependent.

When a jump or branch instruction I is fetched, the two instruction cycles
starting 1 and 2 steps later generate results which would spoil the continuation of
the computation once the jump has been executed (after the stage EX(I) which
updates *PC* to its correct value). In order to separate the correctness proof
for the parallelization of *DLX* from the concern about such control hazards, we
assume in this section that in the transformation P^p of P the compiler places two

empty instructions (formalized by the value *undef*) after each jump or branch instruction occurring in the DLX-program P; we stipulate that these empty instructions do not start an instruction cycle. Without loss of generality we assume that the empty instructions are put into new locations which are linked by the extended *next* function to the old locations in the standard way.

Letting the "compiler" avoid control conflicts by arranging the instructions of P into P^p-code, we have to work with a slightly extended PC-preservation rule. When a jump or branch instruction I is fetched at address $l = reg\,(PC)$, PC is updated to $l' = next\,(l)$ which in P^p is the address of *undef*. But in the $EX(I)$-stage the new value of PC must be computed on the basis of the value of $next\,(next\,(l'))$, i.e. the value of $next\,(l)$ for the DLX-program P. Therefore $PC1$ has to store this value when PC—in case a jump or branch instruction has been fetched— contains the address of the empty instruction. Therefore the PC-preservation rule in DLX^p is $PC1 \leftarrow next\,(next\,(PC))$.

DLX^p **Correctness Theorem.** *Let P be an arbitrary DLX-program, P^p its transformation obtained by inserting two empty instructions after each occurrence of a jump or branch instruction. Let C be the computation of DLX started with program P and C^p the corresponding computation of DLX^p started with P^p. If C^p is data hazard free, then C and C^p have the same result.*

Proof . The decomposition of DLX/DLX^p-computations into instruction cycles allows us to prove the theorem instructionwise, using an induction over the given DLX-computation. For the inductive step we need a stronger inductive hypothesis than what is stated in the theorem. For its formulation we introduce the notion of *relevant locations* which allows us to define locally the relation between sequential states and their pipelined counterparts, avoiding the flushing technique used in [BD94] and [Cy95].

DLX^p**-Lemma.** *Let P, P^p, C, C^p be as in the DLX^p Correctness Theorem. For $n \geq 0$ let IC_n, IC_n^p be the n-th instruction cycle in C, C^p respectively. a) If C^p is data hazard free, then IC_n, IC_n^p are instruction cycles for the same (occurrence of a) DLX-instruction instr and start with the same values for the relevant locations used by instr. b) If IC, IC^p are instruction cycles for instr in C, C^p respectively which start with the same values for the relevant locations used by instr and if instr is not data dependent on any instruction in the pipe, then IC, IC^p compute the same result.*

A location l used by *instr* is called *relevant* except in the following two cases:

Irrelev 1. $l = <reg, IAR>$ **and** $instr = MOVS2I$ enters the pipe 1, 2 or 3 stages after an occurrence of $MOVI2S$ or of $TRAP$; [17]

Irrelev 2. $l = <mem, arg>$ **and** $instr \in LOAD$ enters the pipe 1, 2, or 3 stages after an occurrence of a $STORE$ instruction for the same value arg.[18]

---

[17] No conflict can arise from using $<reg, IAR>$ because $MOVS2I$, the only instruction which uses IAR, can never be in conflict with any preceding instruction. If I writes into IAR, then $I \in \{TRAP, MOVI2S\}$ and I writes into IAR in its third pipe stage; therefore if $I \overset{1,2,3}{<} I'$, then I has already written into IAR when I' uses it.

[18] No conflict can arise from using a memory location because load instructions—the

The projection of relevant and of result locations, out of sequences of computation steps of the pipelined processor, represents the state information which characterizes the sequential execution of the instruction under investigation. As we will see below it is easily shown to be semantically correct in case no potential conflict does occur.[19] Our definition of relevance will be refined in the subsequent upgraded machines by admitting as additional irrelevant locations all those where in a hazardous situation the refined architecture will take care of providing the right values for them when needed. In this way we make it explicit where and how the compiler assumptions can be weakened if the hardware is strenghthened (to solve a given type of conflicts). This illustrates the potential of ASM modelling to deal with hardware/software co-design problems in a rigorous but nevertheless simple and transparent way[20].

The lemma clearly implies the theorem. The proof of the lemma is by induction on n. For $n = 0$ the claim holds by the assumption that C and C^p correspond to each other and therefore are initialized with the same static functions and with the same dynamic functions reg and mem. In the induction step, by inductive hypothesis, for each $i \leq n$, the i-th instruction cycle IC_i^p in C^p starts with the same values for the relevant locations used by $instr\ i$ as does the i-th instruction cycle IC_i in C and they both compute the same result. Therefore IC_{n+1} and IC_{n+1}^p are instruction cycles for the same instruction $instr$ and start with the same values for the relevant locations used by that instruction. Due to the absence of stalls, the $n + 1$-th instruction cycle in C^p starts after the first step of IC_n^p in case the instruction $instr_n$ is neither a branch instruction with

---

only ones which use memory locations—can never be in conflict with preceding store instructions—the only ones which write into memory locations. Indeed if $I \in STORE$ and $I' \in LOAD$, then I updates its result location $<$ *mem, reg (fstop (I)) + ival (I)>* in its fourth pipe stage and I' reads the value of the location $<$ *mem, reg (fstop (I)) + ival (I)>* in its fourth pipe stage too. Therefore if $I \overset{1,2,3}{<} I'$ and I' loads the value of the result location of I as updated by I, then I has already updated this result location when I' loads from there. We remind the reader that DLX does not support self-modifying code.

[19] Our localization constitutes a different way to separate the two concerns which are dealt with in [Taku95:pg.1] by splitting the correctness proof into two independent steps, namely a) showing "that each architectural instruction is implemented correctly by the sequential execution of its pipeline states", and b) showing that "under certain constraints from the actual architecture, no conflicts can occur between the simultaneously executed instructions". A similar separation, into the concern about the correct functionality and the concern about the correct processing of instructions by the pipelining, is suggested also in [Taku93, AL95].

[20] [TaKu95] separate the hardware part EBM from the software constraints SW_Constr for their contribution to imply the pipelining correctness property. The correctness proof can then be split into two steps, namely *a)* EBM implements each instruction correctly by the sequential execution of its pipelined stages, *b)* the software constraints SW_Constr guarantee that in EBM no conflicts can occur between any simultaneously executed instructions. Our stepwise refinements of (ir)relevant locations make the hw/sw-interplay between EBM and SW_Constr directly visible.

Location	Used by *instr* in	Critically in stage
<*reg, nthop*>	$ALU \cup SET \cup$ $BRANCH \cup \{MOVI2S\}$ $JUMP\text{-}\{TRAP\}$	EX(*instr*)
	$LOAD \cup STORE$	MEM(*instr*)
< *reg, IAR*>	$\{MOVS2I\}$	EX(*instr*)
< *mem, arg*>	$LOAD$	MEM(*instr*)
< *reg, PC*>	$JUMP \cup BRANCH$	EX(*instr*)
	$\notin JUMP \cup BRANCH$	IF(*instr*)

Table 2. Critical stages for usage of locations.

true branching condition nor a jump; otherwise the $n+1$–th instruction cycle in C^p starts after the third step of IC_n^p due to the following *Jump Lemma* (which is easily proved by induction on the number of fetched jumps).

Jump Lemma. *If a jump or branch instruction I is fetched in a DLX^p– computation, then the following two fetched instructions are empty and at stage* ID(I) *the register PC1 is updated by the correct value to be used for the computation of the possible new PC–value in stage* EX(I).

Since the other result locations depend on the instruction type we are led to a natural case distinction. For each case it is routine to show that through corresponding updates in IC and IC^p, the same value is computed for the result location. (The details are carried out in [BM96]).

4 Data hazards for non jump/branch instructions

In this section we enrich the architecture so that it can handle data hazards for non jump or branch instructions freeing the compiler from its work to avoid these conflicts; we show how one can weaken the data hazard freeness assumption in the DLX^p correctness theorem and guarantee nevertheless the correctness of the architecture by enriching the rules with three standard features, namely the forwarding technique, new hardware links coming with appropriate additional control logic (multiplexers), and stalling. Technically speaking we refine the DLX^p machine to a machine DLX^{data} which is shown to work correctly also for the execution of non jump or branch instructions I' with data dependence on a previous instruction I in the pipe, i.e. such that condition (i) holds:

$$(i) \quad I \stackrel{1,2,3}{<} I' \wedge (dest\,(I) = fstop\,(I') \vee dest\,(I) = scdop\,(I'))$$

$$\wedge\ I' \notin JUMP \cup BRANCH$$

In DLX no write after write hazard can occur, because writing is allowed only in one pipe stage, namely WB, and because together with any stalled instruction every later instruction in the pipe is also stalled. *DLX* has also no write after read hazard because the read stage, namely ID, precedes the write stage.

We will specify the rule refinements piecemeal, following the case distinctions whether the data hazard to be handled involves a memory access or not and whether the distance between the data dependent instructions in the pipe is 1, 2 or 3. This case analysis will justify the correctness of the refined architecture and therefore establish the following theorem.

DLX^{data} **Correctness Theorem.** *Let C be the computation of DLX started with program P and C^{data} the corresponding computation of DLX^{data} started with P^p. Assume that in C^{data} no occurrence of a jump or branch instruction is in the pipe together with an occurrence of an instruction on which it is data dependent. Then C and C^{data} compute the same result.*

Proof method. We define DLX^{data} by incrementing DLX^p, technically speaking as a conservative extension of DLX^p, so that whenever an instruction I' without data dependence to any previous instruction I satisfying (i) occurs in the pipe, DLX^{data} computes I' the same way as DLX^p does. This conservativity of the refinement allows us to prove the correctness of DLX^{data} by a case analysis in which instructions without data dependency in the pipe are dealt with by reusing the DLX^p–*Lemma* whereas the remaining instructions are dealt with by rule refinements corresponding to the cases under analysis.

Since the value of any result location different from $< reg, PC >$ is determined by the values of the arguments which are used in stage *EX* or *MEM*, it suffices to locally modify the relevant DLX^p–rules in such a way that even in the case of data dependence the correct arguments are provided. One can then weaken the assumption in the DLX^{data}–correctness statement below that corresponding instruction cycles in DLX^p and DLX^{data} start with the same values of the relevant locations used by their instruction; namely we take the hazardous locations out of the set of the relevant ones (exactly because in DLX^{data} they are taken care of by the architecture). This is a typical example how we use conservative refinements together with the localization or projection technique to mimic the way the computer architect proceeds when he enriches the processor.

Proof. As in the preceding section it suffices to prove the following lemma.

DLX^{data}–**Lemma.** *Let P , P^p, C, C^p, C^{data}, IC_n be as in the DLX^p–Lemma and in the theorem and let IC_n^{data} be the n-th instruction cycle in C^{data}.*
a) If C^{data} is free of data hazards for jump or branch instructions, then IC_n and IC_n^{data} are instruction cycles for the same DLX–instruction I' and start with the same values for the relevant locations used by I'.

Let IC, IC^p, IC^{data} be instruction cycles for any I' in C, C^p, C^{data} respectively which start with the same values for the relevant locations used by I'. Then the following two properties hold:
b) If I' is not data dependent on any I in the pipe, then IC^{data} and IC^p, and therefore also IC, compute the same result.

c) If $I' \notin JUMP \cup BRANCH$ is data dependent on some $I \overset{1,2,3}{<} I'$, then IC^{data} and IC compute the same result.

The DLX^{data}–*Lemma* is proved by induction on the number n of instruction cycles. For $n = 0$ the claim is satisfied by the assumption that C, C^p and C, C^{data} correspond to each other. The inductive step for *a)* is proved in the same way as shown for the DLX^p–*Lemma*; the Jump Lemma is true also for DLX^{data} because the same program modification P^p of P is used for C^{data} as for C^p.

For *b)* let I' be data independent of any I which precedes it in the pipe. Then it is easily checked that for each DLX^{data}–rule which is applied in IC^{data} for the execution of (this occurrence of) I', in any of its five pipe stages, the branch is taken which constitutes the DLX^p–part of that rule. Since by assumption IC^p and IC^{data} start with the same values for the relevant locations used by I', the effect of these DLX^{data}–rules applications to I' in C^{data} is the same as that of the DLX^p–rules in IC^p and in particular the values of the result locations of I' computed in IC^p and IC^{data} coincide. From the DLX^p–*Lemma* it follows that also IC and IC^{data} compute the same result.

For *c)* assume we have instructions I, I' in C^{data} satisfying (i). By the assumption on jump/branch instructions we know that the following holds: a) $I \in ALU \cup SET \cup LOAD \cup JLINK \cup \{MOVS2I\}$ and
b) $I' \neq \{MOVS2I\}$, i.e. $I' \in ALU \cup SET \cup LOAD \cup STORE \cup \{MOVI2S\}$. The reason is that only in these cases, *dest (I)*, *fstop (I')*, *scdop (I')* respectively are defined (see table 3 and remember that *dest (I)* = R31 for $I \in JLINK$). Therefore it is natural to distinguish three cases depending on whether the data hazard involves a memory access or not. We distinguish two subcases depending on the distance between data dependent instructions in the pipe. For each case we are going to show that the values of the result locations of I' in IC are the same as the ones produced by executing I' through the refined rules in IC^{data}. Let $MEM = LOAD \cup STORE$ and $REG = INSTRUCTION - MEM$. In going through these cases we explain also the required refinement of DLX^p–rules to DLX^{data}–rules (which are fully spelled out in the appendix).

function	instructions
dest (instr)	*instr* $\in ALU \cup SET \cup LOAD \cup JLINK$ $\cup \{MOVS2I\}$
fstop (instr)	*instr* $\in ALU \cup SET \cup MEM \cup JLINK$ $\cup \{MOVI2S\} \cup BRANCH \cup PLAINJ$
scdop (instr)	*instr* $\in ALU \cup SET \cup STORE$

Table 3. Domain of definition of *dest, fstop, scdop*.

4.1   Case $I \in$ REG

In this case it follows from a) that $I \in ALU \cup SET \cup JLINK \cup \{MOVS2I\}$. *dest (I)* receives its correct value, to be used by I', when it is updated in the

WB–stage of I by the value in $C1$; the latter has been copied in the *MEM*–stage of I from C where it has appeared in the *EX*–stage of I (as the result of an $ALU \cup SET$–operation or as content of $PC1$ or of IAR). If I' enters the pipe 3 or 2 steps after I, then the *ID*–stage of I'—in which the operands of I' are read—overlaps with the *WB*–stage or with the *MEM*–stage of I during which the expected operand value is available in $C1$ or C respectively.

In case I' enters the pipe one step after I, the expected operand value is computed during the *ID*–stage of I' and is available in the *EX*–stage of I' but not before. As a consequence the data hazard can be resolved in those two cases by refining the *ID*–rules $OPERAND$ (for the first case) and the *EX*–rules ALU, $MOVI2S$, MEM_ADDR, $Pass\_B\_to\_MDR$ (for the second case).

Subcase $I \overset{2,3}{<} I'$. In this case the architecture can resolve the data hazard between I' and I by the following refinement of the $DLX^p - OPERAND$ rule which guarantees that in case of conflict the correct value of A or B is taken from $C1$ or C and not from $nthop$ (I'):

> **if** $nthop$ $(IR) \in \{dest$ $(IR3),$ $dest$ $(IR2)\}$
> **then if** $nthop$ $(IR) = dest$ $(IR3) \neq dest$ $(IR2)$ **then** $nthReg \leftarrow C1$
>   **if** $nthop$ $(IR) = dest$ $(IR2)$ **then** $nthReg \leftarrow C$
> **else** $nthReg \leftarrow nthop$ (IR)

In case of two successive updates of *dest* (I), the last one counts (due to the sequentiality of the execution of P in DLX). In the sequel we will refer to the above case distinction in the refined rule $OPERAND$ by the following notation (where $nth \in \{fst, \, scd\}, fstReg = A, \, scdReg = B$):

$$C' = \begin{cases} C1 \text{ if } & nthop \text{ } (IR) = dest \text{ } (IR3) \text{ and } nthop \text{ } (IR) \neq dest \text{ } (IR2) \\[2mm] C \text{ if } & nthop \text{ } (IR) = dest \text{ } (IR2) \end{cases}$$

Reflecting the strengthening of the architecture by the rule refinement, in the $DLX^{data}-Lemma$ we weaken the assumptions by enlarging the set of non relevant I'–used locations by:

Irrelev 3. $< reg, nthop(I') >$ such that $I' \notin JUMP \cup BRANCH$ and for some $I \overset{3,2}{<} I'$ with $I \in REG$ holds $nthop$ $(I') = dest$ (I).

Therefore the DLX^{data} $OPERAND$ rule guarantees that the correct arguments for the *EX*–or *MEM*–stage rules of I' are loaded into A, B in both cases, *a)* when I' has no data dependency from any instruction in the pipe and *b)* in the case of data dependence on $I \in REG \wedge I \overset{3,2}{<} I'$. The price for this hazard resolution is a direct link between the register file–exits A, B and C, $C1$. For the case $I \overset{3}{<} I'$ our solution avoids the introduction of two file accesses (one for writing followed by one for reading [HP90]) per clock cycle.

Subcase $I \overset{1}{<} I'$. If I' immediately follows I in the pipe, then the I'–operand value, to be computed by I, comes out of the ALU and goes into C at the end of the *EX*–stage of I. Thus by *forwarding* the ALU–result as next ALU–input directly without passing through C and A, B, the ALU is enabled to compute the EX–stage of I' with the correct arguments.

The formalization of this forwarding technique consists in a refinement of the EX–rules for the cases with can arise here for I', namely $I' \in ALU \cup SET$, $I' \in MEM$, $I' = MOVI2S$. In each case we add to the corresponding EX–rule of DLX^p a clause which in the data hazard case provides the argument C instead of A or B respectively. This is at the expense of introducing a direct link between C and both ALU ports (for $I' \in ALU \cup SET$) and IAR (for $I' \in = MOVI2S$) and MAR and $SMDR$ (for $I' \in MEM$) together with some control logic (multiplexers) for selecting the forwarded value as the ALU input rather than the value from the register file. For example for $I' \in MEM$ we obtain the following rule refinements (both rules will be furthermore refined by an additional clause below):

MEM_ADDR	Pass_B_to_SMDR
if *opcode* $(IR1) \in LOAD \cup STORE$	if *opcode* $(IR1) \in STORE$
then if *fstop* $(IR1) = dest\ (IR2)$	then if *scdop* $(IR1) = dest\ (IR2)$
then $MAR \leftarrow val_{fst} + ival\ (IR1)$	then $SMDR \leftarrow val_{scd}$
else $MAR \leftarrow A + ival\ (IR1)$	else $SMDR \leftarrow B$

$$\text{where } val_{nth} = \begin{cases} C & \text{if } nthop\ (IR1) = dest\ (IR2) \\ nthReg & \text{otherwise} \end{cases}$$

$$\textbf{and } nth \in \{fst,\ scd\}, fstReg = A,\ scdReg = B.$$

Similarly one proceeds for the refinements of the rules ALU and $MOVI2S$, see the DLX^{data} appendix. Since the refinement of these EX–rules solves the data conflict under study, we add the following non relevant I'–used locations:

Irrelev 4. $< reg, nthop(I') >$ such that for some $I \overset{1}{<} I'$ with $I \in REG$ one of the following holds: a) *opcode* $(I') \in ALU \cup SET$, *iop* $(opcode\ (I')) = true$, *dest* $(I) = fstop\ (I')$, $nth = fst$;
b) *opcode* $(I') \in ALU \cup SET$, *iop* $(opcode\ (I')) = false$, *dest* $(I) = nthop\ (I')$;
c) *opcode* $(I') \in MEM \cup \{MOVI2S\}$, *dest* $(I) = fstop\ (I')$, $nth = fst$;
d) *opcode* $(I') \in STORE$, *dest* $(I) = scdop\ (I')$, $nth = scd$.

Therefore the refined EX–rules of DLX^{data} provide the correct arguments for the EX–stage rules of I' in both cases, through A, B when I' has no data dependency on any instruction in the pipe, and through the forwarded freshly computed I–result in the data dependency case $I \in REG$ and $I \overset{1}{<} I'$.

4.2 Case $I \in$ MEM and $I' \in$ REG

$I \in MEM$, $I' \in REG$ and (i), (a), (b) above yield $I \in LOAD$ **and** $I' \in ALU \cup SET \cup \{MOVI2S\}$. The value *val* loaded by an instruction I is available only at the end of I's MEM–stage, namely in $LMDR$. Therefore non–MEM–instructions I' which enter the pipe 3 or 2 steps later than I and use *val* as operand, can grep it from $LMDR$ in their ID or EX–stage respectively. As for the case $I \in REG$, it suffices to refine the rule $OPERAND$ and the relevant EX–stage rules (here

ALU and *MOVI2S*) furthermore. If however I' enters the pipe immediately after I, then the pipeline has to be stopped for one stage, starting at the latest just before the *EX*–stage of I', in such a way that after the pipeline takes off again, I' can grep from *LMDR* the value I meantime has loaded there.

Subcase $I \overset{3,2}{<} I'$. For the refinement of the *OPERAND*–rule, making use of our abbreviated notation above it suffices to refine C' by adding the case of data dependency of the ante–ante–preceding instruction:

$$
C' =
\begin{cases}
C1 & \text{if } nthop\ (IR) = dest\ (IR3) \quad last\ modification\ in \\
& \text{and } nthop\ (IR) \neq dest\ (IR2)\ ante - ante - preceding \\
& \text{and } opcode\ (IR3) \notin LOAD \quad not\ load\ instr \\[2mm]
\overline{LMDR} & \text{if } nthop\ (IR) = dest\ (IR3) \quad last\ modification\ in \\
& \text{and } nthop\ (IR) \neq dest\ (IR2)\ ante - ante - preceding \\
& \text{and } opcode\ (IR3) \in LOAD \quad load\ instr \\[2mm]
C & \text{if } nthop\ (IR) = dest\ (IR2) \quad last\ modification\ in \\
& \qquad\qquad\qquad\qquad\qquad\qquad ante - preceding\ instr
\end{cases}
$$

where $nth \in \{\ fst,\ scd\ \}$, $fstReg = A$, $scdReg = B$, $\overline{LMDR} = \overline{opcode(IR3)}\ (LMDR)$. Similarly an additional clause is introduced in the preceding refinement for the rules *ALU, MOVI2S* for which we refine the definition of val_{nth} as follows (see the DLX^{data} appendix for details):

$$
val_{nth} =
\begin{cases}
C & \text{if } nthop\ (IR1) = dest\ (IR2) \\[2mm]
\overline{LMDR} & \text{if } nthop\ (IR1) = dest\ (IR3)\ \textbf{and}\ opcode\ (IR3) \in LOAD \\
& \text{and } nthop\ (IR1) \neq dest\ (IR2) \\[2mm]
nthReg & \textbf{otherwise}
\end{cases}
$$

where $nth \in \{\ fst,\ scd\ \}$, $fstReg = A$, $scdReg = B$.

This further refinement of the rules *OPERAND, ALU, MOVI2S* comes together with adding the following nonrelevant locations.

Irrelev 5. $< reg, nthop(I') >$ such that for some $I \in LOAD$ with $I' \in REG - (JUMP \cup BRANCH)$ **and** $nthop\ (I') = dest\ (I)$ one of the following holds:
a) $I \overset{3}{<} I'$; **b)** $I \overset{2}{<} I'$, $iop\ (opcode\ (I')) = true$, $nth = fst$; **c)** $I \overset{2}{<} I'$, $iop\ (opcode\ (I')) = false$.
Therefore the furthermore refined *ID*–rule *OPERAND* of DLX^{data} guarantees that the correct arguments for the *EX*–or *MEM*–stages rules for I' are loaded into A, B in case of non data dependency of I', but also in the data dependency case with an $I \overset{3}{<} I'$, $I' \in REG$, $I \in LOAD$; the refined *EX*–rules *ALU, MOVI2S* provide the correct arguments for the *EX*–stage rule applications through A, B (in case of no data conflict) or through the forwarded value freshly loaded by I in case of data dependence on $I \overset{2}{<} I'$, $I' \in REG$, $I \in LOAD$.

Subcase $I \overset{1}{<} I'$. In this case the pipelined execution of I' (and therefore also of later instructions) has to be stopped at the latest just before the *EX*–stage of I', until the value to be loaded by I becomes available, namely in *LMDR*. It is common practice to add a *pipeline interlock* which detects this situation and stops the pipelining until the situation has been resolved. We formalize this by introducing a new function *load_risk*, defined by:

$$opcode~(IR2) \in LOAD \textbf{ and } reg~(IR1) \in REG ~-~ (JUMP \cup BRANCH)$$
$$\textbf{and } dest~(IR2) \in \{fstop~(IR1)~,~scdop~(IR1)\}.$$

By putting the rules of stage *EX*, *ID* and *IF* under the additional guard $\neg load\_risk$ we obtain that in case of *load_risk* they are not executed whereas the *MEM*–and *WB*–rules are executed. By adding to the *FETCH*–rule the clause *if load_risk then $IR2 \leftarrow undef$*, we obtain that immediately after the execution of this *FETCH*–rule the condition *load_risk* will be *false* (because *opcode (IR2)* $\in LOAD$ is false by *opcode (undef) = undef*) and the full pipelined execution will be resumed. At this point $I' = reg~(IR1)$ still holds but I has been copied by *Preserv IR2* from *IR2* to *IR3*; therefore the subcase of data dependency considered here is reduced to the previous subcase and resolved by the refined EX–rules in DLX^{data}.

By the introduction of the *load_risk* guard to the rules in $IF \cup ID \cup EX$ and of the new *load_risk* rule, the architecture takes care of providing the right arguments for the execution of any $I' \in REG ~-~ (JUMP \cup BRANCH)$ which is data dependent on a load instruction $I \overset{1}{<} I'$, without changing the behavior for instructions without data conflict. This yields the following additional non relevant location:

Irrelev 6. $< reg, nthop(I') >$ such that $I' \in REG ~-~ (JUMP \cup BRANCH)$ and some $I \in LOAD$ satisfies $I \overset{1}{<} I' \wedge nthop~(I') = dest~(I)$.

4.3 Case I, $I' \in$ MEM

Subcase $I \overset{2,3}{<} I'$. The data conflict can be resolved by using the *OPERAND*–rule or the once more refined *EX*–stage rules in order to provide the value loaded by I as operand for I'. The *OPERAND* rule as refined in the previous case already resolves the conflict if $I \overset{3}{<} I'$. If $I \overset{2}{<} I'$, the two *EX*–rules for the *MEM*–instruction I' are *MEM_ADDR* and *Pass_B_to_SMDR*; their refinement is obtained by including into the guard, for the forwarding case, as new disjunct *fstop (IR1) = dest (IR3)* **and** *opcode (IR3)* $\in LOAD$ for *MEM_ADDR* and *scdop (IR1) = dest (IR3)* **and** *opcode (IR3)* $\in LOAD$ for *Pass_B_to_MDR*. This yields the two final *EX*–stage rules of DLX^{data} shown in the appendix. This refinement implies introducing direct links between *LMDR* and *MAR*, *SMDR* and the following new non relevant locations:

Irrelev 7. *a)* $< reg, nthop(I') >$ for $I' \in MEM$ and some $I \in LOAD$ satisfying $I \overset{3}{<} I'$ and $nthop~(I') = dest~(I)$;

b) $< reg, fstop(I') >$ for $I' \in MEM$ and some $I \in LOAD$ satisfying $I \overset{2}{<} I'$ and $fstop\ (I') = dest\ (I)$;

c) $< reg, scdop(I') >$ for $I' \in STORE$ and some $I \in LOAD$ satisfying $I \overset{2}{<} I'$ and $scdop\ (I') = dest\ (I)$.

Subcase $I \overset{1}{<} I'$. The MEM–instruction I' can use the value loaded by the preceding instruction I in two ways, as datum to be stored (case a) or as address for the load or store operation (case b).

Case a. $dest\ (I) = scdop\ (I')$. In this case $I' \in STORE$ and the value loaded by I is needed by I' in its MEM–stage—during which it is available in $LMDR$. Therefore this case can be handled again by forwarding, formalized through refining the $STORE$–rule (see the DLX^{data}–appendix) at the expense of a direct link between $LMDR$ and the memory input port. Since the refined rule resolves the data conflict for the case under study, the claim of the lemma follws if we add the following non relevant locations:

Irrelev 8. $< reg, scdop(I') >$ for $I' \in STORE$ **and** some $I \in LOAD$ satisfying $I \overset{1}{<} I'$ and $scdop\ (I') = dest\ (I)$.

Case b. $dest\ (I) = fstop\ (I')$. In this case I' needs its first operand during its EX–stage when the memory address is computed. But $dest\ (I)$ is loaded into $LMDR$ only during the MEM–stage of I so that the pipeline must be interrupted again for one clock cycle, namely we have to uphold the execution of the rules for the EX–stage of I' and therefore also for the two preceding stages ID and IF. This can be formalized by refining the guard $load\_risk$ through the additional case $dest\ (IR2) = fstop\ (IR1)$ **and** $reg\ (IR1) \in MEM$. Thereby the modified rules resolve the data conflict in this case, establishing the claim of the lemma with the following additional non relevant locations:

Irrelev 9. $< reg, fstop(I') >$ for $I' \in MEM$ **and** some $I \in LOAD$ satisfying $I \overset{1}{<} I'$ and $fstop\ (I') = dest\ (I)$.

5 Handling control hazards

We extend now DLX^{data} to a machine DLX^{pipe} which—as we will show—handles also control hazards correctly without help from the compiler.

Control hazards are those created by jump instructions (under which we subsume also branch instructions). They present two problems, namely

a) to guarantee that after fetching a jump instruction I', the next instruction which will be fetched is the one I' requires to jump to, i.e. the instruction whose address is the value of PC as updated through the execution of I' ,

b) the data dependence of a jump instruction on a preceding instruction in the pipe.

As part of our *divide and conquer* approach we have postponed these two problems up to now by a) assuming, for the correctness proofs, that DLX^{data}–computations are always started with the "compiled" version P^p of P into which two empty instructions are inserted after each jump instruction in P (allowing us to use the *Jump Lemma*), and by *b)* assuming that there are no data dependent jump instructions in DLX^{data}–computations. In this section we transform DLX^{data} first to a model DLX^{ctrl} with the same functionality as DLX^{data} but which does not need any more the compilation of empty instructions after jumps. Then we refine DLX^{ctrl} to DLX^{pipe} and prove that it handles also data dependent jump instructions correctly.

5.1 Computing jump addresses in the ID phase

The problem here is to guarantee at run time that when a *JUMP* or *BRANCH* instruction I is fetched, no other instruction I' is fetched before the computation of the new value of *PC*, to be determined by I, is done. Since after fetching I it needs at least one clock cycle for I to compute the new value for *PC* [HP90], fetching has to be stopped for at least one pipe stage. One can avoid to stall the pipe for a second pipe stage by a special decoding which permits to detect jump instructions immediately after the *IF*–stage, combined with anticipating the computation of the new *PC*–value in the *ID*–stage (instead of the *EX*–stage used in DLX^{data}). As effect we will obtain that in DLX^{ctrl}, one pipe stage after the *IF*–stage of a jump instruction I, the value of *PC* is already the correct *PC* result value of I.

Formally we replace the *EX*–rules *JUMP*, *BRANCH* and the *PC*–updating part of *TRAP* in DLX^{data} by new *ID*–rules which are obtained by substituting *IR1*, *PC1*, *A* by *IR*, *PC*, *fstop (IR)* respectively. The *IAR*–updating part $TRAP_{IAR}$ of *TRAP* and the *LINK*–rule remain in *EX*–stage, because they update result locations different from *PC* whose computation needs not to be changed in going to DLX^{ctrl}. The zero–test in *BRANCH*–instructions can be done without using the ALU by relying upon the usual standard output of registers. (See below for one more addition to the *BRANCH*–rule.)

The *FETCH*–rule of DLX^{data} is refined by introducing an additional guard *pc_risk* [21] which prevents *IR* and *PC* to be updated in case a jump instruction, fetched one clock cycle ago, triggers the correct update of *PC* through one of the new *ID*–stage rules *JUMP*, *BRANCH* or $TRAP_{PC}$. In this case *IR* is set to *undef* so that in the next clock cycle *pc_risk* will be false and the *FETCH*–rule will have again the same effect in DLX^{data} and in DLX^{ctrl}. We define *pc_risk* as *opcode (IR)* $\in$ *JUMP* $\cup$ *BRANCH* and delete the guard ¬*jumps* in the *FETCH*–rule. Since only the rules $TRAP_{IAR}$ and *LINK* in DLX^{ctrl} still use *PC1*, we can replace the DLX^{data}–rule for preservation of *PC* by its else–branch *PC1* ← *PC*. For the complete rule set of DLX^{ctrl} see the appendix.

The DLX^{ctrl} **Correctness Theorem** is the same as for DLX^{data} with C^{data} replaced by the corresponding computation C^{ctrl} of *P* by DLX^{ctrl}. The proof is by reduction to the DLX^{data} correctness theorem and relies upon the fact that corresponding applications of homonymous rules in DLX^{data} and DLX^{ctrl} compute the same result. (We consider the update guarded by ¬*jumps*

---

[21] Using this guard (and similar guards load-update-risk etc. below) is similar to the introduction of SW-constraints in [TaKu95].

in the $FETCH$–rule of DLX^{data} as homonymous to the corresponding new up-date in the $FETCH$–rule of DLX^{ctrl}.) The proof follows by induction on the lenght n of C from the following analogue of the DLX^{data}–Lemma.

DLX^{ctrl}–**Lemma.** *Let P, P^p, C, C^{data}, IC_n, IC_n^{data} be as in the DLX^{data}– Lemma and let IC_n^{ctrl} be the nth instruction cycle in the computation C^{ctrl} of P by DLX^{ctrl}.*

a) If C^{ctrl} (and therefore also C^{data}) is free of occurrences of jump or branch instructions which are data dependent on any instruction in the pipe , then IC_n^{ctrl} and IC_n^{data} (and therefore also IC_n) are instruction cycles for the same DLX- instruction I and start with the same values for the relevant locations used by I.

Let IC, IC^{data}, IC^{ctrl} be instruction cycles for any I in C, C^{data}, C^{ctrl} resp. which start with the same values for the relevant locations used by I. Then the following two properties hold:

b) If I is not data dependent on any instruction in the pipe, then IC^{data} and IC^{ctrl} compute the same result, namely the result of the computation of I in IC.

c) If $I \notin JUMP \cup BRANCH$ is data dependent on some $I' \overset{1,2,3}{<} I$, then IC^{ctrl} and IC^{data} (and therefore IC) compute the same result.

Proof: The proof is by induction on n. For $n=0$ the claim holds because C and C^{data}, C^{ctrl} are initialized correspondingly. For the inductive step of $a)$ the proof for IC_n^{ctrl} and IC_n^{data} goes along the same lines as for the DLX^p–Lemma.

For $b)$ and $c)$ we have only to show that IC^{data} and IC^{ctrl} compute the same result. By the DLX^{data}–Lemma we can then infer that this is the result computed by IC in DLX. Since DLX^{ctrl} has the same rules as DLX^{data} except for those which update the result location $< reg, PC >$, it suffices to check that IC^{ctrl} and IC^{data} compute the same value for $< reg, PC >$.

We distinguish two cases depending on whether the non empty instruction I fetched at the beginning of IC^{ctrl} and IC^{data} is or is not in $JUMP \cup BRANCH$.

Case 1. $I \in JUMP \cup BRANCH$

The *Jump Lemma* guarantees that I is followed in C^{data} by two empty instruc- tions and that $PC1$ is updated in the stage $ID(I)$ by the correct value to be used in the stage $EX(I)$ for the computation of the new PC–value. As a result of the execution of I in C^{data} the new value to be computed for the register PC is ready after the rules for the $EX(I)$ pipe stage have been executed. This value is the correct value because by assumption I is not data dependent on any instruction in the pipe, therefore in stage $ID(I)$ in IC^{data} the correct values are loaded into A and B and then used in stage $EX(I)$ together with reg $(PC1)$ to compute the new value by which PC is updated in this stage. The two empty instructions which follow any jump or branch instruction in P^p guarantee that during the ID–stage of I, $\neg jumps$ holds so that PC is again updated by $next$ (PC) and therefore IR is updated in stage $ID(I)$ and $EX(I)$ with $undef$ (when the $FETCH$ rule is applicable at all), thus "stalling" the pipe for two stages.

In C^{ctrl} the correct next PC–value is ready after the rules for the pipe stage

ID(I) have been executed; indeed the new *JUMP-, TRAP-* and *BRANCH*-rules update the register PC in stage ID(I) by the correct value, due to the assumption that I is not data dependent on any instruction in the pipe. (Remember the assumption made when defining P^p that if l is the value of PC from where I has been fetched, then the value of *next (l)* in *DLX*—which is used in DLX^{ctrl} through PC as basis for the computation of the new value to which PC is then updated in EX(I)—coincides in DLX^{data} with the value *next (next (l'))* where $l' = $ *next (l)* in DLX^{data}. Through $PC1$ this value is used in DLX^{data} as basis for the computation of the same new value of PC).

$TRAP_{IAR}$ and the *LINK* rule are the only ones in DLX^{ctrl} which still use $PC1$. Since the DLX^{ctrl}–computations start with P instead of P^p, $TRAP_{IAR}$ and *LINK* need the value to which PC was updated when I was fetched. Therefore the simple copying rule $PC1 \leftarrow PC$ in DLX^{ctrl} provides the correct value which in DLX^{data} was provided by *next (next (next(PC)))*.

The guard *pc_risk* in the *FETCH*-rule prevents, in the case under consideration, a possibly inconsistent update of PC by *next (PC)* and updates IR by *undef*. This guarantees that except for copying *undef* into IR_i, no rule is applicable in ID(*undef*), EX(*undef*), MEM(*undef*), WB(*undef*).

Case 2. $I \notin JUMP \cup BRANCH$
Since I is not empty, by the *Jump Lemma* in the two previous clock cycles no jump instructions have been fetched in C^{data}; therefore the register PC is correctly updated to *next (PC)* when I is fetched in C^{data} (in which moment *not load_risk* is true). The same effect is obtained in C^{ctrl} by applying the *not pc_risk ? IF*-rule. (Since $I \notin JUMP \cup BRANCH$, in this case we need not consider the effect of the rules *JUMP, BRANCH* and *TRAP*.)

5.2 Data hazards for jump instructions

In this section we refine DLX^{ctrl} to our final model DLX^{pipe} which takes care also of jump instructions $I' \in JUMP \cup BRANCH$ with data dependence—namely *dest (I) = fstop (I')*—on an instruction I preceding I' in the pipe by 1, 2 or 3 steps. From table 3 we know that in this case

$$I \in ALU \cup SET \cup LOAD \cup JLINK \cup \{MOVS2I\}.$$

We distinguish two cases depending on the distance between I and I' in the pipe and on whether I is a *LOAD* instruction or not. For distance 3 and for distance 2 to a non–load instruction I, the *forwarding* technique can be applied; distance 2 to a load–instruction I and distance 1 create a *stall*.

Case $I \overset{3}{<} I'$ or ($I \overset{2}{<} I'$ and $I \notin$ LOAD). If I' is fetched 3 clock cycles later that I, then it reads its first operand in its *ID*–stage when I is in its *WB*–stage and has the new value for *dest (I)* available in $C1$ (if $I \notin LOAD$) or in *LMDR* (if $I \in LOAD$). Therefore it suffices to forward this value—at the expense of direct links between PC and $C1$, $LMDR$—in a refinement of the two rules for $JUMP$ and $BRANCH$ by the following additional clauses; for $JUMP$:

> if $fstop\ (IR) = dest\ (IR3)$
> then if $opcode\ (IR3) \notin LOAD$ then $PC \leftarrow C1$
> if $opcode\ (IR3) \in LOAD$ then $PC \leftarrow \overline{LMDR}$

For a short display of the jump rule we will abbreviate this as follows:

$$\text{if } fstop(IR) = dest(IR3) \text{ then } PC \leftarrow PC'$$

$$PC' = \begin{cases} C1 & \text{if } opcode\ (IR3) \notin LOAD \\ \overline{LMDR} & \text{if } opcode\ (IR3) \in LOAD \end{cases}$$

where $\overline{LMDR} = \overline{opcode(IR3)}(LMDR)$. In the $BRANCH$–rule we add the clause:

> if $fstop\ (IR) = dest\ (IR3)$ then if $reg\ (PC') = 0$
> then $PC \leftarrow PC +_{PC} ival(IR)$

The same forwarding technique allow us to cope with the data hazard in case I' is fetched 2 steps after a non–load instruction I on which it depends. In this case the expected new value of $dest\ (I)$ can be forwarded from C for use in $JUMP$ and $BRANCH$ which are therefore refined once more as follows:

> $\boxed{\text{BRANCH}}$ if $not\ load\_risk$
> then if $opcode\ (IR) \in BRANCH$
> then if $fstop\ (IR) \in \{dest\ (IR3),\ dest\ (IR2)\}$
> then if $reg\ (PC') = 0$
> then $PC \leftarrow PC +_{PC} ival(IR)$
> else if $reg\ (fstop\ (IR)) = 0$
> then $PC \leftarrow PC +_{PC} ival\ (IR)$

> $\boxed{\text{JUMP}}$ if $not\ load\_risk$
> then if $opcode\ (IR) \in PLAINJ \cup JLINK$
> then if $iop\ (opcode\ (IR)) = true$
> then $PC \leftarrow PC +_{PC} ival\ (IR)$
> else if $fstop\ (IR) \in \{dest\ (IR3),\ dest\ (IR2)\}$
> then $PC \leftarrow PC'$
> else $PC \leftarrow fstop\ (IR)$

where $PC' = \begin{cases} C1 & \text{if } fstop\ (IR) = dest\ (IR3) \quad last\ modification\ in \\ & \text{and } fstop\ (IR) \neq dest\ (IR2)\ ante - ante - preceding \\ & \text{and } opcode\ (IR3) \notin LOAD \quad not\ load\ instr \\ \\ \overline{LMDR} & \text{if } nthop\ (IR) = dest\ (IR3) \quad last\ modification\ in \\ & \text{and } fstop\ (IR) \neq dest\ (IR2)\ ante - ante - preceding \\ & \text{and } opcode\ (IR3) \in LOAD \quad load\ instr \\ \\ C & \text{if } fstop\ (IR) = dest\ (IR2) \quad last\ modification\ in \\ & \qquad\qquad\qquad\qquad\qquad ante - preceding\ instr \end{cases}$

These refined rules guarantee that DLX^{ctrl} provides the correct argument for the branching test and also the correct PC–value the machine has to jump to, even in case of the data dependency considered here. This justifies the claim for the corresponding case in the DLX^{ctrl}–lemma below for which we can enlarge the set of irrelevant locations as follows:

Irrelev 10. $< reg,\ fstop(I') >$ *such that* $I' \in JUMP \cup BRANCH$ **and** *fstop* $(I') = dest\ (I)$ *for some* I *satisfying* $I \overset{3}{<} I'$ **or** $(I \overset{2}{<} I'$ *and* $I \notin LOAD)$.

Case $(I \overset{2}{<} I'$ **and** $I \in$ **LOAD**$)$ **or** $(I \overset{1}{<} I'$ **and** $I \notin$ **LOAD**$)$. In this case I' needs its first operand in its ID–stage when I, in its MEM–or EX–stage, is providing the expected new value in $LMDR$ or C respectively. Therefore the pipe has to be stopped for one clock cycle to prevent the ID–stage of I' and the preceding IF–stage from proceeding further. This can be done by putting the IF–and ID–rules under an additional guard *pc_data_risk: BOOL* which formalizes this case[22], and by adding the new rule **if** *pc_data_risk* **then** $IR1 \leftarrow undef$. We incorporate this additional rule into the refined $FETCH$–rule. We prove now that this refinement resolves the data hazard between I' and I.

After I' has been fetched, *pc_data_risk* is true. Therefore during the following clock cycle, the rules for the pipe stage of I (MEM or EX respectively) and in the first case also for the instruction preceding I in the pipe are executed, but $FETCH$ loads *undef* into $IR1$, keeping IR and PC unchanged, and none of the ID–rules can fire; moreover $IR2$ is loaded with $IR1$—which in the case under study is a non–load instruction. We show now that as a result of that, *pc_data_risk* becomes false after one clock cycle: *reg (IR1) = undef* implies *dest (reg (IR1)) = undef ≠ fstop (reg (IR))*, so that the second or–condition of *pc_data_risk* is not satisfied; by $IR2 \notin LOAD$ also the first or–condition of *pc_data_risk* not satisfied. Therefore the pipeline restarts and the data dependency has developed into a conflict which has been dealt with already in the preceding case. This establishes the corresponding case in the proof of the DLX^{ctrl}-lemma below for which we enlarge the set of irrelevant locations as follows, anticipating already the next subcase $I \overset{1}{<} I'$ and $I \in$ LOAD:

Irrelev 11. $< reg,\ fstop(I') >$ *such that* $I' \in JUMP \cup BRANCH$ **and** *fstop* $(I') = dest\ (I)$ *for some* I *satisfying* $(I \overset{2}{<} I'$ **and** $I \in LOAD))$ **or** $I \overset{1}{<} I'$.

Case $I \overset{1}{<} I'$ **and** $I \in$ **LOAD.** In this case I' has to wait two clock cycles during which I can load the needed value. The pipelining is stopped in this case

---

[22] i.e. *pc_data_risk = opcode (IR)* $\in BRANCH \cup JUMP$ **and** *[(fstop (IR) = dest (IR2)* **and** *opcode (IR2)* $\in LOAD)$ **or** *fstop (IR) = dest (IR1)]*. Anticipating the next subcase, we have formulated the condition *fstop (IR) = dest (IR1)* for both subcases, namely *reg (IR1)* $\notin LOAD$ or *reg (IR1)* $\in LOAD$.

for two clock cycles during which pc_data_risk is true (first through its second or–clause, then through its first clause.)

This concludes the upgrade DLX^{ctrl} of DLX^{pipe} , see the appendix. We can prove now our main theorem by induction over the given DLX–computation using the following lemma and the DLX^{ctrl}–Correctness Theorem.

DLX^{pipe}–**Lemma.** *Let* P, C, C^{ctrl}, IC_n, IC_n^{ctrl} *be as in the* DLX^{ctrl} – *Lemma and let* IC^{pipe} *be the* n–*th instruction cycle in the computation* C^{pipe} *of* P *by* DLX^{pipe}.

a) IC_n^{ctrl} *and* IC_n^{pipe} *(and therefore* IC_n*) are instruction cycles for the same* DLX–*instruction* I' *and start with the same values for the relevant locations used by* I'.

Let IC, IC^{ctrl}, IC^{pipe} *be instruction cycles for any* I *in* C, C^{ctrl}, C^{pipe} *respectively which start with the same values for the relevant locations used by* I'. *Then the following two properties hold:*

b) *If* I' *is not data dependent on any* I *in the pipe or* $I' \notin JUMP \cup BRANCH$ *is data dependent on some* $I \overset{1,2,3}{<} I'$, *then* IC^{pipe} *and* IC^{ctrl} *(and therefore* IC*) compute the same result.*

c) *If* $I' \in JUMP \cup BRANCH$ *is data dependent on some* $I \overset{1,2,3}{<} I'$, *then* IC^{pipe} *and* IC *compute the same result.*

Proof. The proof of the lemma is by induction on n. For $n = 0$ the claim follows from the assumption that C, C^p, C^{ctrl} are initialized correspondingly. For the inductive step, $a)$ is proved as in the DLX^p–*Lemma*.

$b)$ follows from the DLX^{ctrl}–*Lemma* and the conservativity of the extension DLX^{pipe} of DLX^{ctrl}; namely the same branches are taken, in the rules of C^{ctrl} and of C^{pipe}, by all instructions $I' \notin JUMP \cup BRANCH$ which are data dependent on some $I \overset{1,2,3}{<} I'$ and also by instructions I' which depend on no other instruction in the pipe. This is the case because in refining DLX^{ctrl} to DLX^{pipe}, only the rules $JUMP$ and $BRANCH$ have been extended, and only for a data dependence case, and because the additional guard pc_data_risk, which has been introduced for the rules in IF and in ID, does concern only instructions in $JUMP \cup BRANCH$ with a data dependence.

For $c)$ we distinguish the three possible cases, namely that for some I, $a)$ $I \overset{3}{<} I'$ or ($I \overset{2}{<} I'$ and $I \notin LOAD$), $b)$ ($I \overset{2}{<} I'$ and $I \in LOAD$) or ($I \overset{1}{<} I'$ and $I \notin LOAD$), $c)$ $I \overset{1}{<} I'$ and $I \in LOAD$. For each case we have shown above that the values of the result locations, as produced by executing I' in IC^{pipe} and IC respectively, are the same; in fact as part of the explanation of the refinement of the DLX^{ctrl}–rules to the DLX^{pipe}–rules we have proved that the data hazard is resolved correctly by the DLX^{pipe}–refinement of the $JUMP$ and $BRANCH$ rules (and the additional guard pc_data_risk for the rules in IF and ID) and that the result of executing I' in DLX^{pipe} coincides with the result of executing I' in IC.

Conclusion. We have developed a practical method to handle aspects of modern processor design which are most susceptible to errors. Our method supports modular design and analysis techniques and provides the possibility to pinpoint design errors at an early stage. The models we define are Abstract State Machines in the sense of Gurevich and therefore can be (implemented and) executed using ASM interpreters, providing the possibility to use the models as prototypes for simulation (see also the discussion of the falsifiability property of ASM prototypes in [Bo95]). Using ASMs is economical and quickly learnt: it requires no special theoretical training and directly supports the designer's operational view at the appropriate level of abstraction.

Our method is applicable to more complex processors than DLX, to more advanced pipelining techniques than the basic ones discussed in this paper, and to more sophisticated memory systems. Applications of the method become really interesting where mechanical tool oriented methods face intrinsic limitations (see for example the *"major bottleneck"* for model checking techniques, identified in [BD94] as the computational efficiency of logical decision procedures). We have given some hints indicating that the approach to hardware design and analysis proposed in this paper and in [BoDC95] can also be turned into a practical framework which can be used by the computer architect to formulate and analyse hardware/software co-design problems in a rigorous yet transparent way.

Acknowledgement. The first author expresses his thanks to $DIMACS$ (Joint Technology Center of Rutgers and Princeton University, Bellcore and ATT Bell Labs) in New Brunswick for the hospitality during the Fall of 1995 and to GMD-$FIRST$ (Institute for Computer Architecture and Software Technology of the German National Research Center for Information Technology) in Berlin for the hospitality during the Fall of 1996 when part of this work was done. We are grateful to the following colleagues for their interest in this research, for their criticism and for many stimulating and illuminating discussions: Rajeev Alur, Jörg Bormann, David van Campenhout, David Cyrluk, David Dill, Hans Eveking, Friedrich von Henke, Thomas Henzinger, Ramayya Kumar, Kent Palmer, Natarajan Shankar, Mandajan Srivas and Klaus Waldschmidt.

References

[AL95] M.Aagaard and M.Leeser. *Reasoning About Pipelines with Structural Hazards*, Springer LNCS 901, 1995.

[A95] W. Ahrendt. *Von PROLOG zur WAM, Verifikation der Prozedurübersetzung mit KIV.* Diploma thesis, University of Karlsruhe, December 1995, pp.115.

[ALLSW94] T.Arora, T.Leung, K.Levitt, T.Schubert, and P.Windley: *Report on the UCD microcoded Viper verification project.* Springer LNCS 780, 1994, 239-252.

[Be93] D.L.Beatty. *A Methodology for Formal Hardware Verification, with Application to Microprocessors.* PhD thesis, School of CS, CMU, Aug. 1993.

[Bo95] E. Börger. *Why use evolving algebras for hardware and software engineering.* in: M. Bartosek, J. Staudek, J. Wiedermann (Eds), SOFSEM'95, Springer LNCS 1012, 1995, 236–271.

[BoDC95] E. Börger and G. Del Castillo. *A formal method for provably correct com-position of a real–life processor out of basic components (The APE100 reverse engineering project).* In: Proc. First IEEE International Conference Engineer-ing of Complex Computer Systems (ICECCS'96). IEEE Comp. Soc. Press, Los Alamitos CA, 1995, 145-148.

[BD95] E. Börger and I. Durdanović. *Correctness of Compiling Occam to Transputer Code.* in: Computer Journal 39, 1996, 52–92.

[BM96] E. Börger and S. Mazzzanti. *A Correctness Proof for Pipelinening in RISC Architectures* DIMACS Technical Report 96-22, July 1996, pp.60.

[BR95] E. Börger and D. Rosenzweig. *The WAM - Definition and Compiler Cor-rectness.* in: *Logic Programming: Formal Methods and Practical Applications* (C.Beierle, L.Plümer, Eds.), Elsevier Science, 1995, 20-90 (chapter 2).

[Bow87] J.P.Bowen. *Formal specifacation and documentation of microprocessor in-struction sets.* In: Microprocessing and Microprogramming 21, 223–230, 1987.

[BB93] O. Buckow and J. Bormann. *Formale Spezifikation und Verifikation eines SPARC-kompatiblen Prozessors mit einem interaktiven Beweissystem,* Siemens Research and Development, München 1993.

[BD94] J.R. Burch and D.L.Dill. *Automatic verification of pipelined microprocessor control.* Conf. on Computer-Aided Verification, 1994.

[C88] J.A. Cohn. *A proof of correctness of the VIPER microprocessor: The first level.* In G.Birtwhistle and P.Subrahmanyam, editors, *VLSI Specification, Verifica-tion, and Synthesis,* pages 27-72. Kluwer Academic Publisher, 1988.

[C89] J.A. Cohn. *Correctness properties of the Viper block model: The second level.* In: G.Birtwistle, editor, *Proceedings of the 1988 Design Verification Confer-ence.* Springer-Verlag, 1989.

[Cy93] D.Cyrluk. *Microprocessor verification in PVS: A methodology and simple ex-ample.* Technical Report SRI-CSL-93-12, SRI CS Lab, 1993.

[Cy95] D.Cyrluk. *A PVS specification, implementation and verification of DLX.* SRI, Palo Alto (Oral Communication).

[Deho94] M. Dehof. *Formale Spezifikation und Verifikation des DLX-RISC-Prozessors.* Diploma Thesis, Inst. f. Technik der Informationsverarbeitung, University of Karlsruhe, August 1994.

[DeTa94] M. Dehof and S. Tahar. *Implementierung des DLX-RISC-Prozessors in einer Standardezellen-Entwurfsumgebung,* Technical Report No. SBF 358-C2-9/94, Institute of Computer Design and Fault Tolerance, University of Karlsruhe, Germany, March 1994.

[G95] Y. Gurevich. *Evolving Algebras 1993: Lipari Guide.* In: Specification and val-idation methods, Ed. E. *Börger,* Oxford University Press, 1995.

[HMC94] E. Harcourt, J.Mauney, and T.Cook. *From processor timing specifications to static instruction scheduling.* In *Static Analysis Symposium,* September 1994.

[Hen93] J.L. Hennessy. *Designing a computer as a microprocessor: Experience and lessons from the MIPS 4000.* A lecture at the Symposium on Integrated Sys-tems, Seattle, Washington, March, 1993.

[HP90] J.L. Hennessy and D.A. Patterson. *Computer Architecture: a Quantitative Ap-proach* Morgan Kaufman Publisher, 1990. Revised second edition 1996.

[Her92] J. Herbert. *Incremental design and formal verification of microcoded micro-processor.* In V. Stavridou, T.F. Melham, and R.T. Boute, editors, *Theorem Provers in Cicruit Design, Proocedings of the IFIP WG 10.2 International Working Conference, Nijmegen, The Netherlands.* North-Holland, June 1992.

[Hug95] J.Huggins. *Kermit: specification and verification.* In: Specification and validation methods, Ed. E. *Börger*, Oxford University Press, 1995.

[Hunt89] W.A. Hunt. *Microprocessor design verification.* Journal of Automated Reasoning, 5:429–460, 1989.

[J88] J.Joyce. *Formal verification and implementation of a microprocessor.* In: G.Birtwhistle and P.A.Subrahmanyan (Eds), VLSI specification, verification, and synthesis. Kluwer Ac. Press 1988.

[J89] J.Joyce. *Multi-level verification of microprocessor-based systems.* PhD thesis, Cambridge Dec. 89.

[JBG86] J.Joyce, G.Birtwistle, and M.Gordon. *Proving a computer correct in higher order logic.* TR 100, Computer Lab., University of Cambridge, 1986.

[La79] L.Lamport. *How to make a multiprocessor computer that correctly executes multiprocess programs.* IEEE Trans. on Computers, 1979, C-28,690-691.

[LC91] M.Langevin and E.Cerny. *Verification of processor-like circuits.* In: P.Prinetto and P.Camurati (Eds), *Advanced research Workshop on Correct Hardware Dwsign Methodologies*, June 1991.

[P96] C.Pusch. *Verification of Compiler Correctness for the WAM.* In: Proc. Theorem Proving in Higher Order Logics (TPHOL'97, Turku), Springer LNCS 1125, 1996.

[Ro92] A.W.Roscoe. *Occam in the specification and verification of microprocessors.* Philosophical Transactions of the Royal Society of London, Series A: Physical Sciences and Engineering, 339(1652): 137-151, Apr.15, 1992.

[SGGH91] J.B.Saxe, S.J.Garland, J.V.Guttag and J.J.Horning. *Using transformations and verification in circuit design.* TR 78, DEC System Res. Center, 1991.

[SB90] M. Srivas and M. Bickford. *Formal verification af a pipelined microprocessor.* IEEE Software, 7(5):52–64, September 1990.

[Ta95] S.Tahar. *Eine Methode zur formalen Verifikation von RISC-Prozessoren.* Fortschrittberichte VDI, Reihe 10: Informatik/Kommunikationstechnik Nr. 350, VDI-Verlag, Düsseldorf 1995, pp.XIV+162.

[TaKu93] S.Tahar and R.Kumar. *Towards a methodology for the formal hierarchical verification of RISC processors.* Proc. IEEE Int.Conf. on Computer Design (ICCD93), Cambridge/Mass; Oct.1993, pp. 58-62.

[TaKu94] S.Tahar and R.Kumar. *Implementational issues for verifying RISC–pipeline conflicts in HOL.* Springer LNCS 859, 1994, pp. 424-439.

[TaKu95] S. Tahar and R.Kumar. *A practical methodology for the formal verification of RISC processors.* FZI TR Sept. 95, Karlsruhe, pp. iii+46.

[Win90] P.J.Windley. *A hierarchical methodology for verifying microprogrammed mircoprocessors.* Proc. IEEE Symp. on Security and Privacy, Oakland, May 1990, pp. 345-357.

[Win94] P.J. Windley. *Formal modelling and verification of microprocessors.* IEEE Transactions on Computers, 1994.

[Windley94] P.J. Windley. *Specifying instruction-set architecture in HOL: a primer.* Springer LNCS 859, 1994, pp. 440-455.

[WC95] P.J. Windley and M.L. Coe. *A Correctness Model for Pipelined Microprocessors.* Springer LNCS 901.

A The sequential machine *DLX*

FETCH

if *mode = FETCH*
then *IR ← mem (PC)*
    PC ← next (PC)
    mode := OPERAND

ALU if *mode = ALU*
then if *iop (opcode) = true*
        then *TEMP ← ival*
        else *TEMP ← B*
    mode := ALU'

ALU'

if *mode = ALU'*
then *C ← $\overline{opcode}$ (A, TEMP)*
    mode := WRITE_BACK

MEM_ADDR

if *mode = MEM_ADDR*
then *MAR ← A + ival*
    if *opcode ∈ STORE*
    then *mode := Pass_B_to_MDR*
    else *mode := MEM_ACC*

STORE

if *mode = MEM_ACC*
∧ *opcode ∈ STORE*
then *mem (MAR) ← MDR*
    mode := FETCH

SUBWORD

if *mode = SUBWORD*
then *C ← $\overline{opcode}$ (MDR)*
    mode := WRITE_BACK

BRANCH

if *mode = JUMPS*
∧*opcode ∈ BRANCH*
then if *reg (A) = 0*
    then *PC ← ival + PC*
    mode := FETCH

MOVS2I

if *mode = IAR ∧ opcode = MOVS2I*
then *C ← IAR*
    mode := WRITE_BACK

OPERAND

if *mode = OPERAND*
then *A ← fstop   B ← scdop*
    mode := new_mode
where *new_mode =*
ALU if *opcode ∈ ALU ∪ SET*
IAR if *opcode ∈ {MOVS2I, MOVI2S}*
JUMPS if *opcode ∈ JUMP ∪ BRANCH*
MEM_ADDR if *opcode ∈ LOAD*
                    ∪ STORE

WRITE_BACK

if *mode = WRITE_BACK*
then *dest ← C*
    mode := FETCH

Pass_B_to_MDR

if *mode = Pass_B_to_MDR*
then *MDR ← B*
    mode := MEM_ACC

LOAD

if *mode = MEM_ACC ∧ opcode ∈ LOAD*
then *MDR ← mem (MAR)*
    mode := SUBWORD

TRAP

if *mode = JUMPS ∧ opcode = TRAP*
then *IAR ← PC*
    PC ← ival
    mode := FETCH

JUMP

if *mode = JUMPS*
and *opcode ∈ PLAINJ ∪ JLINK*
then if *iop (opcode) = true*
        then *PC ← PC + ival*
        else *PC ← A*
    if *opcode ∈ PLAINJ*
        then *mode := FETCH*
        else *C ← PC*
            mode := WRITE_BACK

MOVI2S

if *mode = IAR ∧ opcode = MOVI2S*
then *IAR ← A*
    mode := FETCH

B The parallel machine DLX^p

IF Let $jumps = opcode\ (IR1) \in JUMP \lor (opcode\ (IR1) \in BRANCH \land reg\ (A) = 0)$

$\boxed{\text{FETCH}}$ $\quad IR \leftarrow mem_{instr}\ (PC),$ if $\neg jumps$ then $PC \leftarrow next\ (PC)$

ID $\boxed{\text{Preserv IR}}$ $\boxed{\text{Preserv PC}}$ $\qquad\qquad \boxed{\text{OPERAND}}$

$\quad IR1 \leftarrow IR \qquad PC1 \leftarrow next\ (next\ (PC))\quad A \leftarrow fstop, B \leftarrow scdop$

EX

$\boxed{\text{ALU}}$ **if** $opcode\ (IR1) \in ALU \cup SET$ $\qquad\qquad\qquad \boxed{\text{Preserv IR1}}$
$\qquad$ **then if** $iop\ (opcode\ (IR1)) = true$ $\qquad\qquad\qquad IR2 \leftarrow IR1$
$\qquad\qquad$ **then** $C \leftarrow \overline{opcode\ (IR1)}\ (A,\ ival\ (IR1))$
$\qquad\qquad$ **else** $C \leftarrow \overline{opcode\ (IR1)}\ (A,\ B)$

$\boxed{\text{MEM\_ADDR}}$ $\qquad\qquad\qquad\qquad\qquad\quad \boxed{\text{Pass\_B\_to\_MDR}}$
if $opcode\ (IR1) \in LOAD \cup STORE$ $\qquad$ **if** $opcode\ (IR1) \in STORE$
then $MAR \leftarrow A + ival\ (IR1)$ $\qquad\qquad$ **then** $SMDR \leftarrow B$

$\boxed{\text{MOVS2I}}$ $\qquad\qquad\qquad\qquad\qquad\qquad\quad \boxed{\text{MOVI2S}}$
if $opcode\ (IR1) = MOVS2I$ $\qquad\qquad$ **if** $opcode\ (IR1) = MOVI2S$
then $C \leftarrow IAR$ $\qquad\qquad\qquad\qquad$ **then** $IAR \leftarrow A$

$\boxed{\text{JUMP}}$ $\qquad\qquad\qquad\qquad\qquad\qquad\qquad \boxed{\text{TRAP}}$
if $opcode\ (IR1) \in PLAINJ \cup JLINK$ $\qquad$ **if** $opcode\ (IR1) = TRAP$
then if $iop\ (opcode\ (IR1)) = true$ $\qquad\quad$ **then** $IAR \leftarrow PC1$
$\qquad\quad$ **then** $PC \leftarrow ival\ (IR1) + PC1$ $\qquad\qquad PC \leftarrow ival\ (IR1)$
$\qquad\quad$ **else** $PC \leftarrow A$

$\boxed{\text{BRANCH}}$ **if** $opcode\ (IR1) \in BRANCH$ $\qquad\qquad \boxed{\text{LINK}}$
$\qquad\quad$ **then if** $reg\ (A) = 0$ $\qquad\qquad\qquad\qquad$ **if** $opcode\ (IR1) \in JLINK$
$\qquad\qquad\quad$ **then** $PC \leftarrow PC1 + ival\ (IR1)$ $\qquad$ **then** $C \leftarrow PC1$

MEM

$\boxed{\text{STORE}}$ **if** $opcode\ (IR2) \in STORE$ $\quad \boxed{\text{LOAD}}$ **if** $opcode\ (IR2) \in LOAD$
$\qquad\quad$ **then** $mem\ (MAR) \leftarrow SMDR$ $\qquad\qquad$ **then** $LMDR \leftarrow mem\ (MAR)$

$\boxed{\text{Preserv C}}$ $C1 \leftarrow C$ $\qquad\qquad\qquad\qquad \boxed{\text{Preserv IR2}}$ $IR3 \leftarrow IR2$

WB $\boxed{\text{WRITE\_BACK}}$ **if** $opcode\ (IR3) \in ALU \cup SET \cup \{MOVS2I\} \cup JLINK$
$\qquad\qquad\qquad\qquad$ **then** $dest\ (IR3) \leftarrow C1$

$\qquad\qquad\qquad$ **if** $opcode\ (IR3) \in LOAD$
$\qquad\qquad\qquad$ **then** $dest\ (IR3) \leftarrow \overline{opcode\ (IR3)}\ (LMDR)$

C Data hazards handling machine DLX^{data}

IF

$\boxed{\text{FETCH}}$ if *not load_risk*
        then $IR \leftarrow mem_{instr}\ (PC)$
                if¬*jumps* then $PC \leftarrow next\ (PC)$
        else $IR2 \leftarrow undef$

jumps = *opcode (IR1)* $\in$ *JUMP* or (*opcode (IR1)* $\in$ *BRANCH* and *reg (A) = 0*)
load_risk = *opcode (IR2)* $\in$ *LOAD* and
        [(*dest (IR2)* $\in$ {*fstop (IR1), scdop (IR1)*}
        or (*dest (IR2) = fstop (IR1)* and *IR1* $\in$ *MEM*)].

ID

$\boxed{\text{Preserv IR}}$ $\boxed{\text{Preserv PC}}$ $\boxed{\text{OPERAND}}$
if *not load_risk* if *not load_risk*   if *not load_risk*
then$IR1 \leftarrow IR$ then $PC1 \leftarrow next(next(PC))$ then if *nthop (IR)* $\in$
                                                {*dest (IR3), dest (IR2)*}
                                            then $nthReg \leftarrow C'$
                                            else $nthReg \leftarrow nthop\ (IR)$

where $C' = \begin{cases} C1 & \text{if } nthop\ (IR) = dest\ (IR3) \quad last\ modification\ in \\ & \text{and } nthop\ (IR) \neq dest\ (IR2)\ ante - ante - preceding \\ & \text{and } opcode\ (IR3) \notin LOAD \quad not\ load\ instr \\ \\ \overline{LMDR} & \text{if } nthop\ (IR) = dest\ (IR3) \quad last\ modification\ in \\ & \text{and } nthop\ (IR) \neq dest\ (IR2)\ ante - ante - preceding \\ & \text{and } opcode\ (IR3) \in LOAD \quad load\ instr \\ \\ C & \text{if } nthop\ (IR) = dest\ (IR2) \quad last\ modification\ in \\ & \qquad\qquad\qquad\qquad\qquad\quad ante - preceding\ instr \end{cases}$

and $nth \in$ { *fst, scd* }, *fstReg = A, scdReg= B*, $\overline{LMDR} = \overline{opcode(IR3)}$ *(LMDR).*

EX

$\boxed{\text{ALU}}$
if*not load_risk* and *opcode (IR1)* $\in$ *ALU* $\cup$ *SET*
then if*iop (opcode (IR1)) = true*
        then if *fstop (IR1) = dest (IR2)*
                or [*fstop (IR1) = dest (IR3)* and *opcode (IR3)* $\in$ *LOAD*]
                then $C \leftarrow \overline{opcode\ (IR1)}\ (val_{fst}, ival\ (IR1))$
                else $C \leftarrow \overline{opcode\ (IR1)}\ (A, ival\ (IR1))$
        else if *dest(IR2)* $\in$ {*fstop(IR1), scdop(IR1)*}
                or [*dest(IR3)* $\in$ {*fstop(IR1), scdop(IR1)*}
                    and *opcode(IR3)* $\in$ *LOAD*]
                then $C \leftarrow \overline{opcode\ (IR1)}\ (val_{fst}, val_{scd})$
                else $C \leftarrow \overline{opcode\ (IR1)}\ (A, B)$

$$\text{where } val_{nth} = \begin{cases} C & \text{if } nthop\ (IR1) = dest\ (IR2) \\ \overline{LMDR} & \text{if } nthop\ (IR1) = dest\ (IR3) \text{ and } opcode\ (IR3) \in LOAD \\ & \text{and } nthop\ (IR1) \neq dest\ (IR2) \\ nthReg & \text{otherwise} \end{cases}$$

$\boxed{\text{MEM\_ADDR}}$

if *not load_risk*
then if $opcode\ (IR1) \in LOAD \cup STORE$
    then if $fstop(IR1) = dest(IR2)$
       or $[fstop(IR1) = dest(IR3)$
        and $opcode(IR3) \in LOAD]$
      then $MAR \leftarrow val_{fst} + ival\ (IR1)$
      else $MAR \leftarrow A + ival\ (IR1)$

$\boxed{\text{Pass\_B\_to\_MDR}}$

if *not load_risk* and
  $opcode\ (IR1) \in STORE$
then if $scdop(IR1) = dest(IR2)$
    or $[scdop(IR1) = dest(IR3)$
      $\wedge\ opcode(IR3) \in LOAD]$
  then $SMDR \leftarrow val_{scd}$
  else $SMDR \leftarrow B$

$\boxed{\text{MOVI2S}}$

if *not load_risk*
then if $opcode\ (IR1) = MOVI2S$
    then if $fstop\ (IR1) = dest\ (IR2)$
      or $[fstop\ (IR1) = dest\ (IR3)$
      and $opcode\ (IR3) \in LOAD]$
      then $IAR \leftarrow val_{fst}$
      else $IAR \leftarrow A$

$\boxed{\text{MOVS2I}}$

if *not load_risk*
then if $opcode\ (IR1) = MOVS2I$
    then $C \leftarrow IAR$

$\boxed{\text{Preserv IR1}}$ if *not load_risk* then $IR2 \leftarrow IR1$

$\boxed{\text{TRAP}}$

if *not load_risk*
then if $opcode\ (IR1) = TRAP$
   then $IAR \leftarrow PC1$
      $PC \leftarrow ival\ (IR1)$

$\boxed{\text{JUMP}}$

if *not load_risk*
then if $opcode\ (IR1) \in PLAINJ \cup JLINK$
  then if $iop\ (opcode\ (IR1)) = true$
    then $PC \leftarrow ival\ (IR1) + PC1$
    else $PC \leftarrow A$

$\boxed{\text{LINK}}$

if *not load_risk*
then if $opcode\ (IR1) \in JLINK$
   then $C \leftarrow PC1$

$\boxed{\text{BRANCH}}$

if *not load_risk*
then if $opcode\ (IR1) \in BRANCH$
  then if $reg\ (A) = 0$
    then $PC \leftarrow PC1 + ival\ (IR1)$

MEM

$\boxed{\text{STORE}}$

if $opcode\ (IR2) \in STORE$
then if $opcode\ (IR3) \in LOAD$ and $dest\ (IR3) = scdop\ (IR2)$
    then $mem\ (MAR) \leftarrow \overline{(LMDR)}$
    else $mem\ (MAR) \leftarrow SMDR$

$\boxed{\text{LOAD}}$

$\boxed{\text{WRITE\_BACK}}$

$\boxed{\text{Preserv IR2}}$

$\boxed{\text{Preserv C}}$

as in DLX^P

D Machine DLX^{ctrl} precomputing control code

IF

$\boxed{\text{FETCH}}$ if *not load_risk* then if *not pc_risk*
$\qquad\qquad\qquad\qquad\qquad\qquad$ then $IR \leftarrow mem_{instr}\ (PC)$
$\qquad\qquad\qquad\qquad\qquad\qquad\qquad PC \leftarrow next\ (PC)$
$\qquad\qquad\qquad\qquad\qquad\qquad$ else $IR \leftarrow undef$
$\qquad\qquad\qquad\qquad\qquad$ else $IR2 \leftarrow undef$

where *pc_risk* = *opcode (IR)* $\in$ *JUMP* $\cup$ *BRANCH*
$\qquad$ *load_risk* = as in DLX^{data}.

ID

$\boxed{\text{Preserv PC}}$ if *not load_risk* then $PC1 \leftarrow PC$

$\boxed{\text{BRANCH}}$
if *not load_risk*
then if *opcode (IR)* $\in$ *BRANCH*
$\qquad$ then if *reg (fstop (IR))* = *0*
$\qquad\qquad$ then $PC \leftarrow PC + ival\ (IR)$

$\boxed{\text{JUMP}}$
if *not load_risk* and
$\qquad$ *opcode (IR)* $\in$ *PLAINJ* $\cup$ *JLINK*
then if *iop (opcode (IR))* = *true*
$\qquad$ then $PC \leftarrow PC + ival\ (IR)$
$\qquad$ else $PC \leftarrow fstop\ (IR)$

$\boxed{\text{TRAP}_{PC}}$
if *not load_risk*
then if *opcode (IR)* = *TRAP*
$\qquad$ then $PC \leftarrow ival\ (IR)$

$\boxed{\text{Preserv IR}}$ $\boxed{\text{OPERAND}}$ *as in DLX^{data}*

EX

$\boxed{\text{TRAP}_{IAR}}$ if *not load_risk*
$\qquad\qquad\qquad$ then if *opcode (IR1)* = *TRAP*
$\qquad\qquad\qquad\qquad$ then $IAR \leftarrow PC1$

Other rules of group EX as in DLX^{data}.

MEM

$\boxed{\text{STORE}}$ $\boxed{\text{LOAD}}$ $\boxed{\text{Preserv C}}$ $\boxed{\text{Preserv IR2}}$ as in DLX^{data}

WB

$\boxed{\text{WRITE\_BACK}}$ as in DLX^{data}

E   The fully pipelined machine DLX^{pipe}

IF

FETCH if *not load_risk*
then if *not pc_data_risk*
then if *not pc_risk*
then $IR \leftarrow mem_{instr}\ (PC)$
$PC \leftarrow next\ (PC)$
else $IR \leftarrow undef$
else $IR1 \leftarrow undef$
else $IR2 \leftarrow undef$

$pc\_data\_risk = opcode\ (IR) \in BRANCH \cup JUMP$ and
$[(fstop\ (IR) = dest\ (IR2)$ and $opcode\ (IR2) \in LOAD)$
or $(fstop\ (IR) = dest\ (IR1))\ ]$.
pc_risk $= opcode\ (IR) \in JUMP \cup BRANCH$.
$load\_risk$ $= opcode\ (IR2) \in LOAD$ and
$[(dest\ (IR2) \in \{fstop\ (IR1),\ scdop\ (IR1)\}$
and $IR1 \in REG - (JUMP \cup BRANCH)$
or $(dest\ (IR2) = fstop\ (IR1)$ and $IR1 \in MEM)]$.

ID

Preserv IR
if *not load_risk* and *not pc_data_risk*
then $IR1 \leftarrow IR$

OPERAND
if *not load_risk* and *not pc_data_risk*
then if $nthop\ (IR) = dest\ (IR3)$
or $nthop\ (IR) = dest\ (IR2)$
then $nthReg \leftarrow C'$
else $nthReg \leftarrow nthop\ (IR)$

Preserv PC
if *not load_risk* $\wedge$ *not pc_data_risk*
then $PC1 \leftarrow PC$

$TRAP_{PC}$
if *not load_risk* $\wedge$ *not pc_data_risk*
then if $opcode\ (IR) = TRAP$
then $PC \leftarrow ival\ (IR)$

BRANCH if *not load_risk* and *not pc_data_risk*
then if $opcode\ (IR) \in BRANCH$
then if $fstop\ (IR) \in \{dest\ (IR3),\ dest\ (IR2)\}$
then if $reg\ (PC') = 0$
then $PC \leftarrow PC +_{PC}\ ival(IR)$
else if $reg\ (fstop\ (IR)) = 0$
then $PC \leftarrow PC +_{PC}\ ival\ (IR)$

JUMP if *not load_risk* and *not pc_data_risk*
then if $opcode\ (IR) \in PLAINJ \cup JLINK$
then if $iop\ (opcode\ (IR)) = true$
then $PC \leftarrow PC +_{PC}\ ival\ (IR)$
else if $fstop\ (IR) \in \{dest\ (IR3),\ dest\ (IR2)\}$
then $PC \leftarrow PC'$
else $PC \leftarrow fstop\ (IR)$

$$
\text{where } PC' = C' = \begin{cases}
C1 & \text{if } nthop\ (IR) = dest\ (IR3) \quad \text{last modification in} \\
& \text{and } nthop\ (IR) \neq dest\ (IR2)\ ante - ante - preceding \\
& \text{and } opcode\ (IR3) \notin LOAD \quad \text{not load instr} \\
\\
\overline{LMDR} & \text{if } nthop\ (IR) = dest\ (IR3) \quad \text{last modification in} \\
& \text{and } nthop\ (IR) \neq dest\ (IR2)\ ante - ante - preceding \\
& \text{and } opcode\ (IR3) \in LOAD \quad \text{load instr} \\
\\
C & \text{if } nthop\ (IR) = dest\ (IR2) \quad \text{last modification in} \\
& \qquad\qquad\qquad\qquad\qquad\qquad\quad ante - preceding\ instr
\end{cases}
$$

$nth \in \{fst, scd\}$, $fstReg = A$, $scdReg = B$, $\overline{LMDR} = \overline{opcode\ (IR3)}\ (LMDR)$.

EX

$\boxed{\text{ALU}}$

if $(not\ load\_risk)$ and $opcode\ (IR1) \in ALU \cup SET$
then if $iop\ (opcode\ (IR1)) = true$
      then if $fstop(IR1) = dest(IR2)$
            or $[fstop(IR1) = dest(IR3)$ and $opcode(IR3) \in LOAD]$
            then $C \leftarrow \overline{opcode(IR1)}\ (val_{fst},\ ival\ (IR1))$
            else $C \leftarrow \overline{opcode(IR1)}\ (A,\ ival\ (IR1))$
      else if $dest\ (IR2) \in \{fstop\ (IR1),\ scdop\ (IR1)\}$
            or $[dest(IR3) \in \{fstop(IR1),\ scdop(IR1)\}$
              and $opcode(IR3) \in LOAD]$
            then $C \leftarrow \overline{opcode\ (IR1)}\ (val_{fst},\ val_{scd})$
            else $C \leftarrow \overline{opcode(IR1)}\ (A,\ B)$

$\boxed{\text{MEM\_ADDR}}$

if $(not\ load\_risk)$ and $opcode\ (IR1) \in LOAD \cup STORE$
then if $fstop\ (IR1) = dest\ (IR2)$
         or $[fstop\ (IR1) = dest\ (IR3)$ and $opcode\ (IR3) \in LOAD]$
         then $MAR \leftarrow val_{fst} + ival\ (IR1)$
         else $MAR \leftarrow A + ival\ (IR1)$

$\boxed{\text{Pass\_B\_to\_MDR}}$

if $(not\ load\_risk)$ and $opcode\ (IR1) \in STORE$
then if $scdop\ (IR1) = dest\ (IR2)$
         or $[scdop\ (IR1) = dest\ (IR3)$ and $opcode\ (IR3) \in LOAD]$
         then $SMDR \leftarrow val_{scd}$
         else $SMDR \leftarrow B$

$\boxed{\text{MOVI2S}}$

if $(not\ load\_risk)$ and $opcode\ (IR1) = MOVI2S$
then if $fstop\ (IR1) = dest\ (IR2)$
         or $[fstop\ (IR1) = dest\ (IR3)$ and $opcode\ (IR3) \in LOAD]$
         then $IAR \leftarrow val_{fst}$
         else $IAR \leftarrow A$

$\boxed{\text{MOVS2I}}$ **if** *not load_risk*
           then if *opcode* $(IR1) = MOVS2I$
             then $C \leftarrow IAR$

where $val_{nth} = \begin{cases} C & \text{if } nthop \ (IR1) = dest \ (IR2) \\ \overline{LMDR} & \text{if } nthop \ (IR1) = dest \ (IR3) \text{ and } opcode \ (IR3) \in LOAD \\ & \text{and } nthop \ (IR1) \neq dest \ (IR2) \\ nthReg & \text{otherwise.} \end{cases}$

$\boxed{\text{TRAP}_{IAR}}$
if *not load_risk*
then if *opcode* $(IR1) = TRAP$
       then $IAR \leftarrow PC1$

$\boxed{\text{LINK}}$
if *not load_risk*
then if *opcode* $(IR1) \in JLINK$
       then $C \leftarrow PC1$

$\boxed{\text{Preserv IR1}}$
if *not_load_risk*
then $IR2 \leftarrow IR1$

MEM

$\boxed{\text{Preserv IR2}}$ $IR3 \leftarrow IR2$ $\boxed{\text{Preserv C}}$ $C1 \leftarrow C$

$\boxed{\text{STORE}}$
if *opcode* $(IR2) \in STORE$
then if *opcode* $(IR3) \in LOAD$
       and *dest* $(IR3) = scdop \ (IR2)$
     then *mem* $(MAR) \leftarrow \overline{(LMDR)}$
     else *mem* $(MAR) \leftarrow SMDR$

$\boxed{\text{LOAD}}$
if *opcode* $(IR2) \in LOAD$
then $LMDR \leftarrow mem \ (MAR)$

WB

$\boxed{\text{WRITE\_BACK}}$
if *opcode* $(IR3) \in ALU \cup SET \cup \{MOVS2I\} \cup JLINK$
then *dest* $(IR3) \leftarrow C1$

if *opcode* $(IR3) \in LOAD$
then *dest* $(IR3) \leftarrow \overline{LMDR}$

Integrating VDM^{++} and Real-Time System Design

K. Lano, S. Goldsack, J. Bicarregui,
Dept. of Computing, Imperial College, 180 Queens Gate, London, SW7 2BZ.
S. Kent, Dept. of Computing, University of Brighton.

Abstract. This paper presents work performed in the EPSRC "Object-oriented Specification of Reactive and Real-time Systems" project. It aims to provide formal design methods for real-time systems, using a combination of the VDM^{++} formal method and the HRT-HOOD method.

We identify refinement steps for hard real-time systems in VDM^{++}, together with a case study of a mine-pump control system, involving a combination of VDM^{++} and HRT-HOOD.

We also consider the representation of hybrid systems in VDM^{++}.

1 Introduction

Formalisms for real-time object-oriented specification are still at an early stage of development. The TAM formalism [18] describes real-time systems without module structures, and hence, does not make use of principles of locality or encapsulation to support reasoning and transformation. TAM can be seen as a subset of the Real Time Action Logic (RAL) notation used here. Extensions of object-oriented methods to cover real-time aspects, such as HRT-HOOD [4] or Octopus [1] do not provide a formal semantics or refinement concept, although their design and analysis techniques are useful frameworks for development using formal notations such as VDM^{++}. VDM^{++} provides a number of constructs for specifying real-time constraints, such as the **whenever** statement [7]. However there does not yet exist a systematic method for the development of real-time systems using VDM^{++}.

We will show here that existing methods such as HRT-HOOD can be enhanced and combined with VDM^{++}/RAL, as follows:

- HRT-HOOD object descriptions can be given as VDM^{++} class definitions.

- Timing constraints in HRT-HOOD objects can be formally expressed using RAL formulae.

- Operation request types can be represented by VDM^{++} operation definitions, combined with RAL formulae.

The advantage of carrying out design in HRT-HOOD using VDM^{++} is that precise mathematical constraints can be expressed in the language, and that design can be carried out from an abstract functional and time-specification through to a level close to a final implementation language such as Ada95. VDM^{++} also allows a more abstract description of a system using a "model 0" flat specification, with timing and synchronisation constraints expressed via logical formulae rather than via pseudocode.

In the appendix we describe the RAL formalism, which combines aspects of RTL, LTL and modal action logic formalisms in a coherent framework in order to attempt to meet the criteria and capabilities for real-time formalisms given in [8], and to enable formal treatment of real-time constraints as found in methods such as HRT-HOOD. RAL can be used as an underlying semantics of VDM^{++} to support reasoning about internal consistency and refinement of classes.

2 VDM^{++}

A VDM^{++} specification consists of a set of *class* definitions, where these have the general form:

```
class C
types
  T = TDef
values
  const : T = val
functions
  f : A → B
  f(a) == Defnf(a)
time variables
  input i₁ : T₁;
     ...
  input iₙ : Tₙ;
       o₁ : S₁;
     ...
       oₘ : Sₘ;
  assumption i₁, ..., iₙ  == Assumes;
  effect i₁, ..., iₙ, o₁, ..., oₘ  == Effects
instance variables
  vC : TC;
inv objectstate == InvC;
init objectstate == InitC
methods
  m(x : Xm,C)  value y : Ym,C
    pre Prem,C(x, vC)  ==  Defnm,C;
  ...
sync ...
thread ...
aux reasoning ...
end C
```

The `types`, `values` and `functions` components define types, constants and functions as in conventional VDM (although class reference sets @D for class names D can be used as types in these items – such classes D are termed *suppliers* to C, as are instances of these classes. C is then a *client* of D). The `instance variables` component defines the attributes of the class, and the `inv` defines an invariant over a list of these variables: `objectstate` is used to include all the attributes. The `init` component defines a set of initial states in which an object of the class may be as a result of object creation. Object creation is achieved via an invocation of the operation **C!new**, which returns a reference to the new object of **C** as its result.

The `time variables` clause lists continuously varying attributes which are either inputs (the i_j) with matching `assumption` clauses describing assumptions about how these inputs change over time, or outputs (the o_k), with `effect` clauses defining how these are related to the inputs and to each other.

The methods of **C** are listed in the `methods` clause. The list of method names of **C**, including inherited methods, is referred to as **methods**(**C**).

Methods can be defined in an abstract declarative way, using *specification statements*, or by using a hybrid of specification statements, method calls and procedural code. Input parameters are indicated within the brackets of the method header, and results after a `value` keyword. Preconditions of a method are given in the `pre` clause.

Other clauses of a class definition control how **C** inherits from other classes: the optional `is subclass of` clause in the class header lists classes which are being extended by the present class – that is, all their methods become exportable facilities of **C**.

Dynamic behaviour of objects of **C** is specified in the `sync` and `thread` clauses, which must not conflict. In the `sync` clause, which describes the behaviour of *passive* objects, either an explicit history of an object can be given, as a *trace* expression involving regular expressions in terms of method names, or as a set of *permission* statements of the form:

per Method ⇒ Cond

restricting the conditions under which methods can initiate execution. The guard condition **Cond** can involve event counters #act(m), #req(m) and #fin(m).

Threads describe the behaviour of active objects, and can involve general statements, including a `select` statement construct allowing execution paths to be chosen on the basis of which messages are received first by the object, similar to the `select` of Ada or `ALT` of OCCAM.

A set of internal consistency requirements are associated with a class, which assert that its state space is non-empty, and that the definition of a method maintains the invariant of the class, and that the initialisation predicate implies the invariant.

Refinement obligations, based on theory extension, can also be given.

In discussing the semantics of VDM^{++} and HRT-HOOD we will make use of the following RAL terms (Table 1).

2.1 Examples of Specification

An example of the type of properties that can be specified in an abstract declarative manner using RAL and VDM^{++} is a *periodic* timing constraint: "m initiates every **t** seconds, and in the order of its requests":

$$\forall i : N_1 \cdot \uparrow(m(x), i+1) = \uparrow(m(x), i) + t$$

Symbol	Meaning
@C	Set of possible object identifiers of objects of class **C**
$\overline{\text{C}}$	Set of identifiers of currently existing objects of class **C**
$\uparrow(\mathbf{m(x)},\mathbf{i})$	Activation time of **i**-th invocation of action **m(x)**
$\rightarrow(\mathbf{m(x)},\mathbf{i})$	Request time of **i**-th invocation of action **m(x)**
$\downarrow(\mathbf{m(x)},\mathbf{i})$	Termination time of **i**-th invocation of action **m(x)**
#act(m)	Number of activations of **m** to date
#req(m)	Number of received requests of **m** to date
#fin(m)	Number of terminated invocations of **m** to date
φ©t	φ holds at time **t**
e⊛t	Value of **e** at time **t**
♣(φ := **true**, **i**)	Time at which φ becomes true for the **i**-th time

Table 1. RAL Symbols and Meanings

More generally, in HRT-HOOD, periodic objects involve a periodic action such as **m**, a period **P** for its execution period, and a deadline **D** for its completion within each period:

$$\forall i : N_1 \cdot$$
$$P * (i - 1) \le \uparrow(m, i) \land$$
$$\downarrow(m, i) \le P * (i - 1) + D$$

where $D < P$.

This periodic behaviour can be expressed via a periodic thread and the **whenever** statement of VDM^{++}, provided that the methods of the class concerned are mutex:

```
effects ... ==
  whenever P | now
  also from D ==>
        #fin(m) = #fin(m) + 1  ∧  #act(m) = #act(m) + 1
...
thread
  periodic(P)(m)
```

In words "within **D** time units of any time **t** which is a multiple of **P**, there will have been one more initiated and completed execution of **m** than at **t**". Here $\overleftarrow{\text{att}}$ denotes the value of **att** at the time of occurrence of the most recent trigger event.

Periodic objects may also support high-priority asynchronous actions. These methods, say **interrupt**, will have:

$$\forall i : N_1 \cdot \mathbf{delay}(\mathbf{interrupt}, i) < \delta$$

where $\mathbf{delay}(m, i) = \uparrow(m, i) - \rightarrow(m, i)$, δ is some small time bound for the response time of the periodic object to this interrupt.

Sporadic constraints can also be directly expressed. In HRT-HOOD the sporadic action m may have a deadline **D**, which we can express as a constraint on its duration:

$$\forall i : N_1 \cdot \textbf{duration}(m, i) \leq D$$

where **duration**$(m, i) = \downarrow(m, i) - \uparrow(m, i)$. There may also be interrupts specified as for periodic objects.

Timeouts can be specified for synchronous methods in HRT-HOOD. These require that an operation **exception** is executed if an operation **op** fails to respond to a request within t time units:

$$
\begin{aligned}
&\textbf{timeout\_on\_delay}(\textbf{op}, \textbf{exception}, t) \equiv \\
&\quad (\forall i : N_1 \cdot \\
&\qquad \textbf{delay}(\textbf{op}, i) \geq t \Rightarrow \\
&\qquad\quad \exists! j : N_1 \cdot \uparrow(\textbf{exception}, j) = \rightarrow(\textbf{op}, i) + t) \wedge \\
&\quad (\forall j : N_1 \cdot \\
&\qquad \exists! i : N_1 \cdot \uparrow(\textbf{exception}, j) = \rightarrow(\textbf{op}, i) + t \wedge \\
&\qquad\quad \textbf{delay}(\textbf{op}, i) \geq t)
\end{aligned}
$$

Likewise, a timeout on the duration of **op** can be expressed.

Priority levels for operations and objects can be stated in a number of ways. For example, we could say that a high priority action α being active suppresses any low-priority actions β:

$$\#\textbf{active}(\alpha) > 0 \Rightarrow \#\textbf{active}(\beta) = 0$$

where $\#\textbf{active}(m) = \#\textbf{act}(m) - \#\textbf{fin}(m)$, the number of currently executing invocations of m. Alternatively, if both α and β are waiting to execute, it is always an α instance that is chosen:

$$\textbf{per } \beta \Rightarrow \#\textbf{active}(\alpha) = 0$$

"there are no active invocations of α at initiation of execution of β".

Other forms of prioritisation constraints, permission constraints, timeouts and responsiveness constraints can all be specified in a direct manner using RAL [12]. All the forms of method invocation protocols for concurrent objects described in [4] can be precisely described in this logic in a similar way. The Ada rendez-vous interpretation is the default for VDM++.

3   Models of Time

We wish to model both discrete and hybrid systems, with continuous behaviour specified using *time variables* which may be constrained by general differential and integral calculus formulae. In order that such class specifications are meaningful, we place the following constraints on instance and time variables and the set **TIME** of times.

We require that each (discrete) instance variable $\mathbf{att} : \mathbf{T}$ in a VDM^{++} specification can only take finitely many different values (including the undefined value $\mathbf{nil_T}$) over its lifetime:

$$\{\mathbf{att} \circledast \mathbf{t} \mid \mathbf{t} \in \mathbf{TIME}\} \in \mathbb{F}(\mathbf{T})$$

and the set of times over which $\mathbf{att}$ has a particular value is a finite union of intervals:

$$\forall \mathbf{val} : \mathbf{T};\ \exists \mathbf{I} : \mathbf{N} \to \mathcal{I}(\mathbf{TIME});\ \mathbf{n} : \mathbf{N} \cdot$$
$$\{\mathbf{t} \mid \mathbf{t} : \mathbf{TIME} \wedge \mathbf{att} \circledast \mathbf{t} = \mathbf{val}\} = \bigcup_{\mathbf{i} \leq \mathbf{n}} \mathbf{I}(\mathbf{i})$$

where $\mathcal{I}(\mathbf{TIME})$ is the set of interval subsets of $\mathbf{TIME}$, that is, subsets $\mathbf{J}$ such that

$$\forall \mathbf{t}, \mathbf{t}' : \mathbf{J} \cdot \mathbf{t} < \mathbf{t}' \Rightarrow$$
$$\forall \mathbf{t}'' : \mathbf{TIME} \cdot \mathbf{t} \leq \mathbf{t}'' \leq \mathbf{t}' \Rightarrow \mathbf{t}'' \in \mathbf{J}$$

This is a generalisation of the model of [20]. $\mathbf{e} \circledast \mathbf{t}$ is the value of $\mathbf{e}$ at time $\mathbf{t}$.

In addition, we usually expect that the lifetime of an object is contiguous:

$$\forall \mathbf{obj} : @\mathbf{C} \cdot$$
$$\{\mathbf{t} \mid \mathbf{t} : \mathbf{TIME} \wedge (\mathbf{obj} \in \overline{\mathbf{C}}) \circledast \mathbf{t}\} \in \mathcal{I}(\mathbf{TIME})$$

where $@\mathbf{C}$ is the type of *possible* object identities of objects of class $\mathbf{C}$, whilst $\overline{\mathbf{C}} : \mathbb{F}(@\mathbf{C})$ is the set of identities of objects of $\mathbf{C}$ that currently exist.

The usual model of time in VDM^{++} is that $\mathbf{TIME}$ is the set of non-negative real numbers, with the usual ordering and operations. Because of the above properties of attributes representing instance variables however, they can be considered to use a simpler concept of time: a discrete non-dense set of points from the non-negative reals. Likewise for action symbols denoting methods.

In addition, in order for **whenever** φ ... statements to be well-defined, the times $\clubsuit(\mathbf{a}.\varphi := \mathbf{true}, \mathbf{i})$ at which $\mathbf{a}.\varphi$ becomes true for the i-th time must be well-defined, ie, there must be clear cutoff points for φ. This will be the case, if, for example, φ is of the form **time_variable** $\geq$ **constant**, where **time_variable** is an input time variable with a piecewise continuous graph.

4 Formalising Real-time Refinement

The most interesting forms of refinement step are those which involve a transformation from a continuous model of the world to a discrete model. In VDM^{++} the continuous model is expressed by means of *time variables* of a class, which define attributes that change their value in a manner independent of the methods of the class, and, for real-valued time variables, possibly in a continuous manner.

Figure 1 shows the kinds of model and model transformation which are typically performed in VDM^{++}, starting from a highly abstract continuous model of both the combined controlled system/controller (an *essential* model in the

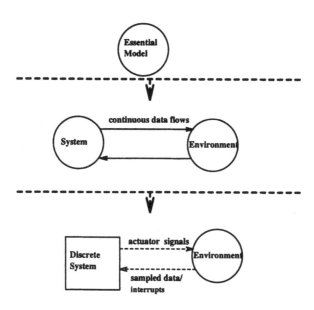

Fig. 1. Model Transformations: Continuous to Discrete

terms of Syntropy [6]), then moving to a model where the controller is separated from the controlled system, and then to a model where the controller is replaced by a discretisation. In the step to the discrete controller, output time variables received by the controlled system are replaced by calls to actuator devices to achieve the required effects. Inputs from the controlled system to the controller are replaced by sampling methods.

This is the approach taken in the paper [7]. Alternatively, the continuous controller model may be divided into a number of continuous classes before discretisation, as is done in the case study of Section 4.5 below.

A fundamental problem in each of these approaches is the transition from the continuous model to the discrete model. The transformation of data in this case does not fit into the usual VDM^{++} style of functional refinement: there is no simple retrieve function that converts the discretisation of a continuous quantity back to that quantity, because some information has been lost. Instead there are two alternatives:

1. accept that this is a case of *relational* data refinement. This complicates the proof theory of refinement and the difficulty of proof and provides no real "measure of closeness" between the continuous and discrete models;

2. use a concept of *approximate* refinement, whereby the retrieve function takes a concrete data item into an approximation of some ideal value (in this case, the retrieve function is some digital-analogue (DA) conversion which tries to recover a continuous function from a sequence of sampled points taken from it).

Other alternatives could include mapping the discrete value to the *set* of possible continuous abstractions of it, which should include the actual abstraction it derives from.

We will examine both 1 and 2 below.

4.1 Periodic Constraint Refinement

A common form of transformation from continuous to discrete views of a system starts from a class which abstractly describes a reaction to an event (the condition C becoming true):

```
class ContinuousController
time variables
  input it :  X;
        ot :  S;
  effect it, ot   ==
    whenever  C(it)
    also from δ₀  ==>  ot = v
end ContinuousController
```

If we can assume that C remains true for at least $t_C > 0$ time units from the points where it becomes true, then we can define a sampling approach:

```
class AbstractController
-- refines ContinuousController
time variables
  input it :  X;
        ot :  S;
  effect it, ot   ==
    whenever  C(it) ∧ (P | now)
    also from δ  ==>  ot = v
end AbstractController
```

where $P \leq t_C$ and $\delta + P \leq \delta_0$. These constraints guarantee that no $C := \text{true}$ events are missed by the **AbstractController**, and that it responds within δ_0 to such events. $P \mid \text{now}$ denotes that P divides the current time.

This class is then implemented by a periodic action:

```
class ConcreteController
-- refines AbstractController
time variables
  input it :  X;
instance variables
  id :  X;
  ot_obj :  @SClass
methods
  react()  ==
    (id := it;
     if  C(id)
     then
       ot_obj!set(v))
```

```
thread
  periodic(P)(react)
aux reasoning
  ∀ i: N₁  ·  ↓(react, i) ≤ P * (i − 1) + δ
end ConcreteController
```

The abstract specification requires that, at each periodic sampling time $\mathbf{P}$, if $\mathbf{C(it)}$ holds, then within a response deadline δ, $\mathbf{ot}$ has the value $\mathbf{v}$:

$$
\begin{aligned}
\forall i : N \cdot \\
\quad C(it) \odot (P * i) \Rightarrow \\
\qquad \exists t : TIME \cdot \\
\qquad\quad P * i \le t \le P * i + \delta \wedge (ot = v) \odot t
\end{aligned}
$$

The form of the test guarantees that the times $\clubsuit(\mathbf{C(it)} \wedge (\mathbf{P} \mid \mathbf{now}) := \mathbf{true}, \mathbf{i})$ are well-defined, regardless of $\mathbf{C}$. $\varphi \odot t$ expresses that φ holds at time t.

The concrete specification attempts to satisfy this specification by an explicit sampling and invocation. The periodic thread implies that:

$$
\begin{aligned}
\forall i : N_1 \cdot \\
\quad P * (i - 1) \le \uparrow(react, i) \wedge \downarrow(react, i) \le P * i
\end{aligned}
$$

and we also know from the deadline specification that in fact $\downarrow(\mathbf{react}, \mathbf{i}) \le \mathbf{P} * (\mathbf{i} - 1) + \delta$.

So assume $\mathbf{C(it)}$ holds at $\mathbf{P} * \mathbf{i}$ for some $\mathbf{i} \in N$. Assume also that it remains true until the point $\uparrow(\mathbf{react}, \mathbf{i}+1)$ where it is sampled by the concrete controller. Generally we will require that the period of sampling is fast enough with respect to the rate of change of it that it does not significantly change its value in the delay from $\mathbf{P} * \mathbf{i}$ to $\uparrow(\mathbf{react}, \mathbf{i} + 1)$.

Then **react** will send the message **set(v)** to the object that replaces the output time variable **ot**, ie, we could have the refinement that **ot** in the abstract class is interpreted by **ot_obj.ot** in the concrete, where **ot** is still a time variable in **ot_obj**.

We can then prove that the effect $\mathbf{ot} = \mathbf{v}$ is achieved by the time $\mathbf{P} * \mathbf{i} + \delta$, if **set** is synchronous, because then it terminates and achieves $\mathbf{ot} = \mathbf{v}$ before $\downarrow(\mathbf{react}, \mathbf{i} + 1) \le \mathbf{P} * \mathbf{i} + \delta$ as required.

If **set** is asynchronous, with maximum delay plus duration ϵ, say, then we must use $\delta - \epsilon$ as the deadline for **react** in **ConcreteController**.

A consistency check that $\mathbf{duration(react)} \le \gamma$ for whichever deadline γ is chosen must be made[1].

At the abstract specification level, a corresponding check must be made that multiple **whenever** statements do not require conflicting situations to occur at the same time.

Thus we can, in principle, prove such a refinement step correct.

---

[1] Under the assumption that the processor is adequate and not overloaded.

4.2 Discrete Sampling

The difficulty comes in the conversion from analogue to digital quantities –
such as the relationship between **id** and **it** in the **ConcreteController** class.
Consider the general case where we have an original abstract system which just
contains a time variable (an example in the case study would be the operator
console, containing **alarm**, or the module containing a sampled version of the
CH4 level).

```
class Output
time variables
  ot : S
end Output
```

A refinement could instead contain a sampled version of the same data:

```
class Output_1
instance variables
  d : seq of S
methods
  set(v : S)  ==  d := d ⌢ [v];

  access() value S
       pre len(d) > 0  ==   return d(len(d))
end Output_1
```

with the refinement relation being that **d** is some sampling of the abstract variable:

$$\mathbf{sampled}(\mathbf{d}, \mathbf{ot}, \mathbf{P}, \mathbf{D})$$

This is an abbreviation for:

$$\forall k : N \cdot \exists t : \mathbf{TIME} \cdot$$
$$k * P < t \leq k * P + D \ \land \ k + 1 \in \mathrm{dom}(d \circledast (k * P + D)) \ \land$$
$$d(k + 1) \circledast (k * P + D) = ot \circledast t$$

That is, **ot** is sampled in each interval of the form $(k * P, k * P + D]$ and its
value assigned to the $k + 1$-th element of **d**.

More generally, we could sample an expression **e** involving various time
variables and instance variables, including previous values of **d**. The predicate
sampled(**d**, **ot**, **P**, **D**) does describe a refinement *relation* between the abstract
and concrete data, but not a retrieve *function* from the concrete state to the
abstract state: for a given **d** there is not necessarily a unique **ot** such that
sampled(**d**, **ot**, **P**, **D**) is true. An alternative is to somehow characterise the error between a functional refinement mapping from **d** back to the continuous
domain, and the original **ot** value (Figure 2 shows the general situation here).
For example, let **ot** be a boolean quantity, modelled by a $\{0, 1\}$ value set. Assume
that the minimum inter-arrival time of events that change the value of **ot** is **I**.
Then the abstract model can be regarded as a sequence of intervals $I_0, I_1, \ldots,$

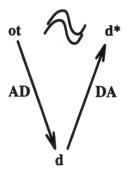

ot   d*

AD   DA

d

Fig. 2. Approximate Refinement

which are demarcated by changes in the value of **ot**. Each of the I_j has length $\text{len}(I_j) \geq I$.

Define the DA conversion function from **d** to a time series **d*** as follows:

$$
\begin{aligned}
\mathbf{d^*(t)} \quad &= \mathbf{d}(1) \quad \text{if } \mathbf{t} \leq \mathbf{t_0} \\
&= \mathbf{d}(2) \quad \text{if } \mathbf{t_0} < \mathbf{t} \leq \mathbf{t_1} \\
&= \dots \\
&= \mathbf{d}(i+2) \quad \text{if } \mathbf{t_i} < \mathbf{t} \leq \mathbf{t_{i+1}}
\end{aligned}
$$

where t_i is the time at which the i+1-th sample of **ot** is taken, so $\mathbf{d^*(t_i)} = \mathbf{ot(t_i)}$.

Other DA functions are possible, but the error estimations are similar in each case. Here, we are concerned to estimate the integral

$$\int_0^T \mid \mathbf{ot(t)} - \mathbf{d^*(t)} \mid \, \mathbf{dt}$$

for a given time **T**. This is our measure of error in the approximate retrieve function that takes **d** to **d***. We can break this integral down into a sum:

$$\Sigma_{i \in N_1, j \in N} \mid \mathbf{ot(\uparrow I_j)} - \mathbf{d^*(t_i)} \mid * \mathbf{len(I_j \cap (t_{i-1}, t_i])}$$

over all the segments $I_j \cap (t_{i-1}, t_i]$ where the values of both variables are constant, and $t_i \leq T$. $\uparrow I_j$ refers to the starting time point of the j-th interval.

If we then assume that $\mathbf{P} + \mathbf{D} < \mathbf{I}$, we can infer that:

$$\int_0^{k*I} \mid \mathbf{ot(t)} - \mathbf{d^*(t)} \mid \, \mathbf{dt} \ \leq \ 2k * (\mathbf{P} + \mathbf{D})$$

This enables a crude upper bound to be placed on the error in the approximation. Clearly, the smaller we make the period and the deadline **P** and **D** involved in the sampling, the more accurate is the approximate refinement. $\mathbf{D} < \mathbf{P}$ is necessary, and **D** must be greater than the duration of the sampling method involved (in a system object such as **ConcreteController** above).

Similar reasoning can be applied to other discrete-valued time variables.

4.3 Approximate Refinement

Given some DA conversion function **R** as described in the previous section, we require the following conditions for an "approximate refinement" based on this function.

1. **R** should be the left inverse of an adequate (surjective) function from the continuous to the discrete space;

2. the error between the continuous approximated variable **c** and **R(d)** where **d** is the corresponding discrete variable, should be boundable in terms which can be engineered by the system designer;

3. each axiom φ of the continuous system should be provable in the interpreted form $\varphi[\mathbf{R(d)}/\mathbf{c}]$ in the discrete system.

These three conditions have the following justifications:

1. We should be able to see the transformation from a continuous to the discrete model as an abstraction step, removing domain detail that is irrelevant to the implemented system, similar to the transition from the *essential* models of Syntropy [6], which describe the "real world" domain, and the *specification* models, which describe the required software (Figure 3).

2. The developer should be able to control the quality of the discrete approximation, and trade off this quality against other aspects, such as efficiency.

3. Every property required by the abstract continuous class should be true in interpreted form in the new class – this is the usual meaning of refinement and subtyping as theory extension.

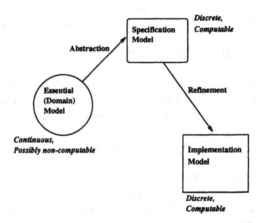

Fig. 3. Essential, Specification and Implementation Model Relationships

Examples of this process are given in the papers [16, 17].

4.4 Sporadic Constraint Refinement

The final major case of refinement from continuous to discrete views of a system involves the implementation of sporadic response requirements by sporadic, interrupt-driven classes. This can be treated by replacing a boolean input time variable by an interrupt [15].

4.5 Case Study

An example of combined analysis and design using VDM^{++}/RAL and HRT-HOOD is as follows, based on the mine pump example used in [4].
The requirements are as follows:

1. the system should respond to the water low and water high conditions within 20 seconds – switching the pump on if the water goes high with the methane level below the danger level, and switching it off if the water level goes low. The minimum inter-arrival time for these events is 100 seconds;

2. the system should respond to high methane conditions within 1 second, switching off the pump and raising a "methane high" alarm. The methane is sampled at 5 second intervals;

3. the system should respond to critically high levels of CO within 1 second, raising a "CO high" alarm. The CO level is sampled at 60 second intervals;

4. the system should respond to a critically low air flow reading within 2 seconds, raising a "low air flow" alarm. The air flow is sampled at 60 second intervals;

5. the system should respond to a low water flow reading when the pump is on within 3 seconds, raising a "pump failed" alarm. The water flow is sampled at 60 second intervals.

Based on the analysis of the requirements, we could specify the system as a single class with input time variables representing the measured CH4 and CO levels, the water level and water and airflow levels, together with the outputs – the data logger, motor state and alarm. This class can be derived from a context diagram and the detailed requirements of functionality:

```
class Model_0
types
  Alarm_status  = < high_methane > | < low_air_flow > | < high_water > |
                  < high_co > | < pump_failed > | < safe >
values
  ch4_high :  N  =  undefined;
  co_high :  N  =  undefined;
  jitter_range :  N  =  undefined;
  water_low :  N  =  undefined
time variables
  input co_reading :  N;
```

```
input ch4_reading :  N;
input low_sensor :  bool;
input high_sensor :  bool;
input water_flow :  N;
input air_flow :  N;
      motor_on :  bool;
      alarm :  Alarm_status;
```

/* Requirement 1: */

```
effect high_sensor, ch4_reading, motor_on  ==
  whenever high_sensor ∧ (ch4_reading ≤ ch4_high − jitter_range)
  also from 20000 ==> motor_on;
```

```
effect low_sensor, motor_on  ==
  whenever low_sensor
  also from 20000 ==> ¬ (motor_on);
```

/* Requirement 2: */

```
effect ch4_reading, alarm  ==
  whenever ch4_reading ≥ ch4_high ∧ (5000 | now)
  also from 1000 ==> alarm = < high_methane >;
```

```
effect ch4_reading, motor_on  ==
  whenever ch4_reading ≥ ch4_high ∧ (5000 | now)
  also from 1000 ==> ¬ (motor_on);
```

/* Requirement 3: */

```
effect co_reading, alarm  ==
  whenever co_reading ≥ co_high ∧ (60000 | now)
  also from 1000 ==> alarm = < high_co >;
```

/* Requirement 4: */

```
effect air_flow, alarm  ==
  whenever air_flow ≤ flow_low ∧ (60000 | now)
  also from 2000 ==> alarm = < low_air_flow >;
```

/* Requirement 5: */

```
effect water_flow, motor_on, alarm  ==
  whenever (water_flow ≤ water_low) ∧
           motor_on ∧ (60000 | now)
  also from 3000 ==> alarm = < pump_failed >
```

end Model_0

We ignore data logging for simplicity in this initial model. The above class could also contain differential/integral calculus expressions relating the CH4 level to the air-flow rate. **jitter_range** provides some hysteresis for the system.

Notice that many of the above effect clauses give the same sample points $60000 * k$ for the relevant tests on sensor values to be made. It may be infeasible for these tests to be made at exactly this time in the actual system, if the sampling tasks share a processor. However, any minor time-displacement of this kind should not make any difference to the truth or falsity of the test concerned. This is an auxilliary proof requirement which would require knowledge of the relevant rate of changes involved.

We can clearly factor this model into parts which contain smaller subsets of the data. The first subsystem is an environment monitor which contains the various gas monitors:

```
class EnvironmentMonitor
  is subclass of Basic_types
time variables
  input co_reading : N;
  input ch4_reading : N;
  input air_flow : N;
        motor_on : bool;
        alarm : Alarm_status;

effect ch4_reading, alarm ==
  whenever ch4_reading ≥ ch4_high ∧ (5000 | now)
  also from 1000 ==>
                alarm = < high_methane >;

effect co_reading, alarm ==
  whenever co_reading ≥ co_high ∧ (60000 | now)
  also from 1000 ==> alarm = < high_co >;

effect ch4_reading, motor_on ==
  whenever ch4_reading ≥ ch4_high ∧ (5000 | now)
  also from 1000 ==> ¬ (motor_on);

effect air_flow, alarm ==
  whenever air_flow ≤ flow_low ∧ (60000 | now)
  also from 2000 ==> alarm = < low_air_flow >;

methods
  check_safe() value bool ==
    return (ch4_reading ≤ ch4_high − jitter_range)
end EnvironmentMonitor
```

We need the method **check_safe** to support the implementation of the first requirement.

This can be further decomposed into individual monitors plus a protected object to support external queries to the CH4 status:

```
class Ch4_Sensor
```

```
  is subclass of Basic_types
time variables
  input ch4_reading :  N;
       motor_on :  bool;
       alarm :  Alarm_status;
effect ch4_reading, alarm  ==
  whenever ch4_reading  ≥  ch4_high  ∧  (5000  |  now)
  also from 1000 ==>
               alarm  =  < high_methane >;


effect ch4_reading, motor_on  ==
  whenever ch4_reading  ≥  ch4_high  ∧  (5000  |  now)
  also from 1000 ==>   ¬ (motor_on)
end Ch4_Sensor


class Ch4Status
  is subclass of Basic_types
instance variables
  environment_status :  Ch4_status;
init objectstate ==
  environment_status  :=  < motor_safe >
methods
  read() value Ch4_status  ==
    return environment_status;

  write(v :  Ch4_status)  ==
    environment_status  :=  v
end Ch4Status
```

Similarly for the **Air_flow_Sensor** and **CO_Sensor** classes.

Together these can be aggregated to refine the original subsystem specification:

```
class EnvironmentMonitor_1
-- refines EnvironmentMonitor
  is subclass of Basic_types, Ch4_Sensor,
     Air_flow_Sensor, CO_Sensor, Ch4Status
methods
  check_safe() value bool  ==
    return (environment_status  =  < motor_safe >)
end EnvironmentMonitor_1
```

The refinement relation is that the abstract time variables in **EnvironmentMonitor** are implemented by the corresponding time variables in the individual classes, whilst the attribute **environment_status** is a discretisation of the test for motor safety, with a sampling time within 1 second of the times $5000 * k$ for $k \in N$:

```
  sampled(environment_status,
    if ch4_reading  ≤  ch4_high  −  jitter_range
    then  < motor_safe >
```

else
   if ch4_reading $\geq$ ch4_high
   then $<$ motor_unsafe $>$
   else environment_status, 5000, 1000)

Here **environment_status** refers to the value at the previous sampling interval.

This approach corresponds to the architecture given in the paper [4]. Each of the objects with continuous variables will eventually be implemented as cyclic objects using a periodic thread. In the case of the CO sensor and air flow sensor objects the thread will have a periodicity of 60 seconds, whilst the CH4 sensor has a periodicity of 5 seconds.

Because we are using multiple inheritance to put together the components, which corresponds to aggregation in a object-oriented method such as Fusion [5], we must identify priorities for objects (generally, the objects with the shortest deadlines and periods will have the highest priority, etc). In a simple system such as this one, we can interpret the priorities as determining the precedence of execution of two operations: if class **C** has a higher priority than class **D**, then each object **a** of **C** can always commence execution of an operation **m** of **C** in preference to any object **b** of **D** waiting to execute an operation **n** of **D**:

$$\forall j : N_1 \cdot (a.\#\mathbf{waiting(m)} = 0)\circledcirc\uparrow(b!n, j)$$

An internal consistency check must be made, that all the separate periodic and deadline constraints specified in the individual classes can still be satisfied, given these priorities, when they are aggregated into a mutex container class.

We take the priorities from [4]:

Class	Priority
CH4 status	7
motor	6
CH4 sensor	5
CO sensor	4
airflow sensor	3
water flow sensor	2
HLW handler	1

Formally we must have that

$$\forall i,j : N_1 \cdot \mathbf{duration(opcs\_periodic\_code}, i) \leq$$
$$1000 - \mathbf{duration(co\_periodic\_code}, j)$$

where **co_periodic_code** is the periodic action of the CO sensor. Likewise:

$$\forall i,j,k : N_1 \cdot$$
$$\mathbf{duration(opcs\_periodic\_code}, i) +$$
$$\mathbf{duration(co\_periodic\_code}, j) \leq$$
$$2000 - \mathbf{duration(air\_flow\_code}, k)$$

in order that all three periodic codes can achieve their deadlines, assuming that they must be executed in a mutually exclusive and uninterrupted manner.

We can obtain slightly more refined requirements by examining the different periods concerned: **opcs_periodic_code** and **co_periodic_code** only possibly conflict, for example, in the time intervals of length 5000 beginning with a multiple of 60000. Thus we could require just:

$$\forall \ell : \mathbb{N} \cdot$$
$$\text{duration}(\text{opcs\_periodic\_code}, 1 + 12\ell) \le$$
$$1000 - \text{duration}(\text{co\_periodic\_code}, 1 + \ell)$$

The **Basic_types** class encapsulates shared type and constant definitions:

```
class Basic_types
types
  Rate  =  N;
  Ch4_reading  =  N;
  Ch4_status  =  < motor_safe >  |  < motor_unsafe >;
  Alarm_status  =
      < high_methane >  |  < low_air_flow >  |  < high_co >  |
      < high_water >  |  < pump_failed >  |  < safe >
values
  water_low :  N  =  undefined;
  ch4_high :  Ch4_reading  =  undefined;
  co_high :  N  =  undefined;
  jitter_range :  Ch4_reading  =  undefined;
  ch4_sensor_period :  N  =  5000
end Basic_types
```

The implementation of the CH4 sensor is a cyclic object:

```
class Ch4_Sensor_1
-- refines Ch4_Sensor
is subclass of Basic_types
time variables
  input ch4dbr :  Ch4_reading
instance variables
  ch4_present :  Ch4_reading;
  ch4_status :  Ch4_status;
  ch4status :  @Ch4Status;
  pump_controller :  @PumpController;
  operator_console :  @OperatorConsole;
  data_logger :  @DataLogger;
init objectstate ==
  (ch4_present  :=  0;
    ch4_status  :=  < motor_safe >)
methods
  opcs_periodic_code()  ==
    (ch4_present  :=  ch4dbr;
      ch4_status  :=  ch4status!read();
```

```
    if ch4_present ≥ ch4_high
    then
      (if ch4_status = < motor_safe >
       then
         (pump_controller!not_safe();
          operator_console!set_alarm(< high_methane >);
          ch4status!write(< motor_unsafe >) ) )
    else
      if (ch4_present ≤ ch4_high - jitter_range) ∧
          ch4_status = < motor_unsafe >
      then
         (pump_controller!safe();
          ch4status!write(< motor_safe >));
      data_logger!ch4_status(ch4_status) )
thread
  periodic(ch4_sensor_period)(opcs_periodic_code)
aux reasoning
  ∀ i : N₁ ·
     ↓(opcs_periodic_code, i) ≤ (i - 1) * ch4_sensor_period + 1000
end Ch4_Sensor_1
```

The deadline (1000ms) of the **opcs_periodic_code** operation is expressed in the aux reasoning section of the class.

The refinement relation of this object compared to its specification is that **ch4dbr** implements **ch4_reading**, and that **ch4_present** is a sampled copy of this with period 5 seconds and deadline 1 second:

$$\text{sampled}(\text{ch4\_present}, \text{ch4\_reading}, 5000, 1000)$$

We can recast this as an approximate function-based refinement as in Section 4.2.

In the above class the **motor_on** output time variable has been implemented by the internal state **pump_controller.motor_on**. Notice that we use the *abstract specification* of this class within the declarations of **Ch4_Sensor_1**, this is valid because any refinement of **PumpController** must be polymorphically compatible with its specification, and must provide some expression that implements **motor_on**. Likewise, **operator_console.alarm** is a sampled implementation of the abstract attribute **alarm**.

Ch4Status is already in a form that can be directly implemented as a protected (mutex) object.

The high/low water sensor object has the abstract specification:

```
class HighLowWater_Sensor
  is subclass of Basic_types
time variables
  input low_sensor : bool;
  input high_sensor : bool;
  input ch4_reading : N;
        motor_on : bool;
```

```
assumption low_sensor, high_sensor  ==
                  ¬ (low_sensor ∧ high_sensor);

effect high_sensor, ch4_reading, motor_on  ==
   whenever high_sensor ∧ (ch4_reading ≤ ch4_high − jitter_range)
   also from 20000 ==> motor_on;

effect low_sensor, motor_on  ==
   whenever low_sensor
   also from 20000 ==> ¬ (motor_on);
end HighLowWater_Sensor
```

It is refined using the "sporadic constraints refinement" strategy:

```
class HighLowWater_Sensor_1
   is subclass of Basic_types
instance variables
   motor : @Motor
methods
   low_sensor_interrupt()  ==  skip;

   high_sensor_interrupt()  ==  skip
thread
   while true
   do
     sel
       answer high_sensor_interrupt -> motor!operate(),
       answer low_sensor_interrupt -> motor!turn_off()
end HighLowWater_Sensor_1
```

The **operate** method of **Motor** attempts to switch the motor on, checking first that the methane level is safe, using the **check_safe** method. Duration constraints are that the period of 100000ms that is the minimum inter-arrival time between these interrupts must be greater than the maximum duration of the select body, as described in Section 4.4 above. Likewise, this duration must be less than the deadline of 20000ms in the abstract requirements.

The structure of part of the development is shown in Figure 4. **Basic_types** is not shown, for simplicity. Notice that the inheritance of **EnvironmentMonitor** into **Model_1** must hide the method **check_safe** as this method does not appear in the external interface of the system. Likewise the operations of **Ch4Status** should be hidden in **EnvironmentMonitor_1**. The latter class is *hybrid*, as it contains both time variables and ordinary discrete variables. The refinement of **EnvironmentMonitor** by **EnvironmentMonitor_1** involves a *discrete sampling* refinement step on the **environment_status** variable with respect to the expression which tests if the motor is safe to operate.

The refinement from **Ch4_Sensor** to **Ch4_Sensor_1** is a *periodic constraint* refinement step, and also introduces discrete sampling.

The refinement of the high/low water sensor is a *sporadic constraint* refinement step.

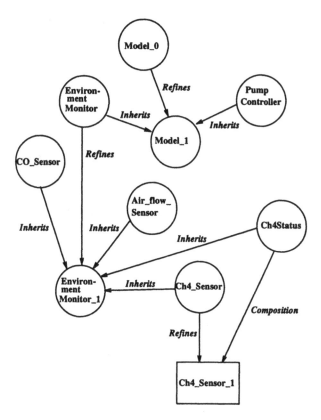

Fig. 4. Partial Structure of Mine System Development

In a complete development we would also refine the other sensors and classes to implementation-level descriptions.

5 Hybrid Systems

The analysis and verification of hybrid systems – systems containing both continuous and discrete aspects – is currently one of the most challenging areas in process control [3, 9]. We can extend the RAL formalism to treat some hybrid systems by adopting the model of *phase transition systems* from [19]. In this model, a hybrid system passes through a set of *phases*, which generalise the concept of a state in a statechart. Within each phase, time variables can modify their value in accordance with a set of differential/integral calculus equations. Transitions between phases occur as a result of some critical exit condition being reached.

We can relate this to RAL by defining a new form of actions termed *phases*:

$\alpha =$
> **wr** ℓ
> **pre G**
> **do** θ
> **exit E**
> **post Q**

These have the intended meaning that **G** always holds at initiation of α:

$$\forall i : N_1 \cdot G \odot \uparrow(\alpha, i)$$

Q holds at termination of α: $\forall i : N_1 \cdot Q \odot \downarrow(\alpha, i)$ and that the phase equations θ hold during each execution of α:

$$\forall i : N_1 ; t : \textbf{TIME} \cdot$$
$$\uparrow(\alpha, i) \leq t < \downarrow(\alpha, i) \Rightarrow \theta \odot t$$

Hooked attributes $\overleftarrow{\textbf{var}}$ occurring in θ refer to the value of **var** at $\uparrow(\alpha, i)$.

We assume that **TIME** is the set of non-negative real numbers in the following.

The **exit** condition controls when α terminates – ie, at the first point in each execution interval where **E** becomes true:

$$\forall i : N_1 ; t : \textbf{TIME} \cdot$$
$$\uparrow(\alpha, i) \leq t < \downarrow(\alpha, i) \Rightarrow \neg E \odot t \wedge$$
$$E \odot \downarrow(\alpha, i)$$

Thus for example, in the "mouse and cat" example of [19], we can represent the activity of the mouse running, terminated by the **safe** state by:

$\textbf{mrun}_{\textbf{norm}} =$
> **wr** x_m, **mstate**
> **pre mstate** $= <$ running $>$
> **do** $\frac{dx_m}{dt} = -v_m$
> **exit** $x_m = 0$
> **post mstate** $= <$ safe $>$

The mouse velocity is v_m and its distance to the hole is x_m.

A similar action is used to represent the cat running and achieving the "mouse caught" state:

$\textbf{crun}_{\textbf{norm}} =$
> **wr** x_c, **cstate**
> **pre cstate** $= <$ running $>$
> **do** $\frac{dx_c}{dt} = -v_c$
> **exit** $x_c = x_m \wedge x_c > 0$
> **post cstate** $= <$ mouse_caught $>$

A "running" activity of the complete system is then a parallel combination of these two activities. It is not a simple || combination however, because the cat achieving the "mouse caught" state must "pre-empt" the mouse running action. Likewise, the mouse achieving the "safe" state before the cat reaches it must pre-empt the cat running action.

In order to represent the abortion or pre-emption of an action, we use the choice combinator ⊓:

$$\forall\, i : N_1 \cdot$$
$$\exists\, j : N_1 \cdot$$
$$(\uparrow(\alpha, i) = \uparrow(\beta, j) \;\wedge\; \downarrow(\alpha, i) = \downarrow(\beta, j)) \;\vee$$
$$(\uparrow(\alpha, i) = \uparrow(\gamma, j) \;\wedge\; \downarrow(\alpha, i) = \downarrow(\gamma, j))$$

where $\alpha = \beta \sqcap \gamma$.

If we want α to be capable of being aborted, ie, to terminate without its normal postcondition holding, then we define a normal behaviour action α_{norm} and an action for abnormal behaviour cases α_{abort}.

α itself is defined as the ⊓ combination of these:

$$\alpha = \alpha_{norm} \sqcap \alpha_{abort}$$

For example, if we consider the cat and mouse problem, the activity of the mouse running has the abort termination:

mrun$_{abort}$ =
    wr x_m, mstate
    pre mstate = < running >
    do $\frac{dx_m}{dt} = -v_m$
    exit $x_m = x_c \wedge x_m > 0$
    post mstate = < caught >

Similarly the cat running has an abnormal termination:

crun$_{abort}$ =
    wr x_c, cstate
    pre cstate = < running >
    do $\frac{dx_c}{dt} = -v_c$
    exit $x_c = 0$
    post cstate = < failed >

This means that the composed action can be combined with the cat running action using the pre-emption operator ↯:

$$\forall\, i : N_1 \cdot$$
$$\exists\, j, k : N_1 \cdot$$
$$\uparrow(\alpha \text{↯} \beta, i) = \uparrow(\alpha, j) = \uparrow(\beta, k) \;\wedge$$
$$\downarrow(\alpha \text{↯} \beta, i) = \downarrow(\alpha, j) = \downarrow(\beta, k)$$

The overall system action is then:

$$\textbf{running} \; = \; (\textbf{mrun} \natural (\textbf{wait}(\delta); \; \textbf{crun}))$$

At termination of an instance $(\textbf{running}, i)$ with duration greater than δ we know that both the **mrun** and **crun** actions have terminated. If **mrun** has terminated normally, ie, the instance of **mrun** is an instance of $\textbf{mrun}_{norm}$, then we know that $\textbf{mstate} \; = \; < \textbf{safe} >$ at $\downarrow(\textbf{running}, i)$, and that $x_m = 0$, and hence, that the **crun** action has terminated abnormally: the instance of **crun** involved is an instance of $\textbf{crun}_{abort}$.

Conversely, if the instance of **crun** is an instance of $\textbf{crun}_{norm}$ then $\textbf{cstate} =$ $< \textbf{mouse\_caught} >$ at $\downarrow(\textbf{running}, i)$, and that $x_m = x_c \neq 0$, so that the instance of **mrun** involved must be an instance of $\textbf{mrun}_{abort}$.

It is not possible for both actions to fail – in that case $x_m = x_c > 0$ and $x_c = 0$, a contradiction.

Discretisation of this problem will require sampling of the v_c and v_m variables, and calculation of the x_c and x_m variables. We must choose a fine enough granularity of sampling so that the point where the cat catches the mouse is not missed. This requires some tolerance $| \; x_m - x_c \; | < \; \epsilon$ in the distances as the criterion for "catches", instead of equality. The sampling period τ must then satisfy

$$\tau \; < \; \frac{2*\epsilon}{v_c - v_m}$$

6   Conclusions

We have described some techniques for combining VDM^{++} with HRT-HOOD, and how real-time refinement can be formalised in RAL. All the forms of constraint described in [8] can be expressed in RAL, except required internal nondeterminism. External required non-determinism (the capability to respond to several different messages) is expressed via the **enabled** predicate.

The formalism possesses a sound semantics, and it is therefore consistent relative to ZF set theory. The advantage of the formalism over other real-time and concurrency formalisms is the conciseness of the core syntax and axiomatisation, and its ability to express the full range of reactive and real-time system behaviour via derived constructs. The TAM formalism of [18] can be regarded as a subset of RAL, and could be used to transform specification and code fragments that are purely local to one class and that are within its language. For practical development, we also need higher-level design transformations such as design patterns, and a systematic, tool-supported combination of formal and diagrammatic notations.

We have argued that the concept of approximate (functional) refinement is preferable to the use of relational refinement in carrying out the step from a continuous or hybrid specification of a system to a discrete specification. This is because it provides a simpler formulation of refinement in terms of theory extension, and enables us to measure the degree to which information about the continuous world can be recovered from the discrete refinement.

Animation of VDM^{++} specifications can be performed at the abstract continuous description level, using tools such as gPROMS [2], in order to validate the formal model of the real-world situation expressed in terms of predicates and time variables. This is in contrast to implementation-level simulation as described in [4], which is in terms of threads and processes and may be unconnected to the real-world model. Tool support for proof obligation generation for internal consistency, refinement and subtyping obligations, and for animation of event sequences against VDM^{++} classes is being developed in the "Object-oriented Specification of Reactive and Real-time Systems" project.

Examples of using the logic to express properties of distributed and concurrent systems can be found in the papers [13, 14]. Similar techniques could be applied using the formal language Z^{++}, although VDM^{++} is more suited to the later design and implementation stages.

References

1. M Awad, J Kuusela, and Jurgen Ziegler. *Object-oriented Technology for Real-time Systems*. Prentice Hall, 1996.
2. P I Barton, E Smith and C C Pantelides. Combined Discrete/Continuous Process Modelling Using gPROMS, 1991 AIChE Annual Meeting: Recent Advances in Process Control, Los Angeles, 1991.
3. P Barton and T Park. *Analysis and Control of Combined Discrete/Continuous Systems: Progress and Challenges in the Chemical Processing Industries*, in proceedings of *Chemical Process Control - V: Assessment and New Directions for Research*, January, 1996.
4. A Burns and A Wellings. HRT-HOOD: A structured design method for hard real-time systems. *Real-Time Systems*, 6(1):73–114, January 1994.
5. D Coleman, P Arnold, S Bodoff, C Dollin, H Gilchrist, F Hayes, and P Jeremaes. *Object-oriented Development: The FUSION Method*. Prentice Hall Object-oriented Series, 1994.
6. S Cook and J Daniels. *Designing Object Systems: Object-Oriented Modelling with Syntropy*. Prentice Hall, Sept 1994.
7. E Durr, S Goldsack, and J van Katjwick. Specification of a cruise controller in VDM^{++}. In *Proceedings of Real Time OO Workshop, ECOOP 96*, 1996.
8. S M Celiktin. Interval-Based Techniques for the Specification and Analysis of Real-Time Requirements, PhD thesis, Catholic University of Louvain, September 1994.
9. S Engell and S Kowalewski. *Discrete Events and Hybrid Systems in Process Control*, Proceedings of *Chemical Process Control - V: Assessment and New Directions for Research*, January, 1996.
10. J Fiadeiro and T Maibaum. *Describing, Structuring and Implementing Objects*, in de Bakker *et al.*, *Foundations of Object Oriented languages*, LNCS 489, Springer-Verlag, 1991.
11. F Jahanian and A K Mok. Safety Analysis of Timing Properties in Real-time Systems, *IEEE Transactions on Software Engineering*, SE-12, pp. 890–904, September 1986.

12. S Kent and K Lano. *Axiomatic Semantics for Concurrent Object Systems*, AFRODITE Technical Report AFRO/IC/SKKL/SEM/V1, Dept. of Computing, Imperial College, 180 Queens Gate, London SW7 2BZ.

13. K Lano. *Distributed System Specification in VDM⁺⁺*, FORTE '95 Proceedings, Chapman and Hall, 1995.

14. K Lano, J Bicarregui and S Kent. *A Real-time Action Logic of Objects*, ECOOP 96 Workshop on Proof Theory of Object-oriented Systems, Linz, Austria, 1996.

15. K Lano. *Semantics of Real-Time Action Logic*, Technical Report GR/K68783-3, Dept. of Computing, Imperial College, 1996.

16. K Lano, S Goldsack and A Sanchez. *Transforming Continuous into Discrete Specifications with VDM⁺⁺*, IEE C8 Colloquium Digest on Hybrid Control for real-time Systems, 1996.

17. K Lano. *Refinement and Simulation of Real-time and Hybrid Systems using VDM⁺⁺ and gPROMS*, ROOS project report GR/K68783-13, November 1996, Dept. of Computing, Imperial College.

18. G Lowe and H Zedan. Refinement of complex systems: A case study. *The Computer Journal*, 38(10):785–800, 1995.

19. Z Manna and A Pnueli. Time for concurrency. Technical report, Dept. of Computer Science, Stanford University, 1992.

20. B Mahony and I J Hayes. Using continuous real functions to model timed histories. In P A Bailes, editor, *Proceedings of 6th Australian Software Engineering Conference*. Australian Computer Society, July 1991.

21. J S Ostroff. *Temporal Logic for Real-Time Systems*. John Wiley, 1989.

22. A Pnueli. Applications of temporal logic to the specification and verification of reactive systems: A survey of current trends. In J de Bakker, W P de Roever, and G Rozenberg, editors, *Current Trends in Concurrency*, LNCS vol. 224, Springer-Verlag, 1986.

Real-time Action Logic

Logic

RAL is an extension of the Object Calculus of Fiadeiro and Maibaum [10] to cover durative actions and real-time constraints. The syntactic elements of an RAL theory are: *action symbols*, *attribute symbols*, plus the usual type, function and predicate symbols of typed predicate calculus, including the operators $\in$, set comprehension, $\cup$, $\mathbb{F}$, etc, of ZF set theory. These aspects are as for the standard object calculus.

For each action α, there are function symbols $\rightarrow(\alpha, i)$ the time of request of the i-th invocation of action α, $\uparrow(\alpha, i)$ the time of activation of this invocation, and $\downarrow(\alpha, i)$ the time of termination of this invocation. i ranges over $\mathbb{N}_1$.

Modal operators are $\odot$ "holds at a time" and $\circledast$ "value at a time".

The type **TIME** is assumed to be totally ordered by a relation $<$, with a least element 0, and with $\mathbb{N} \subseteq$ **TIME**. It satisfies the axioms of the set of non-negative elements of a totally ordered ring, with addition operation $+$ and unit 0, and multiplication operation $*$ with unit 1.

The following operators can be defined in terms of the above symbols: (i) the modal action formulae $[\alpha]\mathbf{P}$ "α establishes $\mathbf{P}$". $\mathbf{P}$ may contain references

$\overleftarrow{e}$ to the value of **e** at commencement of the invocation of α being considered; (ii) the operator $\supset$ representing the calling relation between two actions; (iii) the RTL [11] event-time operators $\clubsuit(\varphi := \mathbf{true}, i)$ and $\clubsuit(\varphi := \mathbf{false}, i)$ giving the times of the i-th occurrences of the events of a predicate φ becoming true or false, respectively; (iv) counters $\#\mathbf{req}(\alpha)$, $\#\mathbf{act}(\alpha)$ and $\#\mathbf{fin}(\alpha)$ for request, activation and termination events; (v) the temporal logic operators $\square$, $\diamond$, $\bigcirc$; (vi) action combinators ; , $\|$ (parallel non-interfering execution), assignment, etc.

Specific to the object-oriented view are types @**Any** of all possible object identifiers, and subsorts @**C** of this type which represent the possible object identifiers of objects of class **C**.

A predicate added for concurrent object-oriented systems is a test for *enabling* of an action α (whether a request for execution of α will be serviced or not). This is expressed by **enabled**(α).

Attributes and Actions For a specification **S** consisting of a set of classes, the attribute symbols are as follows: **x.att** for **x** : @**C** and **att** an instance or time variable of a class **C** of **S**. The attribute $\overline{\mathbf{C}}$ for each class **C** represents the set of *existing* objects of **C**. This is of type $\mathbb{F}(@\mathbf{C})$.

Derived attributes of a class will include *event counters* $\#\mathbf{act}(\mathbf{m})$, $\#\mathbf{fin}(\mathbf{m})$ as defined below.

The action symbols are: $\mathbf{new_C}(\mathbf{c})$ for **C** a class of **S** and **c** : @**C**; **x**!**m**(**e**) for **x** : @**C** and **m** a method of **C**, with **e** : $\mathbf{X_{m,C}}$ a term in the type of the input parameters of **m** in **C**.

pre Guard post Post where **Guard** is an expression over a set of attributes, and **Post** can additionally contain expressions of the form $\overleftarrow{e}$ referring to the value of the expression **e** at commencement of execution of the action.

We write $\mathbf{x}.\uparrow(\mathbf{m}(\mathbf{e}), \mathbf{i})$ for $\uparrow(\mathbf{x}!\mathbf{m}(\mathbf{e}), \mathbf{i})$ etc to make the notation used for objects more uniform.

Derived Actions and Attributes For an object **x** : @**C** event occurrence times $\clubsuit(\varphi := \mathbf{true}, i)$ and $\clubsuit(\varphi := \mathbf{false}, i)$ can be defined from the above language.

Event counters are also derived operators:

$$\mathbf{x}.\#\mathbf{act}(\mathbf{m}(\mathbf{e})) = \\ \mathbf{card}(\{\mathbf{j} : \mathbb{N}_1 \mid \mathbf{x}.\uparrow(\mathbf{m}(\mathbf{e}), \mathbf{j}) < \mathbf{now}\})$$

This definition involves $<$ because we consider $\#\mathbf{act}(\mathbf{m})$ to be incremented indivisibly *just after* the moment at which **m** initiates execution. Similarly we can define $\mathbf{x}.\#\mathbf{req}(\mathbf{m}(\mathbf{e}))$ and $\mathbf{x}.\#\mathbf{fin}(\mathbf{m}(\mathbf{e}))$.

The actions **pre G post P** name actions α with the following properties:

$$\forall \mathbf{i} : \mathbb{N}_1 \cdot \mathbf{now} = \uparrow(\alpha, \mathbf{i}) \Rightarrow \mathbf{G} \odot \uparrow(\alpha, \mathbf{i})$$
$$\forall \mathbf{i} : \mathbb{N}_1 \cdot \mathbf{now} = \uparrow(\alpha, \mathbf{i}) \Rightarrow \mathbf{P}[\mathbf{att} \circledast \uparrow(\alpha, \mathbf{i}) / \overleftarrow{\mathbf{att}}] \odot \downarrow(\alpha, \mathbf{i})$$

Formulae $\square_{a,C}\phi$ denotes that ϕ holds at each future initiation time of a method invocation a!m on an object a : @C, where m is a method of the class **C**. In other words it abbreviates

$$\forall i : N_1 \cdot a.\uparrow(m_1, i) \geq now \Rightarrow \phi \odot a.\uparrow(m_1, i)$$
$$\wedge \ldots \wedge$$
$$\forall i : N_1 \cdot a.\uparrow(m_n, i) \geq now \Rightarrow \phi \odot a.\uparrow(m_n, i)$$

where $\underline{methods}(C) = \{m_1, \ldots, m_n\}$.

The calling operator $\supset$ is defined by:

$$\alpha \supset \beta \equiv$$
$$\forall i : N_1 \cdot now = \uparrow(\alpha, i) \Rightarrow$$
$$\exists j : N_1 \cdot \uparrow(\beta, j) = \uparrow(\alpha, i) \wedge \downarrow(\beta, j) = \downarrow(\alpha, i)$$

In other words: every invocation interval of α is also one of β.

The MAL operator $[\alpha]P$ is defined as:

$$[\alpha]P \equiv$$
$$\forall i : N_1 \cdot now = \uparrow(\alpha, i) \Rightarrow P[att \circledast \uparrow(\alpha, i)/\overleftarrow{att}] \odot \downarrow(\alpha, i)$$

where the same substitution is used as for the definition of **pre G post P** above.

We can then show $[\mathbf{pre\ G\ post\ P}](\overleftarrow{G} \wedge P)$ and that

$$(\alpha \supset \beta) \Rightarrow ([\beta]P \Rightarrow [\alpha]P)$$

for any **P** in the language concerned.

Conditionals have the expected properties:

$$E \Rightarrow (\text{if } E \text{ then } S_1 \text{ else } S_2 \quad \supset \quad S_1)$$
$$\neg E \Rightarrow (\text{if } E \text{ then } S_1 \text{ else } S_2 \quad \supset \quad S_2)$$

Similarly, **while** loops can be defined.

A synchronous method invocation a!m(e) is interpreted as an **invoke** statement:

invoke a!m(e)

An instance (S, i) of this statement has the properties:

$$\forall i : N_1 \cdot \exists j : N_1 \cdot$$
$$\uparrow(S, i) = a.\rightarrow(m(e), j) \wedge$$
$$\downarrow(S, i) = a.\downarrow(m(e), j)$$

Axioms The axioms of predicate calculus and ZF set theory are adopted, with some modifications.

The core logical axioms include:

(C1): $\forall i : N_1 \cdot \rightarrow(m(e), i) \leq \rightarrow(m(e), i + 1)$

"the $\rightarrow(m(e), i)$ times are enumerated in order of their occurrence."

$$(C2): \forall i : N_1 \cdot \rightarrow(m(e), i) \leq \uparrow(m(e), i) < \downarrow(m(e), i)$$

"every invocation must be requested before it can initiate, and initiates before it terminates."

The *compactness* condition is that for every $p \in N_1$ there are only finitely many values $\uparrow(\alpha, i) < p$, for each action α. Similar conditions are required for the $\rightarrow$ and $\downarrow$ times.

Of key importance for reasoning about objects is a *framing* or *locality* constraint [10], which asserts that over any interval in which no action executes, no attribute representing an instance variable changes in value.

This locality principle reduces to that of the object calculus in the case that all actions have duration 1 and $\mathbf{TIME} = \mathbf{N}$.

The usual inference rules of predicate logic are taken. In addition the following rule is adopted:

$$\frac{\Gamma \vdash \varphi}{\Gamma \vdash \forall t : \mathbf{TIME} \cdot \varphi \odot t}$$

Interpretations of Class Features

The theory Γ_S of a system is the union of the theories Γ_C of the separate classes within it, which are defined as follows.

If we have a method definition in class $\mathbf{C}$ of the form:

$$m(x : X_{m,C}) \text{ value } y : Y_{m,C}$$
$$\text{pre } Pre_{m,C} == Code_{m,C};$$

then the action $a!m(e)$ has the properties:

$$a.Pre_{m,C}[e/x] \wedge a \in \overline{C} \Rightarrow$$
$$a!m(e) \supset a.Code_{m,C}[e/x]$$

where each attribute **att** of $\mathbf{C}$ occurring in $\mathbf{Pre}_{m,C}$ is renamed to a.att in $a.\mathbf{Pre}_{m,C}$ and similarly for $\mathbf{Code}_{m,C}$. Additionally, invocations of actions $b!n(f)$ within $\mathbf{Code}$ are explicitly written as **invoke** $b!n(f)$ statements.

The initialisation of a class $\mathbf{C}$ can be regarded as a method $\mathbf{init}_C$ which is called automatically when an object c is created by the action $\mathbf{new}_C$:
$$\mathbf{new}_C(c) \supset c!\mathbf{init}_C.$$

$\mathbf{new}_C$ itself has the property: $c \notin \overline{C} \Rightarrow [\mathbf{new}_C(c)](\overline{C} = \overset{\leftarrow}{\overline{C}} \cup \{c\}).$

A method must be enabled when it initiates execution:

$$\forall x : @C; \ i : N_1; \ e : X_{m,C} \cdot$$
$$\text{enabled}(x!m(e)) \odot x.\uparrow(m(e), i)$$

for all methods $\mathbf{m}$ of $\mathbf{C}$.

The invariant of a class is true at every method initiation and termination time: $\Box_{a,C}\mathbf{Inv_C} \wedge \forall i : N_1 \cdot \mathbf{Inv_C}@a.\!\downarrow\!(m_j, i)$ for each method $\mathbf{m_j}$ of $\mathbf{C}$ and $a :$ @$\mathbf{C}$. However, the typing constraints for attributes are *always* true: $\Box^\tau(a.att \in \mathbf{T})$ for each attribute declaration $\mathbf{att : T}$ of $\mathbf{C}$.

Permission guards for a method $\mathbf{m}$ give conditions which must be implied by $\mathbf{enabled(m)}$:

per m $\Rightarrow$ **G**

yields the axiom $\mathbf{enabled(m)} \Rightarrow \mathbf{G}$.

The **whenever** construct of VDM^{++} is interpreted as follows. A statement

whenever χ **also from** δ **==>** φ

asserts that φ must be true at some point in each interval of the form $[t, t + \delta]$ where t is a time at which χ becomes true.

Thus it can be expressed directly as:

$$\forall i : N_1; \ \exists t : \mathbf{TIME} \cdot \varphi@t \ \wedge$$
$$\clubsuit(\chi := \mathbf{true}, i) \leq t \leq \clubsuit(\chi := \mathbf{true}, i) + \delta$$

This definition yields a transitivity principle.

We can extend this interpretation to classes involving time variables, provided that we restrict **TIME** to be the set of non-negative real numbers. The representation of **assumption** and **effect** clauses then uses *phase actions*, which have ongoing activities terminated by critical conditions. These activities only change a certain subset of the time variables (ie, for each assumption clause, the variables listed in the header, and for effect clauses, those output variables listed in the clause header).

In detail, for each **assumption** clause

assumption it$_1$, ..., it$_p$ **==** $\mathbf{A(it_1, ..., it_p)}$

we have a phase action

wr $\mathbf{it_1, \ldots, it_p}$
pre self $\in \overline{\mathbf{C}}$
do $\mathbf{A(it_1, \ldots, it_p)}$
exit self $\notin \overline{\mathbf{C}}$
post true

which continues for the lifetime of the current object, allows only $\mathbf{it_1, \ldots, it_p}$ to change, and requires that their changes obey the formula $\mathbf{A}$ at all times in this lifetime.

If an input time variable **it** does not appear in an **assumption** clause, then there is a default action for **it** with $\mathbf{A}$ being **true**. Similarly for output time variables without effect clauses. The **now** attribute is treated in this way, except that its activity clause is $\frac{d\,\mathbf{now}}{dt} = 1$.

Likewise, an **effect** clause

effect it$_1$, ..., it$_p$, ot$_1$, ..., ot$_q$ == E(it$_1$, ..., it$_p$, ot$_1$, ..., ot$_q$)

has an interpretation as an action:

wr ot$_1$,...,ot$_q$
pre self ∈ $\overline{C}$
do E(it$_1$,...,it$_p$,ot$_1$,...,ot$_q$)
exit self ∉ $\overline{C}$
post true

All of these actions are lifted to be actions at the class level by substitution of particular object references a : @C for **self**, and a.it$_i$ for it$_i$, etc.

Notice that since these actions execute over the entire lifetime of an object of the class, a more refined concept of locality, involving write frames for actions, is necessary in order to reason about changes to attributes over intervals. More precisely, if an attribute **att** of a : @C changes in value between times $t_1 < t_2$, then there is some t : **TIME** with $t_1 \leq t \leq t_2$ such that some action m of C is executing on a at t, and has **att** in its write frame.

Finally, the formulae listed in the **aux reasoning** part of a class are conjoined together, and lifted to refer to particular objects, in order to obtain their meaning in the class theory.

If class C inherits class D, the theory of D is included in that of C, except that methods m of D defined in both classes are renamed to D'm in the theory of C.

Subtyping and Refinement Concepts

Theory Morphisms

The concept of a theory morphism for RAL is similar to that for the object calculus. A morphism σ : **Th1** → **Th2** maps each type symbol **T** of **Th1** to a type symbol $\sigma(T)$ of **Th2**, each function symbol of **Th1** to a function symbol of **Th2**, and each attribute of **Th1** to an attribute of **Th2**. Actions of **Th1** are mapped to actions of **Th2**.

The type **TIME** is always mapped to itself.

We can construct a category of theories with theory morphisms as categorical arrows as usual. Theory morphisms can be used to decompose the description of a class or system theory into theories for individual objects, and theories of the individual classes.

Refinement

The concepts of subtyping and refinement in VDM^{++} correspond to a particular form of theory morphism. Class C is a supertype of class D if there is a *retrieve function* **R** : **T$_D$** → **T$_C$** between the respective states, and a renaming ϕ of methods of **C** to those of **D**, such that for every $\varphi \in \mathcal{L}_\mathbf{C}$, $\Gamma_\mathbf{C} \vdash \varphi$ implies that $\Gamma_\mathbf{D} \vdash \phi(\varphi[\mathbf{R}(\mathbf{v})/\mathbf{u}])$ where **v** is the tuple of attributes of **D**, **u** of **C**.

ϕ must map internal methods of **C** to internal methods of **D**, and external methods to external methods. The notation $\mathbf{C} \sqsubseteq_{\phi,\mathbf{R}} \mathbf{D}$ is used to denote this relation.

D is a refinement of **C** if it is a subtype of **C** and the retrieve function **R** satisfies the condition of *adequacy*:

$$\forall \mathbf{u} \in \mathbf{T}_\mathbf{C} \cdot \mathbf{Inv}_\mathbf{C}(\mathbf{u}) \Rightarrow$$
$$\exists \mathbf{v} \in \mathbf{T}_\mathbf{D} \cdot \mathbf{Inv}_\mathbf{D}(\mathbf{v}) \wedge \mathbf{R}(\mathbf{v}) = \mathbf{u}$$

That is, **R** is onto. In addition, no new external methods can be introduced in **D**.

Refinement proofs can be decomposed into modular proofs of stronger but more local obligations, such as that preconditions can be weakened and post-conditions strengthened, etc.

Reactive Systems

An Approach to the Design of
Distributed Systems with B AMN

Michael Butler

Dept. of Electronics & Computer Science, Univ. of Southampton,
Highfield, Southampton SO17 1BJ, United Kingdom,
M.J.Butler@ecs.soton.ac.uk

Abstract. In this paper, we describe an approach to the design of distributed systems with B AMN. The approach is based on the action-system formalism which provides a framework for developing state-based parallel reactive systems. More specifically, we use the so-called CSP approach to action systems in which interaction between subsystems is by synchronised message passing and there is no sharing of state. We show that the abstract machines of B may be regarded as action systems and show how reactive refinement and decomposition of action systems may be applied to abstract machines. The approach fits in closely with the stepwise refinement method of B.

1 Introduction

In the B method [1], a system is specified as an abstract machine consisting of some state and some operations acting on that state. This is essentially the same structure as an action system [5] which describes the behaviour of a parallel reactive system in terms of the atomic actions (i.e., operations) that can take place during its execution. The operations of both B machines and action systems are described using notations based on Dijkstra's guarded command language [11]. As with B machines, action systems may be refined in a stepwise manner. Techniques for refining the atomicity of operations and for composing systems in parallel have been developed for action systems and such techniques are important for the development of parallel/distributed systems.

Different views as to what constitutes the observable behaviour of an action system may be taken. In the state-based view, the evolution of the state during execution is observable but not the identity of the operations that cause the state transitions. In the event-based view, the execution of an operation is regarded as an event, but only the the identity of the event is observable and the state is regarded as being internal and not observable. The event-based view corresponds to the way in which system behaviour is modelled in various process algebras such as ACP [7], CCS [14] and CSP [12]. An exact correspondence between action systems and CSP was made by Morgan [15]. Using this correspondence, techniques for event-based refinement and parallel composition of action systems have been developed in [8, 9]. In this paper, we shall use the event-based view of action systems, applying the techniques of [8, 9] to B abstract machines. For a description of the state-based view of action systems see [6].

```
MACHINE M
SETS SS
CONSTANTS C
PROPERTIES P
VARIABLES v
INVARIANT I
INITIALISATION init
OPERATIONS
    y₁ ⟵ op₁(x₁) ≙ S₁
    y₂ ⟵ op₂(x₂) ≙ S₂
    ...
END
```

Fig. 1. Abstract machine outline.

The paper is organised as follows: In Sections 2 and 3 respectively, we show how action systems may be described using B AMN and compare action system refinement with refinement in B. In Section 4, we describe the notion of internal actions that are outside the control of a system's environment and show how such actions may be introduced to a system in a refinement step. In Section 5, we describe a technique for combining action systems in parallel in which interaction between subsystems is based on shared actions. Section 6 outlines the correspondence between CSP and the event-based view of action systems.

2 Abstract Machines and Actions Systems

B Abstract Machine Notation (AMN)

B AMN is a model-oriented formal notation and is part of the B-method developed by Abrial [1]. A system in B is specified as an abstract machine which has the form outlined in Fig. 1. An abstract machine consists of some sets (SS), constants (C) and variables (v) which are modelled using standard set-theoretic constructs. Properties (P) and invariants (I) are first-order predicates. Operations act on the variables while preserving the invariant and have input parameters (x_i) and output parameters (y_i). The initialisation (*init*) and operations (S_i) are written in the generalised substitution notation of B AMN which includes constructs such as *assignment* ($v := E$), *precondition statements* (PRE Q THEN S END), *guarded statements* (SELECT Q THEN S END) and *unbounded choice* (ANY x WHERE Q THEN S END).

The semantics of generalised substitutions is defined by weakest-precondition formulae: for statement S and postcondition P, the formula $[S]P$ (weakest precondition of S w.r.t. P) characterises those initial states from which S is guaranteed to terminate in a state satisfying P. The semantics of several AMN constructs are specified in Fig. 2. Note that this is only a subset of the full language.

$$[v := E]P \;\hat{=}\; P[E/v]$$
$$[\text{PRE } Q \text{ THEN } S \text{ END}]P \;\hat{=}\; Q \wedge [S]P$$
$$[\text{SELECT } Q \text{ THEN } S \text{ END}]P \;\hat{=}\; Q \Rightarrow [S]P$$
$$[\text{ANY } x \text{ WHERE } Q \text{ THEN } S \text{ END}]P \;\hat{=}\; (\forall x \bullet Q \Rightarrow [S]P)$$
$$[S; T]P \;\hat{=}\; [S]([T]P)$$

Fig. 2. AMN semantics.

Actions and Guarding

Actions will be specified as statements in the generalised substitution notation of B AMN. For action S, the formula $[S]$ *false* represents those initial states from which S is guaranteed to establish any postcondition; to see this, we have that for any predicate P,

false $\Rightarrow P$,

and, since $[S]$ is monotonic, we get

$[S]$ *false* $\Rightarrow [S]P$.

That is, S behaves miraculously in an initial state satisfying $[S]$ *false* since it can establish any postcondition P. For example, the statement

SELECT *false* THEN T END

is miraculous in any initial state since

$[\text{SELECT } false \text{ THEN } T \text{ END}]$ *false* $=$ *true*.

We take the view that a statement is "enabled" only in those initial states in which it behaves non-miraculously. The condition under which a statement S is enabled is called its *guard*, written $gd(S)$, where

$$gd(S) \;\hat{=}\; \neg ([S] \textit{false}).$$

From this we get the following rules for calculating the guards of guarded statements, unbounded choice statements and assignment statements:

$$gd(\text{ SELECT } G \text{ THEN } S \text{ END }) = G \wedge gd(S)$$
$$gd(\text{ ANY } x \text{ WHERE } P \text{ THEN } S \text{ END }) = (\exists x \cdot P \wedge gd(S))$$
$$gd(\ x := E\ ) = \textit{true}.$$

For example, we get

$$gd(\text{ ANY } x \text{ WHERE } x \in a \text{ THEN } a := a \setminus \{x\} \text{ END })$$
$$= (\exists x \cdot x \in a \wedge \textit{true})$$
$$= a \neq \{\},$$

```
MACHINE
    VM1
VARIABLES
    n
INVARIANT
    n ∈ {0,1}
INITIALISATION
    n := 0
OPERATIONS
    coin  ≙  SELECT n = 0 THEN n := 1 END

    choc  ≙  SELECT n = 1 THEN n := 0 END
END
```

Fig. 3. Simple vending machine.

which means this statement is enabled only when $a \neq \{\}$.

Note the difference between PRE and SELECT statements: a PRE statement aborts (i.e., is not guaranteed to terminate) when Q is not satisfied, while a SELECT statement is disabled and hence cannot be executed when Q is not satisfied.

Action Systems

An action system consists of some state variables, a set of actions, each with its own unique name, and an initialisation statement. An action system proceeds by firstly executing the initialisation. Then, repeatedly, an enabled action is selected and executed. An action system deadlocks if no action is enabled, and diverges (behaves chaotically) whenever some action aborts.

Fig. 3 contains an action system, called *VM1*, specified as a B abstract machine. This is intended to represent a simple vending machine. The state of the machine is represented by the variable n. The machine has two actions represented by the operations *coin* and *choc*. Initially n is set to 0 so that only the *coin* action is enabled. When the *coin* action is executed, n is set to 1, and only the *choc* action is enabled. Execution of the *choc* action then results in *coin* being enabled again and so on. Thus *VM1* describes a system that alternatively engages in an *coin* action then a *choc* action forever.

As mentioned already, we are taking an event-based view of action systems. This means that the environment of an abstract machine can only interact with the machine through its actions and has no direct access to a machine's state. The environment of a machine can also control the execution of actions by blocking them. This will be seen clearly in Section 5, where parallel composition of action systems is described.

For any abstract machine M, we write $\alpha(M)$ for the set of action names in M. For example,

$$\alpha(VM1) \;=\; \{\ coin, choc\ \}.$$

We write $M.a$ for the action named a in machine M. For example,

$$VM1.coin \;=\; \text{SELECT } n = 0 \text{ THEN } n := 1 \text{ END}.$$

Parameter Passing

The actions of an action system can be *input* actions, with associated input parameters, or *output* actions, with associated output parameters. An input action will be represented by a B AMN operation of the form

$$name(x) \;\hat{=}\; S,$$

where x represents the input parameter(s). An input action models a channel through which a machine is willing to accept an input value whenever that action is enabled. An output action will be represented by a B AMN operation of the form

$$y \longleftarrow name \;\hat{=}\; S,$$

where y represents the output parameter(s). An output action models a channel through which a machine is willing to deliver an output value whenever that action is enabled.

We shall assume that no action can be both an input action and and output action. See Section 7 for a discussion of this issue.

Fig. 4 specifies an action system representing an ordered buffer. It is always ready to accept values of type T on the *left* channel, and to output on the *right* channel a value that has been input but not yet output. Values are output in the order in which they are input.

3 Refinement

Specification machines usually contain abstract data structures that are not directly implementable in a programming language. *Data refinement* is used in order to bring abstract specifications towards implementations by replacing abstract variables with concrete variables that are more easily implemented.

```
MACHINE

    Buffer1

VARIABLES

    s

INVARIANT

    s ∈ seq T

INITIALISATION

    s := ⟨⟩

OPERATIONS

    left(x)  ≘  SELECT x ∈ T THEN s := s ⌢ ⟨x⟩ END

    y ⟵ right  ≘  SELECT s ≠ ⟨⟩ THEN y,s := head(s), tail(s) END
END
```

Fig. 4. Ordered buffer.

Data Refinement of Actions

An abstraction invariant AI relating the abstract variables a and concrete variables c is used to replace abstract statements with concrete statements. If S is a statement that acts on variables a, T is a statement that acts on variables c, and AI is an abstraction invariant then we write

$$S \sqsubseteq_{AI} T$$

for "S is data-refined by T under abstraction invariant AI".

The weakest-precondition definition of data refinement [1, 3, 16, 17] is as follows:

Definition 1 (Data Refinement) $S \sqsubseteq_{AI} T$ *if for each postcondition* P *independent of concrete variable* c,

$$AI \wedge [S]P \Rightarrow [T](\exists a \bullet AI \wedge P).$$

Aside: Definition 1 involves quantification over predicates (P). Abrial [1] shows that the following conditions, which do not involve quantification over predicates, are sufficient to ensure data refinement:

$$AI \wedge pre(S) \Rightarrow pre(T)$$
$$AI \wedge pre(S) \Rightarrow [T](\neg [S]\neg AI).$$

Here, $pre(S) \ ≘ \ [S]true$, i.e., $pre(S)$ represents the initial condition under which S is guaranteed to terminate. Obligations such as these can be checked using the theorem-proving environments associated with the B-method [1, 18].

Data Refinement of Initialisations

$S \sqsubseteq_{AI} T$ means that T data-refines S in such a way that the abstraction invariant AI is preserved from initial to final states. However, no assumption is made about the state of action systems before initialisation. This means that a concrete initialisation must refine an abstract initialisation in such a way that it establishes the abstraction invariant rather than simply preserving it. We write $S \sqsubseteq_{AI}^{init} T$ to deal with this, where

Definition 2 (Data Refinement of Initialisations) $S \sqsubseteq_{AI}^{init} T$ *if for each post-condition P independent of concrete variable c,*

$$[S]P \;\Rightarrow\; [T](\exists a \bullet AI \land P).$$

Simulation

A *simulation* is a relation between action systems M and N with the same alphabet but possibly different state-spaces:

Definition 3 (Simulation) *For abstract action system M and concrete action system N, where $\alpha(M) = \alpha(N)$, M is simulated by N with abstraction invariant AI, denoted $M \sqsubseteq_{AI} N$, provided each of the following conditions hold:*

1. $M.init \sqsubseteq_{AI}^{init} N.init$
2. $M.a \sqsubseteq_{AI} N.a$,   each $a \in \alpha(M)$
3. $AI \land gd(M.a) \Rightarrow gd(N.a)$,   each $a \in \alpha(M)$.

Conditions 1 and 2 ensure that each action of N is a refinement of its counterpart in M, and are referred to as data-refinement conditions. These are precisely the conditions that define refinement of machines in B AMN [1]. Condition 3 ensures that N may only refuse an action when M may refuse it, and is referred to as a progress condition. Intuitively, $M \sqsubseteq_{AI} N$ means that any observable behaviour of N is also an observable behaviour of M. See Section 6 for a more precise definition of what this means.

Example: Unordered Buffer

We specify and refine a buffer that does not guarantee to output values in the order in which they are input. An unordered buffer is described by an action system that has a *bag* of values as its state variable. A bag is a collection of elements that may have multiple occurrences of any element. We write *bag T* for the set of finite bags of type T. Bags will be enumerated between bag brackets $\prec$ and $\succ$. *Addition* of bags b, c, is written $b + c$, while *subtraction* is written $b - c$.

The action system *UBuffer1* of Fig. 5 describes an unordered buffer that communicates values of type T. The initialisation statement of *UBuffer1* sets the bag to be empty. The input action *left* accepts input values of type T, adding them to the bag a. Provided a is non-empty, the output action *right* nondeterministically chooses some element from a, removes it from a and outputs it as y.

MACHINE

    UBuffer1

VARIABLES

    a

INVARIANT

    $a \in bag\ T$

INITIALISATION

    $a := \prec\succ$

OPERATIONS

    $left(x) \;\hat{=}\;$ SELECT $x \in T$ THEN $a := a + \prec x \succ$ END

    $y \longleftarrow right \;\hat{=}\;$ ANY y' WHERE $y' \in a$ THEN $a, y := a - \prec y' \succ, y'$ END

END

Fig. 5. Unordered buffer.

It can be shown that *UBuffer1* is refined by *Buffer1* of Fig. 4. As an abstraction invariant, we use

$$AI \;\hat{=}\; a = bag(s),$$

where $bag(s)$ represents the bag of elements in sequence s. The proof obligations generated by Definition 3 are as follows:

- $UBuffer1.init \sqsubseteq_{AI}^{init} Buffer1.init$
- $UBuffer1.left \sqsubseteq_{AI} Buffer1.left$
- $UBuffer1.right \sqsubseteq_{AI} Buffer1.right$
- $AI \wedge gd(UBuffer1.left) \Rightarrow gd(Buffer1.left)$
- $AI \wedge gd(UBuffer1.right) \Rightarrow gd(Buffer1.right)$

4 Internal Actions

In this section, action systems are extended to include internal actions. Internal actions are not visible to the environment of a machine and are thus outside the control of the environment. Any number of executions of an internal action may occur in between each execution of a visible action. If the action system reaches a state where internal actions can be executed infinitely, then the action system diverges. Internal actions do not have input or output parameters.

An example of an action system with internal actions is given in Fig. 6. *UBuffer2* represents an unordered buffer with an input channel *left* and an output channel *right*. However, instead of having a single bag as its state variable,

```
MACHINE
    UBuffer2
VARIABLES
    b,c
INVARIANT
    b ∈ bag T  ∧  c ∈ bag T
INITIALISATION
    b,c := ≺≻, ≺≻
OPERATIONS
    left(x)  ≙  SELECT x ∈ T THEN b := b + ≺x≻ END

    y ⟵ right  ≙  ANY y′ WHERE y′ ∈ c THEN c,y := c − ≺y′≻, y′ END
INTERNAL OPERATIONS
    mid  ≙  ANY z WHERE z ∈ b THEN b,c := b − ≺z≻, c + ≺z≻ END
END
```

Fig. 6. Unordered buffer with internal action.

UBuffer2 has two bags, b and c. The *left* action places input values in bag b, while the *right* action takes output values from bag c. Values are moved from b to c by the internal action *mid*, which is enabled as long as b is non-empty. Since b is finite, *mid* will eventually be disabled, so it cannot cause divergence.

We write $\beta(M)$ for the set of internal actions in system M.

Refinement with Internal Actions

Intuitively it can be seen that *UBuffer2* behaves in the same way as *UBuffer1* of Fig. 5. We shall introduce a proof rule that allows us to verify that *UBuffer1* ⊑ *UBuffer2*. This rule is a special form of simulation in which the concrete system has some internal actions, and the abstract system has no internal actions.

To ensure that the internal actions do not introduce divergence, a well-foundedness argument is used. A set *WF*, with irreflexive partial order $<$, is *well-founded* if each non-empty subset of *WF* contains a minimal element under $<$. For example, the natural numbers with the usual ordering, or the cartesian product of two or more well-founded sets with lexicographic ordering, all form well-founded sets. The well-foundedness argument requires the use of a well-founded set *WF* and a *variant*, which is an expression in the state-variables. The variant should always be an element of *WF*, and it should be decreased by each internal action of the concrete system.

The simulation rule is as follows:

Definition 4 *Let M and N be action systems where $\alpha(M) = \alpha(N)$ and $\beta(M) = \{\}$. M is simulated by N with abstraction invariant AI, well-founded set WF, and variant E, denoted $M \sqsubseteq_{(AI,WF,E)} N$, provided each of the following conditions hold:*

1. $M.init \sqsubseteq_{AI}^{init} N.init$
2. $M.a \sqsubseteq_{AI} N.a$,   each $a \in \alpha(M)$
3. $skip \sqsubseteq_{AI} N.h$,   each $h \in \beta(N)$
4. $AI \Rightarrow E \in WF$
5. $AI \wedge E = e \Rightarrow [N.h](E < e)$,   each $h \in \beta(N)$
6. $AI \wedge gd(M.a) \Rightarrow gd(N.a) \vee (\exists h \in \beta(N) \bullet gd(N.h))$,   each $a \in \alpha(M)$

Conditions 1, 2, and 3 are data-refinement conditions. Conditions 1 and 2 are the same as in Definition 3. Condition 3 ensures that each internal action of N causes no change to the corresponding abstract state. Conditions 4 and 5 are referred to as non-divergence conditions. Condition 4 ensures that the variant E is an element of WF, while Condition 5 ensures that the internal actions of N always decrease E when executed. Together, Conditions 4 and 5 ensure that the internal actions of N are eventually disabled and so cannot introduce divergence. Condition 6 is a progress condition and ensures that, whenever an action of M is enabled, either the corresponding action of N is enabled, or some internal action of N is enabled.

Example

To show that *UBuffer1* $\sqsubseteq$ *UBuffer2*, we use the abstraction invariant

$$AI \ \hat{=} \ a = b + c.$$

We use the size of bag b, written $\#b$, as a variant, with $\mathbb{N}$ as a well-founded set. Note that *UBuffer2.mid* is a refinement of *skip* under this abstraction invariant since the bag sum $b + c$ is unchanged by its execution. Also *UBuffer2.mid* decreases the variant $\#b$.

Hiding Operator

Let M be an action system, and C be a set of operation names, with $C \subseteq \alpha(M)$. We write $M \backslash C$ for the machine M with each action named in C converted into an internal action. The input/output parameters of an internalised action should be localised using the VAR $x \bullet S$ END construct of B AMN [1]. Note that action hiding is simply a syntactic transformation of M.

Action hiding is monotonic: if M is refined by N, then $M \backslash C$ is refined by $N \backslash C$.

5 Parallel Composition

In this section, we describe a parallel composition operator for action systems. The parallel composition of action systems M and N is written $M \parallel N$. M and N must not have any common state variables. Instead they interact by synchronising over shared actions (i.e., actions with common names). They may also pass values on synchronisation. We look first at basic parallel composition and later look at parallel composition with value passing.

Basic Parallel Composition of Actions

To achieve the synchronisation effect, shared actions are 'fused' using a parallel operator for actions $(S \parallel T)$. This operator satisfies the following properties:

- $x := E \parallel y := F = x, y := E, F$
- SELECT G THEN S END $\parallel$ SELECT H THEN T END $=$
  SELECT $G \wedge H$ THEN $S \parallel T$ END.

The parallel operator models simultaneous execution of statements and the composite action is enabled exactly when both component actions are enabled. Note that the variables assigned to by constituent actions must be independent. The weakest-precondition semantics of this operator is described in [1, 4].

Basic Parallel Composition of Action Systems

The parallel composition of action systems M and N is an action system constructed by fusing shared actions of M and N and leaving independent actions independent. The state variables of the composite system $M \parallel N$ are simply the union of the variables of M and N.

As an illustration of this, consider $N1$ and $N2$ of Fig. 7. $N1$ alternates between an a-action and a c-action, while $N2$ alternates between an b-action and a c-action. The system $N1 \parallel N2$ is shown in Fig. 8. The a- and b-actions of $N1 \parallel N2$ come directly from $N1$ and $N2$ respectively, while the c-action is the fusion of the c-actions of $N1$ and $N2$. The initialisations of $N1$ and $N2$ are also fused to form the initialisation of $N1 \parallel N2$. The effect of $N1 \parallel N2$ is that, repeatedly, the a- or the b-actions can occur in either order, then both systems must synchronise on the c-action.

Parallel Composition with Value-Passing

We extend the parallel operator to deal with parameterised actions and value-passing. An output action from one system is composed with a similarly labelled input action form another in such a way that the output value generated by the first is passed on as the input value for the second. For example, given an output action of the form

$$y \longleftarrow name \; \hat{=} \; \text{SELECT } G \text{ THEN } u, y := U, Y \text{ END}$$

```
MACHINE                           MACHINE

   N1                                N2

VARIABLES                         VARIABLES

   m                                 n

INVARIANT                         INVARIANT

   m ∈ {0,1}                         n ∈ {0,1}

INITIALISATION                    INITIALISATION

   m := 0                            n := 0

OPERATIONS                        OPERATIONS

   a  ≙                              b  ≙
      SELECT                            SELECT
         m = 0                            n = 0
      THEN                              THEN
         m := 1                           n := 1
      END                               END


   c  ≙                              c  ≙
      SELECT                            SELECT
         m = 1                            n = 1
      THEN                              THEN
         m := 0                           n := 0
      END                               END

END                               END
```

Fig. 7. Action systems with common actions.

and an output action if of the form

$$name(x) \;\;\hat{=}\;\; \text{SELECT } x \in A \wedge H \text{ THEN } v := F(x) \text{ END},$$

their value-passing fusion is represented as:

$$y \longleftarrow name \;\;\hat{=}\;\; \text{SELECT } H \wedge G \text{ THEN } u,y,v := U,Y,F(Y) \text{ END}.$$

Notice how $F(x)$ becomes $F(Y)$, modelling the passing of the output value from the output action to the input action. Notice also that the fused action is itself an output action. This allows us to fuse further input actions with the fused action, thereby modelling broadcast communications.

More generally, let $M.name$ be an output action of machine M and $N.name$

```
MACHINE

    N1 ∥ N2

VARIABLES

    m, n

INVARIANT

    m, n ∈ {0, 1}

INITIALISATION

    m, n := 0, 0

OPERATIONS

    a  ≙  SELECT m = 0 THEN m := 1 END

    b  ≙  SELECT n = 0 THEN n := 1 END

    c  ≙  SELECT m = 1 ∧ n = 1 THEN m, n := 0, 0 END

END
```

Fig. 8. Parallel composition of action systems.

be an input action of N. We shall assume[1] that $M.name$ has the form:

$$y \longleftarrow name \ \ \hat{=} \ \ \text{ANY } u', y' \text{ WHERE } P \text{ THEN } u, y := u', y' \text{ END}$$

and that $N.name$ has the form:

$$name(x) \ \ \hat{=} \ \ \text{SELECT } x \in A \wedge H \text{ THEN } v := F(x) \text{ END},$$

where H is independent of x. The value-passing fusion of these two action is defined by:

Definition 5 (Value-passing Fusion)

$$y \longleftarrow name \ \ \hat{=} \ \ \text{ANY } u', y' \text{ WHERE } P \wedge H \text{ THEN } u, y, v := u', y', F(y') \text{ END}.$$

Furthermore, the composition of $M.name$ an $N.name$ is only permitted provided

$$I_M \ \Rightarrow \ [M.name](y \in A),$$

where M is the system to which the output action belongs and I_M is the invariant of M.

---

[1] We only make these assumptions on actions that are to be composed in parallel with other actions.

This restriction ensures that the output value generated by the output action is always acceptable by the input action. Since this restriction guarantees that $y \in A$, the predicate $x \in A$ in the guard of the input action is dropped from the composite action.

The composition of two systems M and N is then constructed by fusing commonly named input-output pairs of actions as described by Definition 5. As before, independently named actions remain independent. The fusion of input-input pairs of actions is also permitted: assume $M.name$ has the form

$$name(x) \ \hat{=} \ \text{SELECT } x \in A \land G \text{ THEN } u := F(x) \text{ END,}$$

and that $N.name$ has the form:

$$name(x) \ \hat{=} \ \text{SELECT } x \in B \land H \text{ THEN } v := G(x) \text{ END,}$$

The fusion of these two action is defined by:

Definition 6

$$name(x) \ \hat{=} \ \text{SELECT } x \in (A \cap B) \land G \land H \text{ THEN } u, v := F(x), G(x) \text{ END.}$$

Fusion of output-output pairs of actions is not permitted. This avoids the introduction of deadlock in situations where two output actions are not willing to output the same value.

Fig. 9 describes the action systems *UBufferL* and *UBufferR*. *UBufferL* is simply an unbounded buffer with *right* renamed to *mid*, while *UBufferR* has *left* renamed to *mid*. When *UBufferL* and *UBufferR* are placed in parallel, they interact via the *mid* channel, with values being passed from *UBufferL* to *UBufferR*. This can be seen by constructing the composite action system *UBufferL* || *UBufferR* as described above (see Fig. 10). The only proof obligation (from Definition 5) associated with this composition is that the *UBufferL.mid* is guaranteed to output a value of type T, i.e.,

$$b \in bag \ T \ \Rightarrow \ [UBufferL.mid](y \in T).$$

If the *mid* action of *UBufferL* || *UBufferR* is hidden, then the resultant action system is the same as *UBuffer2* of Fig. 6:

$$UBuffer2 = (UBufferL \ || \ UBufferR) \backslash \{mid\}.$$

Now, since *UBuffer1* $\sqsubseteq$ *UBuffer2*, we have that:

$$UBuffer1 \sqsubseteq (UBufferL \ || \ UBufferR) \backslash \{mid\}.$$

```
MACHINE

    UBufferL

VARIABLES

    b

INVARIANT

    b ∈ bag T

INITIALISATION

    b := ≺≻

OPERATIONS

    left(x)  ≙
        SELECT
            x ∈ T
        THEN
            b := b + ≺x≻
        END


    y ⟵ mid  ≙
        ANY y' WHERE
            y' ∈ b
        THEN
            b,y := b − ≺y'≻, y'
        END

END
```

```
MACHINE

    UBufferR

VARIABLES

    c

INVARIANT

    c ∈ bag T

INITIALISATION

    c := ≺≻

OPERATIONS

    y ⟵ right  ≙
        ANY y' WHERE
            y' ∈ c
        THEN
            c,y := c − ≺y'≻, y'
        END


    mid(x)  ≙
        SELECT
            x ∈ T
        THEN
            c := c + ≺x≻
        END

END
```

Fig. 9. Buffers.

Design Technique

The derivation of the system $(UBufferL \parallel UBufferR) \backslash \{mid\}$ illustrates a design technique that may be used to decompose an action system into parallel subsystems: refine the state variables so that they may be partitioned amongst the subsystems, introducing internal actions representing interaction between subsystems, then partition the system into subsystems using the parallel operator in reverse. The refinement of the single system can always be performed in a number of steps rather than a single step.

Most importantly, the parallel composition of action systems is monotonic: if M is refined by M' and N is refined by N', then $M \parallel N$ is refined by $M' \parallel N'$. This means that when we decompose a system into parallel subsystems, the subsystems may be refined and decomposed independently.

```
MACHINE

    UBufferL || UBufferR

VARIABLES

    b,c

INVARIANT

    b ∈ bag T  ∧  c ∈ bag T

INITIALISATION

    b,c := ⟨⟩, ⟨⟩

OPERATIONS

    left(x)  ≙
        SELECT
            x ∈ T
        THEN
            b := b + ⟨x⟩
        END


    y ⟵ right  ≙
        ANY y' WHERE
            y' ∈ c
        THEN
            c,y := c − ⟨y'⟩, y'
        END


    y ⟵ mid  ≙
        ANY y' WHERE
            y' ∈ b
        THEN
            b,c,y := b − ⟨y'⟩, c + ⟨y'⟩, y'
        END

END
```

Fig. 10. Parallel buffers.

6 CSP Correspondence

In CSP [12], the behaviour of a process is viewed in terms of the events it can engage in. Value-passing is modelled by grouping events into input channels and output channels. Each process P has an alphabet of events A, and its behaviour is modelled by a set of *failures* F and a set of *divergences* D. A failure is a pair (t,X), where t is a trace of events and X is a set of events; $(t,X) \in F$ means that P may engage in the trace of events t and then refuse all the events in X. A divergence is a trace of events d, and $d \in D$ means that, after engaging the trace d, P may diverge (behave chaotically). Process (A,F,D) is refined by process (A,F',D'), written $(A,F,D) \sqsubseteq (A,F',D')$, if

$$F \supseteq F' \text{ and } D \supseteq D'.$$

In [15], a correspondence between CSP and an event-based view of action systems is described. This involves giving a failures-divergence semantics to action systems, with the execution of actions corresponding to the occurrence of CSP-like communication events. Let $\{M\}$ represent the failures-divergence semantics of action system M. The definition of $\{M\}$ may be found in [8, 15]. Previously we claimed that if M is simulated by N (Definitions 3 and 4), then any observable behaviour of N is an observable behavior of M. The observable behaviour of an action system is represented by its failures-divergence semantics and it can be shown [8, 20] that if M is simulated by N, then

$$\{M\} \sqsubseteq \{N\}.$$

CSP has both a hiding operator $(P \setminus C)$ for internalising events and a parallel composition operator $(P \parallel Q)$ for composing processes based on shared events. Both operators are defined in terms of failures-divergence semantics: Let $[\![P]\!]$ be the failures-divergence semantics of a process P. Then $[\![P \setminus C]\!]$ is defined by $HIDE([\![P]\!], C)$ and $[\![P \parallel Q]\!]$ is defined by $PAR([\![P]\!], [\![Q]\!])$, where $HIDE$ and PAR are described in [12]. It can be shown [8] that the hiding and parallel operators for action systems correspond to the CSP operators; that is, for action systems M and N:

$$\{M \setminus C\} = HIDE(\{M\}, C)$$
$$\{M \parallel N\} = PAR(\{M\}, \{N\}).$$

Since $HIDE$ and PAR are monotonic w.r.t. (failures-divergence) refinement [12], our earlier claim that the hiding and parallel operators for action systems are monotonic is justified.

7 Concluding

Although operations in B AMN can have both input and output parameters, it was stated earlier that actions can either be input actions or output actions, but not both. Consider an AMN action of the form

$$y \longleftarrow name(x) \; \hat{=} \; S.$$

In the implementation of this operation, we would expect a delay between receipt of x and the delivery of y. In particular, we may want to push the computation of y into some internal actions. In order to do this using simulation (Definition 4), the operation should be broken into an input action, representing receipt of x, and an output action, representing delivery of y. In this way, we can introduce internal actions that are executed in between receipt of x and delivery of y, contributing towards the computation of y. It also allows us to interleave other visible actions between receipt of x and delivery of y.

Abrial has proposed an approach to the design of protocols using the B method [2]. With this approach, a protocol is specified as a single operation which is subsequently decomposed into a sequence of steps through a series of refinements. The introduction of each new step in the protocol is justified by showing that it is a data-refinement of the *skip* action. This is the same as our data-refinement condition on internal actions being introduced by a simulation step (Condition 3, Definition 4). Of course, we also require that the nondivergence and progress conditions be satisfied (Conditions 4, 5, 6, Definition 4).

In this paper, we have taken an event-based view of action systems. In [10, 19], B is combined with a state-based view of action systems. With the state-based view, refinement is similar to the definitions presented here, but the parallel composition of machines is somewhat different; actions are not fused, but machines may share variables through which they interact. The choice between an event-based and a state-based view will depend on the nature of the application being developed. The event-based view is more suited to the design of message-passing distributed systems, while the state-based view is more suited to the design of parallel algorithms. Lano and Dick have also developed an approach to dealing with concurrency in B by combining it with a form of temporal logic [13].

We have seen the close correspondence between action systems and the abstract machines of B and seen the similarity between their notions of refinement. Because of this close correspondence, we are able to apply action system techniques such as internalisation of actions and parallel composition to abstract machines. These techniques provide a powerful abstraction mechanism since they allow us to abstract away from the distributed architecture of a system and the complex interactions between its subsystems; a system can be specified as a single abstract machine and only in later refinement steps do we need to introduce explicit subsystems and interactions between them. In [9], the techniques are applied to the design of an electronic mail system[2]; the state of the abstract mail service is a single bag in which messages for all the users are stored; this is then refined and decomposed into a store-and-forward network. The reasoning required to use these techniques involves refinement arguments and variant arguments, which is the sort of reasoning already used in B. The techniques are also very modular since the parallel components of a distributed system can be refined and decomposed separately without making any assumptions about the rest of the system.

---

[2] This uses the notation of [16] rather than B AMN.

References

1. J.R. Abrial. *The B-Book: Assigning Programs to Meanings*. Cambridge University Press, 1996.
2. J.R. Abrial. Extending B without changing it (for developing distributed systems). In H. Habrias, editor, *First B Conference*, November 1996.
3. R.J.R. Back. *Correctness Preserving Program Refinements: Proof Theory and Applications*. Tract 131, Mathematisch Centrum, Amsterdam, 1980.
4. R.J.R. Back and M.J. Butler. Exploring summation and product operators in the refinement calculus. In B. Möller, editor, *Mathematics of Program Construction, 1995*, volume LNCS 947, pages 128–158. Springer–Verlag, 1995.
5. R.J.R. Back and R. Kurki-Suonio. Decentralisation of process nets with centralised control. In *2nd ACM SIGACT-SIGOPS Symp. on Principles of Distributed Computing*, pages 131–142, 1983.
6. R.J.R. Back and K. Sere. Stepwise refinement of parallel algorithms. *Sci. Comp. Prog.*, 13:133–180, 1989.
7. J.A. Bergstra and J.W. Klop. Algebra of communicating processes with abstraction. *Theoret. Comp. Sci.*, 37:77–121, 1985.
8. M.J. Butler. *A CSP Approach To Action Systems*. D.Phil. Thesis, Programming Research Group, Oxford University, 1992.
9. M.J. Butler. Stepwise refinement of communicating systems. *Science of Computer Programming*, 27(2):139–173, September 1996.
10. M.J. Butler and M. Waldén. Distributed system development in B. In H. Habrias, editor, *First B Conference*, November 1996.
11. E.W. Dijkstra. *A Discipline of Programming*. Prentice-Hall, 1976.
12. C.A.R. Hoare. *Communicating Sequential Processes*. Prentice–Hall, 1985.
13. K. Lano and J.Dick. Development of concurrent systems in B AMN. In He Jifeng, editor, *7th BCS-FACS Refinement Workshop*, Workshops in Computing. Springer–Verlag, 1996.
14. R. Milner. *Communication and Concurrency*. Prentice–Hall, 1989.
15. C.C. Morgan. Of wp and CSP. In W.H.J. Feijen, A.J.M. van Gasteren, D. Gries, and J. Misra, editors, *Beauty is our business: a birthday salute to Edsger W. Dijkstra*. Springer–Verlag, 1990.
16. C.C. Morgan and T. Vickers, editors. *On the Refinement Calculus*. Formal Approaches to Computing and Information Technology. Springer, 1994.
17. J.M. Morris. Laws of data refinement. *Acta Informatica*, 26:287–308, 1989.
18. D.S. Nielson and I.H. Sorensen. The B-technologies: a system for computer aided programming. In U.H. Engberg, K.G. Larsen, and P.D. Mosses, editors, *6th Nordic Workshop on Programming Theory*. BRICS, October 1994.
19. M. Waldén and K. Sere. Refining action systems within B-Tool. In *Formal Methods Europe (FME'96)*, volume LNCS 1051, pages 85 – 104. Springer–Verlag, March 1996.
20. J.C.P. Woodcock and C.C. Morgan. Refinement of state-based concurrent systems. In D. Bjørner, C.A.R. Hoare, and H. Langmaack, editors, *VDM '90*, volume LNCS 428, pages 340–351. Springer–Verlag, 1990.

Specifying Reactive Systems in B AMN

K. Lano

Dept. of Computing, Imperial College, 180 Queens Gate, London SW7 2BZ

Abstract. This paper describes techniques for specifying and designing reactive systems in the B Abstract Machine (AMN) language, using concepts from procedural process control. In addition, we consider what forms of concurrent extensions to B AMN would make it more effective in representing such systems.

1 Reactive System Specification

A *reactive* system is a system that requires description of allowed patterns of behaviour for its specification, rather than a simple input/output relation [20].

Reactive systems often involve concurrent execution of processes (in order to achieve responsiveness requirements, or because of inherent distribution in the application) and requirements on system states that must not arise (*safety* constraints), the reachability of certain states (*liveness* constraints) and *periodic* and *sporadic* timing constraints between system responses to events.

Compared to formalisms such as Statecharts [8], SDL, VDM^{++} [4] and CSP which are oriented to describing reactive systems, the B AMN formalism appears quite weak in its capabilities, particularly as concerns liveness or timing constraints [11]. This is unfortunate, as B is one of the most industrially successful and best-supported formal methods, with successful uptake in the field of rail transport in particular [2]. It has advantages over statecharts and SDL in providing a fully formal specification language for abstractly describing complex data types and state transitions in a model-based style, together with a precise definition of refinement. As we will show here, there are techniques and simple language extensions which can overcome the inadequacies of B in many cases.

We will assume knowledge of the B notation. Comprehensive introductions to B can be found in the books [1, 12].

In Section 2 we discuss the various forms of requirement which need to be formalised in order to specify reactive systems, in Section 3 we introduce a case study of a gas burner ignition system [19, 21], which illustrates timing and safety constraints. In Section 4 we present a part of the production cell case study [18] in order to show how the above approach generalises to cover a complex system involving several levels of control and safety and liveness requirements. Section 5 discusses some extensions to B which could make specification of reactive systems in the language more effective.

2 Specifying Reactive Systems

The paper [22] presents a methodology for defining controllers for reactive systems, specifically in the process control domain. This method distinguishes:

- A model of the *system to be controlled*, as a finite state machine (specified, for example, using statechart notation) with *controllable* transitions (resulting from commands from the controller) distinguished from *uncontrolled* transitions (resulting from system responses to controller commands, or spontaneous events). A unique initial state, and a set of *marked* (significant) states are also identified.

- A specification of the desired behaviour of the *controlled system*, including safety, liveness and timing constraints.

- A design of a *controller* which, when combined with the uncontrolled system, will realise the specification.

Here, we can formalise each of these models as B specifications, with additional documentation for some liveness and safety constraints, and carry out internal consistency proofs to verify that the controller achieves the specification, and a refinement process to implement the controller design in an executable system.

The model of the system to be controlled can be expressed as a number of machines (often, one machine for each component, such as a valve, pressure gauge, etc). The specification of the controlled behaviour is split into a number of controller specifications, depending upon the relevant level of granularity of the requirements: safety constraints which require preservation of properties at the finest level of action of the system must be expressed as invariants of a low-level controller which places restrictions on the atomic actions of the unconstrained system.

The controller itself can also be designed as a hierarchy of controllers, with high-level procedures invoking a series of actions from lower-level controllers. For example, in the production cell, a single cycle of robot and press activity can be broken down into a sequence of sub-procedures (such as "extend arm 1 to the table, pick up item, and retract"), which themselves are sequences of low-level control actions (extending arm 1 to the table involves an iteration of an action which monitors this extension).

This division into layers of control is partly enforced by the limitations of B, because the sequencing of operations is not allowed at the abstract specification level, and cyclic dependencies between modules are not allowed. Such a division is nonetheless itself a good design approach, allowing separation of concerns.

2.1 Safety Requirements

These can be divided into: (i) Safety invariants; (ii) Permission guards; (iii) Life histories.

Safety invariants are usually of the form "φ is always true" or "$\neg\ \varphi$ never happens". For example "the robot arms never collide with other robot components" or "gas never flows without air flowing". The invariant of a B machine can *approximately* express such invariants. As part of the proof obligations of a machine M with invariant I, we have the requirements

$$[T]I$$

where **T** is the initialisation, and

$$I \wedge P \Rightarrow [S]I$$

for each operation defined by

$$y \longleftarrow op(x) \; \widehat{=}$$
$$\text{PRE } \mathbf{P}$$
$$\text{THEN } \mathbf{S}$$
$$\text{END}$$

That is, the invariant must be established by the initialisation and must be preserved by each operation[1].

Therefore, the invariant will be true at all time points between operation invocations. Separate reasoning is needed in order to ensure that the invariant is true *during* operations. If it is possible to model the finest significant level of granularity of a system using B operations, then a B machine invariant preserved by such operations can be viewed as asserting that the predicate is true at all times.

Permission guards are of the form "op can only be executed if φ holds". For example "a robot arm can only deposit a piece on the deposit belt if it is clear" or "the gas valve can only open if the air valve is open".

Technically these should be formalised using the SELECT construct:

```
deposit_blank_at_belt =
  SELECT dbelt_clear = TRUE
  THEN    /* Turn off the arm2 magnet and start to retract arm2: */
    arm2_release || arm2_start_retract
  END
```

"Block the deposit action until **dbelt_clear** becomes **TRUE**". However this only makes sense in a concurrent environment where there are two or more clients of the component that manages the **dbelt_clear** variable – otherwise it represents a permanent deadlock.

Instead, we can use preconditions to obtain proof obligations that operations are only called when their guards are true:

```
deposit_blank_at_belt =
  PRE dbelt_clear = TRUE
  THEN  arm2_release || arm2_start_retract
  END
```

---

[1] [S]P is the weakest precondition operator in B and is read as "S establishes P".

Life histories specify that only certain sequences of behaviour (operation executions) are allowed for a component. For example the press in the robot system goes through a cycle of actions **close; open_fully; to_middle**, and the valves in the gas burner example go through a cycle of the form (**open; close**)*, where **t*** indicates that the trace **t** can be performed 0 or more times in succession.

These constraints can be specified using preconditions. For example the behaviour **start_extend; continue_extend*; stop_extend** of the robot arm when extending can be expressed as:

```
arm1_start_extend(dest) =
  PRE arm1_state = retracted & ...
  THEN
    ... || arm1_state := extending
  END;

arm1_continue_extend =
  PRE arm1_state = extending & ...
  THEN  ...
    /* no change of arm1_state */
  END;

arm1_stop_extend =
  PRE arm1_state = extending & ...
  THEN
    arm1_state := extended || ...
  END
```

However this is an indirect and unclear means of expression, and a more abstract approach is to use a regular expression language:

- **op** – accept execution of **op**

- **t**$_1$; **t**$_2$ – accept execution of trace **t**$_1$ followed by one of **t**$_2$

- **t*** – accept execution of **t** repeatedly

- **t**$_1$ | **t**$_2$ – accept either **t**$_1$ or **t**$_2$

- **t**$_1$ **w t**$_2$ – accept interleaved execution of **t**$_1$ and **t**$_2$ (with common actions synchronised)

This is based on the *trace* notation of VDM^{++} [23].

2.2 Liveness Requirements

These are typically of the form "if φ holds, eventually the property ϕ will hold". For example, if the robot arm is extending, eventually it will reach a pre-set destination distance.

This can be partly formalised by using a *variant*, a $\mathbb{N}$-valued expression, strictly decreased by each invocation of some collection of operations op$_i$ which are not the desired operation **op**. Implicitly these operations are called within a higher-level control loop terminated by the variant reaching 0, and which is followed by a call of **op**. For example, see the arm extension process in Section 4.

2.3 Timing Requirements

Timing requirements can be categorised as duration constraints, timeout constraints, periodic or sporadic timing constraints. A typical duration constraint is that the **arm1_extend_process(dest)** operation on the robot should have a duration bounded by **dest/10** seconds. A typical sporadic constraint is that a startup operation on the gas burner should not be invoked within 60 seconds of a preceding failed startup.

We can indicate intended durations using a **tick** action of a clock:

```
MACHINE Clock(maxtime)
VARIABLES now
INVARIANT now: TIME
DEFINITIONS
  TIME == 0..maxtime
INITIALISATION now := 0
OPERATIONS

  tick(lower,upper) =
    PRE lower: TIME & upper: TIME & lower <= upper &
        now + upper <= maxtime
    THEN
      now :: now+lower..now+upper
    END;  ...

END
```

This is used as follows:

```
  arm1_extend_process(dest) =
    PRE dest: DISTANCE & extension <= dest
    THEN
      stop_position := dest || extension := dest ||
      tick(0,dest/10)
    END
```

This can be verified at the code level by appeal to known assembler op code durations [7, 10].

Timeouts can be specified by using a related **Timer** component, as described in Section 3.

3 Case Study 1: Gas Burner

3.1 Requirements

This case study is described in [19, 21]. It involves the control of a simple gas burner (Figure 1) in order to satisfy the following safety requirements:

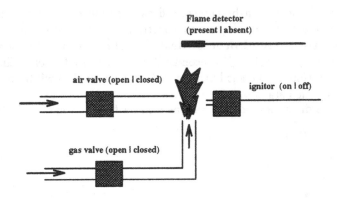

Fig. 1. Gas Burner Devices

- **S1** The gas valve must never be open unless the air valve is open;

- **S2** The flame should never appear if the air valve is closed.

The controlled system must also satisfy the following (sporadic) timing requirements:

- **T1** The gas valve should not be continuously open for more than 30 seconds without the flame being present

- **T2** The system should not be restarted within 60 seconds of a failed startup procedure.

An efficiency requirement **E1**: "the ignitor must not be on unless necessary" is formalised by requiring that the ignitor is only on if the air valve is open. **E2** is that the duration of **startup** must be no more than 35 seconds.

The operator will have operations to start and shut down the system, but otherwise the controller is responsible for autonomously ensuring the above constraints. The data and control flow diagram of the system is given in Figure 2.

States which fail to satisfy **S1** and **E1** can be eliminated by constraining the occurrence of controllable transitions, whilst states failing **S2** cannot be eliminated. Instead the controller must react immediately to the occurrence of such states by initiating a suitable controllable transition.

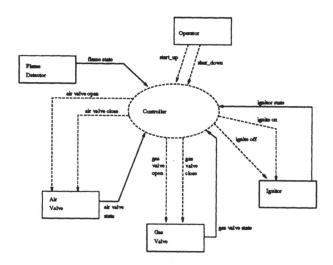

Fig. 2. Gas Burner System DCFG

3.2 Domain Model

The statechart of the uncontrolled system is given in Figure 3. Notice that no constraints are specified here (we could have restricted this state space of 16 states by physical laws such as that flame without the gas valve being open is impossible). The dashed lines in the statechart of Figure 3 indicate uncontrollable transitions – the flame may spontaneously appear or disappear regardless of the state of other components.

The B model of the unconstrained system will be split into a model of the actuator devices and a model of the input flame detector device:

```
MACHINE GB_Actuators  /* Represents unconstrained system */
SEES GB_data  /* Contains type definitions */
VARIABLES avstate, gvstate, istate
INVARIANT
  avstate: AVState & gvstate: GVState &
  istate: IState
INITIALISATION
  avstate := av_closed || gvstate := gv_closed ||
  istate := off
OPERATIONS
  open_av = PRE avstate = av_closed
            THEN avstate := av_open
            END;

  close_av = PRE avstate = av_open
             THEN avstate := av_closed
             END;

  open_gv = PRE gvstate = gv_closed
```

Gas Burner: Domain Model

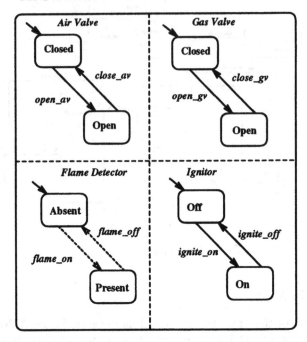

Fig. 3. Gas Burner Statechart – Uncontrolled System

```
            THEN gvstate := gv_open
            END;

    close_gv = ...;

    ignite_on = ...;

    ignite_off = ...
END

and:

MACHINE FlameDetector
SEES GB_data
VARIABLES flame
INVARIANT flame: FDState
INITIALISATION flame := absent
OPERATIONS
    fd <-- sample_fd =
            ANY ff
            WHERE ff: FDState
            THEN flame,fd := ff,ff
            END;
```

```
  becomes_present =
      flame := present;

  becomes_absent =
      flame := absent
END
```

The **sample_fd** operation is for the use of controllers using periodic sampling –
the non-deterministic result represents the changes in the flame state between
sampling points.

GB_data contains type definitions for the state spaces of the components:

```
MACHINE GB_data
SETS
  AVState = {av_open, av_closed};
  GVState = {gv_open, gv_closed};
  IState = { on, off };
  FDState = { present, absent }
END
```

3.3 Specification of Required Behaviour

Considering the constraints **S1** and **E1**, these restrict the state space of the
system to avoid states where the air valve is closed but the gas valve open, or
where the air valve is closed but the ignitor is on. Therefore, they result in a
subset of behaviour of the uncontrolled system shown in Figure 4.

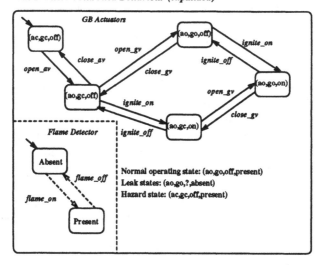

Fig. 4. Controlled Behaviour

The corresponding B machine represents a low-level controller:

```
MACHINE GB_Control
INCLUDES GB_Actuators
PROMOTES open_av, close_gv, ignite_off
SEES GB_data
INVARIANT
  (avstate = av_closed  =>  istate = off) &      /* E1 */
  (gvstate = gv_open  =>  avstate = av_open)      /* S1 */
OPERATIONS
 c_close_av =
    PRE avstate = av_open & istate = off &
        gvstate = gv_closed
    THEN
      close_av
    END;

 c_open_gv =
    PRE avstate = av_open & gvstate = gv_closed
    THEN
      open_gv
    END;

 c_ignite_on =
    PRE avstate = av_open & istate = off
    THEN
      ignite_on
    END
END
```

The PROMOTES clause lists actions which are being promoted to be operations of GB_Control without further constraints. Other operations of GB_Actuators need to be constrained by additional preconditions in order to ensure that E1 and S1 are preserved.

As proposed in [9], *control* relationships are typically represented by the INCLUDES or IMPORTS mechanisms of B, and *monitoring* relationships are represented by the SEES mechanism.

We can formalise S1 and E1 – which refer to *all* times – as invariants of this controller if we make the assumption that its operations represent the lowest level of granularity of action in the system. This would not be the case if we modelled valves that could be in intermediate states between open and closed states, for example [17].

The controller is automatically proven to be correct by the B Toolkit. This shows that if clients of this controller only call its actions within their preconditions, then the invariants S1 and E1 will hold for the controlled system.

In particular, we do not need to repeat these invariants in higher level controllers, because the internal consistency or refinement proofs of these controllers will include the obligations that operations of GB_Control are called within their preconditions [12].

The timing constraints **T1** and **T2** are expressed in a machine which defines the responses of the system to the operator **startup** and **shutdown** actions:

```
MACHINE Gas_Burner(maxtime)
INCLUDES Clock(maxtime)
SEES GB_data, Bool_TYPE
PROMOTES tick
VARIABLES avstate, gvstate, istate, fdstate,
  last_startup_failed, last_startup_time
DEFINITIONS TIME == 0..maxtime
INVARIANT
  avstate: AVState & gvstate: GVState &
  istate: IState & fdstate: FDState &
  last_startup_failed: BOOL & last_startup_time: TIME
INITIALISATION
  avstate := av_closed || gvstate := gv_closed ||
  istate := off || fdstate :: FDState ||
  last_startup_failed := FALSE || last_startup_time := 0
OPERATIONS
  ok <-- startup =
    PRE
      (last_startup_failed = TRUE  =>
         now >= last_startup_time + 60) &
      now + 35 <= maxtime &
      istate = off & avstate = av_closed & gvstate = gv_closed
    THEN
      CHOICE
        tick(30,35) ||    /* Timeout + 0..5 seconds */
        fdstate := absent ||
        gvstate := gv_closed || avstate := av_closed ||
        ok := FALSE ||
        last_startup_failed := TRUE ||
        last_startup_time := now+30
      OR
        tick(0,35) ||
        fdstate := present || gvstate := gv_open ||
        avstate := av_open || ok := TRUE ||
        last_startup_failed := FALSE
      END
    END;

  shutdown =
    PRE gvstate = gv_open & avstate = av_open &
        istate = off
    THEN
      avstate := av_closed || gvstate := gv_closed
    END
END
```

T1 is expressed by the **tick** invocations within the CHOICE in **startup** – the first side of the choice represents the case where the timeout is passed without the flame becoming present. Thus the duration of the operation in this case must be at least 30 seconds, and we place an upper bound of 35 seconds on the duration for efficiency reasons (**E2**). In the case where the flame does appear before the timeout, the duration can be from 0 up to 35 seconds.

T2 is expressed by the precondition of **startup**, which asserts that **startup** should only be invoked if there has not been a failed **startup** within the last 60 seconds.

In a formalism such as VDM^{++} we could more directly express that a leak is only allowed for at most 30 seconds by a calculus expression:

$$\int_a^b \text{leak dt} = b - a \quad \Rightarrow \quad b \leq a + 30$$

where **leak**(t) = 1 if **gvstate** = **gv_open** at t and **fdstate** = **absent**, and 0 otherwise.

We can perform validation of the above specification by animation: running test scenarios of the desired behaviour against the specification to see if what it asserts as the required behaviour actually agrees with the original expectations. For example, a test scenario for requirement **T2** could be:

1. Initialise system,

2. failed **startup** at **now** = 0,

3. attempt another **startup** at **now** = 50 (precondition should be false).

3.4   Design

We can use the concepts of procedural control from [21] to define actions in higher-level controllers for reactive systems such as the gas burner. A procedural controller maps each state in the constrained state space of the system (with forbidden states eliminated) to exactly one action which can be forced to occur to correct or pre-empt undesirable behaviour. In this case the remaining undesirable states are:

1. Leak states – the gas valve open and the flame absent (for more than 30 seconds during startup, or for any duration during normal operation).

2. Uncontrolled ignition states – the flame present but the air valve closed.

The action in the first case is to close the gas valve, and in the second to open the air valve.

Here we will focus on the timing properties and the handling of failed startups, rather than on the control cycles during the idle or active states of the system. The implementation of the outer level gas burner controller is as follows:

```
IMPLEMENTATION GasBurnerI
REFINES GasBurner
SEES Bool_TYPE, GB_data
```

```
IMPORTS Timer(maxtime), GB_Control, FlameDetector,
  lstime_Nvar(maxtime), lsfailed_Vvar(BOOL)
PROMOTES tick
INVARIANT fdstate = flame & lstime_Nvar = last_startup_time &
  lsfailed_Vvar = last_startup_failed & tau = now
INITIALISATION
  lstime_STO_NVAR(0); lsfailed_STO_VAR(FALSE)
OPERATIONS
  ok <-- startup =
   VAR flame_state, time_elapsed, start_time
   IN
     start_time <-- current_time;
     flame_state <-- sample_fd;
     time_elapsed := 0;
     set_timer(30);
     open_av; c_ignite_on;
     c_open_gv;
     WHILE flame_state = absent & time_elapsed <= 30
     DO
       flame_state <-- sample_fd;
       timer_tick(1,5);
       time_elapsed <-- expired
     INVARIANT elapsed <= 35 &
       elapsed: NAT & time_elapsed = elapsed &
       start_time = now & start_time + time_elapsed = tau &
       flame_state = flame &
       istate = on & avstate = av_open & gvstate = gv_open &
       lstime_Nvar = last_startup_time
     VARIANT 35 - elapsed
     END;
     IF flame_state = absent
     THEN /* reset everything */
       close_gv;
       ignite_off; c_close_av;
       ok := FALSE;
       lstime_STO_NVAR(start_time+30);
       lsfailed_STO_VAR(TRUE)
     ELSE
       ignite_off;  ok := TRUE;
       lsfailed_STO_VAR(FALSE)
     END
   END;

  shutdown =
     BEGIN close_gv; c_close_av END; ...
END
```

The **Timer** component allows us to set a timeout (here it is 30 seconds), and to periodically sample the elapsed time, until the timeout has been reached (or until the flame has lit). **timer_tick**$(1, 5)$ acts like a **before elapsed + 5** statement in the formalism of [10]. The **Timer** is an enhancement of the **Clock**:

```
MACHINE Timer(maxtime)
SEES Bool_TYPE
VARIABLES elapsed, timeout, tau
DEFINITIONS
  TIME == 0..maxtime
INVARIANT
  elapsed: TIME & timeout: TIME &
  tau: TIME &
  elapsed <= tau
INITIALISATION
  elapsed := 0 || timeout := 0 || tau := 0
OPERATIONS
  set_timer(xx) =
      PRE xx: TIME
      THEN
         timeout := xx || elapsed := 0
      END;

  timer_tick(lower,upper) =
    PRE lower: TIME & upper: TIME & lower <= upper &
        tau+upper <= maxtime
    THEN
      ANY tt
      WHERE tt: lower..upper
      THEN
         elapsed := elapsed+tt ||
         tau := tau+tt
      END
    END;

  ee <-- expired =
    ee := elapsed;

  tt <-- current_time =
    tt := tau
END
```

A more natural approach to the specification of a timeout would be to have a **Timer** component that signals the controller when the timeout has been reached [14]. However this would require the **Timer** to be able to invoke operations of the controller, which breaks the strict hierarchy of design structures in B. Thus we must adopt instead the above polling approach, which complicates the proof of correctness.

Informally we can reason that the above implementation of **startup** is correct – either the timeout happens or the flame lights within the timeout period, and therefore the loop terminates.

Either the flame is absent at the loop end, and the termination state corresponds to that in the specification (everything is off, delay must be at least the timeout); or the flame is present and the delay may be anything up to time-out + 5s duration – checking that this additional duration is observed requires assembler-level reasoning as in [7].

There are 35 proof obligations for the correctness of the **GasBurner**, of which 19 require detailed interactive proof using properties of time.

3.5 Implementation

We can create a simulator for the system by implementing the actuators by C programs which simply write out the action they are carrying out, and by implementing the timer and flame detector by components which query the user for the appropriate input values. Thus we can test scenarios at the implementation level as well as at the specification level. For example, a simulator for the actuators machine would be:

```
IMPLEMENTATION GB_ActuatorsI
REFINES GB_Actuators
SEES GB_data, basic_io, String_TYPE
IMPORTS
  avstate_Vvar(AVState), gvstate_Vvar(GVState), istate_Vvar(IState)
INVARIANT
  avstate_Vvar = avstate & gvstate_Vvar = gvstate & istate_Vvar = istate
INITIALISATION
  gvstate_STO_VAR(gv_closed);
  istate_STO_VAR(off); avstate_STO_VAR(av_closed)
OPERATIONS
  open_av =
    BEGIN
      PUT_STR("Opening Air Valve"); NWL(1);
      avstate_STO_VAR(av_open)
    END;

  close_av =
    BEGIN
      PUT_STR("Closing Air Valve"); NWL(1);
      avstate_STO_VAR(av_closed)
    END; ....
END
```

For the actual control system, the actuators and sensors would be implemented by interfaces to valve, ignitor and flame detector devices, whilst the timer would be implemented by a system clock – invocations of **timer_tick** are removed from the controller code, but we have to verify that the relevant segments of code whose duration they model actually do have execution times bounded by this duration.

4 Case Study 2: Production Cell

This case study has been used as a means of comparing several approaches to formal specification and development [18].

4.1 Requirements

The components of the system are:

- a press, which responds to signals to close and open, and whose position can be determined;

- a robot, which can rotate clockwise or anti-clockwise, and which has two arms which can be separately extended or retracted. Each arm has an electromagnetic gripper at its end;

- a feed belt, which conveys metal blanks to the rotating table. It can be started and stopped, and it can be determined if a blank is at the end of the belt;

- a deposit belt, which transports work pieces unloaded by the robot to a travelling crane. The belt has a sensor to detect if a piece has reached the point where it can be picked up by the crane;

- a travelling crane, which picks up metal plates from the deposit belt, moving them to the feed belt and unloading them onto this belt;

- a rotating table, which is capable of vertical and rotary movement.

We assume that there is an additional sensor which detects if the deposit belt is clear (the need for this is recognised in several of the papers of [18]) and if a blank is on the rotating table.

In this paper we will focus on the internal control of the robot, and the interaction between the robot and press. We will assume that the rotating table operates correctly and that the presence of a blank at the table may occur 'spontaneously' without being controllable from the robot/press subsystem. Thus the data and control flow diagram of the subsystem is as shown in Figure 5. Event flows are denoted by dashed arrows whilst discrete data flows are indicated by solid arrows.

The safety requirements for the robot and press interaction and individual behaviour are as follows:

1. the robot must not be rotated further than necessary;

2. the robot arms must not be extended more than necessary for picking up or releasing items;

3. the press must not be moved downwards from its lower position, or upwards from its top position;

4. the press may only close when the robot arm is not positioned inside it;

5. the robot arms must not drop metal blanks apart from onto the deposit belt and into the press when the press is open in its middle position;

6. no blank can be deposited in the press if the press already holds a blank;

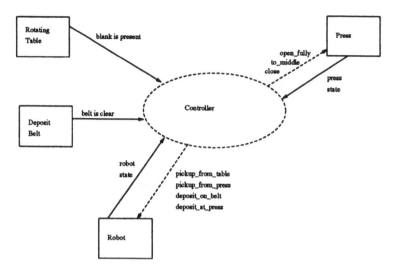

Fig. 5. DCFG of Robot/Press System

7. a new blank may only be deposited by the robot on the deposit belt if this belt is 'clear' in a certain sense.

All of these constraints can be satisfied by constraining controllable transitions (although a way of sensing whether a blank is in the press is needed).

A liveness constraint is that every blank present at the end of the feed belt will eventually arrive, forged, at the beginning of the deposit belt. An efficiency requirement is that concurrency within the system should be maximised.

The processing of the robot/press subsystem is as follows:

– when a blank appears on the rotary table, the robot picks it up with its first arm, then rotates anticlockwise until the second arm is pointing to the press (which must be in the lower position);

– the robot picks up a forged piece from the press with its second arm, then rotates anticlockwise until its second arm is pointing towards the deposit belt;

– the robot deposits the forged piece on the deposit belt, then rotates anti-clockwise until the first arm is pointing to the press (which must be in the middle position – because the first and second arms are at different heights), and deposits the unforged piece in the press. It then rotates clockwise until its first arm is pointing towards the table.

4.2 Domain Model

We can identify 8 significant states for the robot: (i) positioned at the table; (ii) moving from the table to pickup from the press; (iii) positioned at the press for pickup; (iv) moving from the press to the deposit belt; (v) positioned at the deposit belt; (vi) moving from the deposit belt to deposit at the press; (vii) positioned at the press for deposit; (viii) moving from the press deposit position to pickup at the table. Its behaviour is a cycle between these states in this order.

The initial specification of the unconstrained behaviour is therefore as follows:

```
MACHINE ProductionCell
SETS
  RState  =  { positioned_at_table, positioned_press_pickup,
               positioned_press_deposit, positioned_at_deposit,
               rotating_table_press, rotating_press_deposit,
               rotating_deposit_press, rotating_press_table };
  PState  =  { upper, moving_to_close, moving_to_open,
               moving_to_middle, middle, lower };
  RAState =  { extending, retracting, extended, retracted };
  MagState = { holding, released };
  RDirection =  { clockwise, anticlockwise }
CONSTANTS
  delta_dist   /* The maximum arm distance change that
                  can occur in a sampling cycle */
PROPERTIES delta_dist: 1..200
DEFINITIONS
  ANGLE == 0..359;     /* Degrees */
  DISTANCE == 0..1000  /* cm      */
VARIABLES
  rstate, rotation, arm1_state, arm2_state, arm1_extension,
  arm2_extension, arm1_magstate, arm2_magstate, pstate,
  arm1_stop_position, arm2_stop_position,
  blank_present, dbelt_clear
INVARIANT
  rstate: RState & rotation: ANGLE &
  arm1_state: RAState & arm2_state: RAState &
  arm1_extension: DISTANCE & arm2_extension: DISTANCE &
  arm1_magstate: MagState & arm2_magstate: MagState &
  pstate: PState & arm1_stop_position: DISTANCE &
  arm2_stop_position: DISTANCE &
  blank_present: BOOL & dbelt_clear: BOOL
INITIALISATION
  rstate := positioned_at_table || rotation := 0 ||
  arm1_state := retracted || arm2_state := retracted ||
  arm1_extension := 0 || arm2_extension := 0 ||
  arm1_magstate := released || arm2_magstate := released ||
  pstate := middle ||
  arm1_stop_position := 0 || arm2_stop_position := 0 ||
  blank_present :: BOOL || dbelt_clear :: BOOL
OPERATIONS
  blank_arrives =
    PRE blank_present = FALSE
    THEN blank_present := TRUE
    END; ...
```

```
press_start_close =
  PRE pstate = middle
  THEN pstate := moving_to_close
  END; ...

arm1_pick_up =
  PRE arm1_magstate = released
  THEN arm1_magstate := holding
  END;

arm1_start_extend(dest) =
  PRE dest: DISTANCE & arm1_extension <= dest &
    arm1_state = retracted
  THEN
    arm1_stop_position := dest ||
    arm1_state := extending
  END;

arm1_continue_extend =
  PRE arm1_state = extending &
    arm1_extension + delta_dist < arm1_stop_position
  THEN
    arm1_extension :: arm1_extension+1..arm1_extension+delta_dist
  END;

arm1_stop_extend =
  PRE arm1_state = extending &
    arm1_extension <= arm1_stop_position &
    arm1_stop_position <= arm1_extension + delta_dist
  THEN
    arm1_state := extended ||
    arm1_extension := arm1_stop_position
  END; ....
END
```

This can be decomposed into machines for each of the components:

```
MACHINE Press
SEES PC_Types
VARIABLES pstate
INVARIANT pstate: PState
INITIALISATION pstate := middle
OPERATIONS
  press_start_close =
    PRE pstate = middle
    THEN pstate := moving_to_close
    END; ....
END
```

and so forth, where **PC_Types** encapsulates the type and constant definitions of **ProductionCell**.

4.3 Specification of Required Behaviour

The required safety invariants can be expressed in a low-level controller for the production cell, which combines operations of the press and robot in order to provide operations that higher level controllers will interface with:

```
MACHINE PC_Controller
INCLUDES ProductionCell
SEES Bool_TYPE, PC_Types
PROMOTES
  get_arm1_extension, /* & other queries */
  start_rotate_table_press, stop_rotate_table_press,
  start_rotate_press_deposit, stop_rotate_press_deposit,
  start_rotate_deposit_press, stop_rotate_deposit_press,
  start_rotate_press_table, stop_rotate_press_table
DEFINITIONS
  ANGLE == 0..359;      /* degrees */
  DISTANCE == 0..1000; /* cm */
  table_dist == 200;
  press_dist == 150;
  deposit_dist == 100
INVARIANT
  (rstate = positioned_at_table  =>  (arm1_extension <= table_dist &
                                      arm2_extension <= table_dist) ) &
        /* No collision of arm1 or arm2 with table */
  (rstate = rotating_press_table  =>  (arm1_extension = 0 &
                                       arm2_extension = 0)) & ...
        /* + other safety invariants */
OPERATIONS
  start_extend_to_table =
    PRE rstate = positioned_at_table &
      arm1_state = retracted & arm1_magstate = released &
      blank_present = TRUE & pstate = middle
    THEN
      arm1_start_extend(table_dist) || press_start_close
    END;

  continue_extend_to_table =
    PRE rstate = positioned_at_table & arm1_state = extending &
      arm1_extension + delta_dist < table_dist &
      arm1_stop_position = table_dist
    THEN
      arm1_continue_extend
    END;

  complete_extend_to_table =
    PRE rstate = positioned_at_table & arm1_state = extending &
      arm1_magstate = released & blank_present = TRUE &
```

```
          pstate = moving_to_close &
          arm1_extension <= table_dist &
          table_dist <= arm1_extension + delta_dist &
          arm1_stop_position = table_dist
        THEN
          arm1_stop_extend || arm1_pick_up ||
          blank_removed || press_complete_close
        END;

     start_retract_from_table =
        PRE rstate = positioned_at_table & arm1_state = extended &
          pstate = upper
        THEN
          arm1_start_retract || press_start_open
        END;

     continue_retract_from_table = ...

     /* Plus other process steps for the
        robot/press system. */
END
```

At a higher level of granularity we can consider each of the main processes of the robot and press as a single operation, such as "**pickup_from_table_proc**". The specification of this level simply gives the expected effect of such a complete procedure:

```
MACHINE HL_Control
SEES Bool_TYPE, PC_Types
DEFINITIONS
  ANGLE == 0..359;     /* degrees */
  DISTANCE == 0..1000; /* cm */
  table_dist == 200;
  press_dist == 150;
  deposit_dist == 100
VARIABLES
  hl_rstate, hl_arm1_state, hl_arm1_extension,
  hl_arm1_magstate, hl_pstate,
  hl_arm1_stop_position, ...., hl_blank_present
INVARIANT
  hl_rstate: RState & hl_arm1_state: RAState &
  hl_arm1_extension: DISTANCE &
  hl_arm1_magstate: MagState & ... & hl_pstate: PState &
  hl_arm1_stop_position: DISTANCE & hl_blank_present: BOOL
INITIALISATION
  hl_rstate := positioned_at_table || hl_arm1_state := retracted ||
  hl_arm1_extension := 0 || hl_arm1_magstate := released ||
  hl_pstate := middle || hl_arm1_stop_position := 0 ||
  hl_blank_present:: BOOL || ...
OPERATIONS
  pickup_from_table_proc =
```

```
      PRE hl_blank_present = TRUE &
          hl_rstate = positioned_at_table &
          hl_arm1_state = retracted &
          hl_arm1_magstate = released & hl_pstate = middle
      THEN
        hl_blank_present := FALSE ||
        hl_arm1_magstate := holding ||
        hl_pstate := lower
      END;

    rotate_table_press_proc = ...;
END
```

Finally at the top level we specify the activity of the controller purely in terms of the number of blanks processed:

```
MACHINE PC_Outer
SEES Bool_TYPE
VARIABLES processed, active
INVARIANT processed: 0..1000000 & active: BOOL
INITIALISATION
  processed := 0 || active := FALSE
OPERATIONS
  start =
    PRE active = FALSE
    THEN
      active := TRUE || processed := 0
    END;

  pp <-- stop =
    PRE active = TRUE
    THEN
      active := FALSE || pp := processed
    END;

  cycle =
    PRE active = TRUE & processed < 1000000
    THEN
      processed := processed + 1
    END
END
```

In a more detailed specification we would consider the possibility of failures of certain operations and actions, and the propagation upwards of these exceptions. Probably the safest course of action in the case of failure of any component is to bring all the components to a halt.

4.4 Design

We implement the **ProductionCell** machine as an aggregate of the constituent components:

```
IMPLEMENTATION ProductionCellI
REFINES ProductionCell
SEES  Bool_TYPE, PC_Types
IMPORTS
  Robot, Press, FeedBelt, a1_RobotArm, a2_RobotArm
PROMOTES blank_arrives, blank_removed, press_start_close,
 press_complete_open, press_start_middle, press_complete_middle,
 start_rotate_table_press, stop_rotate_table_press,
 stop_rotate_deposit_press, press_start_open,
 start_rotate_press_deposit, stop_rotate_press_deposit,
 start_rotate_deposit_press, start_rotate_press_table,
 stop_rotate_press_table, arm2_pick_up,
 arm2_release, arm1_pick_up, arm1_release, arm1_start_extend,
 arm1_stop_extend, arm1_start_retract, arm1_continue_retract,
 arm1_stop_retract, get_arm1_extension, arm2_start_retract, ...
END
```

We can refine || to ; in the operations of **PC_Controller**, ordering the operation calls to **ProductionCell** to ensure their preconditions hold:

```
IMPLEMENTATION PC_ControllerI
REFINES PC_Controller
SEES Bool_TYPE, PC_Types
IMPORTS ProductionCell
OPERATIONS
  complete_extend_to_table =
    BEGIN
      arm1_stop_extend;
      arm1_pick_up;
      blank_removed;
      press_complete_close
    END; ....
END
```

The refinement relation is the identity (no change in variables).

The high-level controller uses polling loops to implement the high-level processes of the robot and press:

```
IMPLEMENTATION HL_ControlI
REFINES HL_Control
IMPORTS PC_Controller
SEES PC_Types
DEFINITIONS
  ANGLE == 0..359;    /* degrees */
  DISTANCE == 0..1000; /* cm */
  table_dist == 200;
  press_dist == 150;
  deposit_dist == 100;
  delta_dist == 100
INVARIANT
  hl_rstate = rstate & hl_arm1_state = arm1_state &
```

```
     hl_arm1_extension = arm1_extension &
     hl_arm1_magstate = arm1_magstate &
     hl_blank_present = blank_present & hl_pstate = pstate & ...
OPERATIONS

  pickup_from_table_proc =
    VAR arm1ext
    IN
      arm1ext <-- get_arm1_extension;
      start_extend_to_table;
      WHILE arm1ext + delta_dist < table_dist
      DO
        continue_extend_to_table;
        arm1ext <-- get_arm1_extension
      INVARIANT
        arm1_extension <= table_dist &
        rstate = positioned_at_table &
        arm1_state = extending & arm1ext = arm1_extension
      VARIANT
        table_dist - arm1_extension
      END;
      complete_extend_to_table;
      start_retract_from_table;
      WHILE delta_dist < arm1ext
      DO
        continue_retract_from_table;
        arm1ext <-- get_arm1_extension
      INVARIANT arm1_extension > 0  &
        rstate = positioned_at_table &
        arm1_state = retracting & arm1ext = arm1_extension
      VARIANT arm1_extension
      END;
      complete_retract_from_table
    END;

  rotate_table_press_proc = ....;
END
```

The outer level is implemented by:

```
IMPLEMENTATION PC_OuterI
REFINES PC_Outer
SEES Bool_TYPE
IMPORTS HL_Control, proc_Nvar(1000000), act_Vvar(BOOL)
INVARIANT
  proc_Nvar = processed & act_Vvar = active
INITIALISATION
  act_STO_VAR(FALSE)
OPERATIONS
  start = BEGIN
            proc_STO_NVAR(0);
```

```
                act_STO_VAR(TRUE)
            END;

pp <-- stop = BEGIN
                  pp <-- proc_VAL_NVAR;
                  act_STO_VAR(FALSE)
              END;

cycle =
  BEGIN
    pickup_from_table_process;
    rotate_table_press_process;
    pickup_from_press_process;
    rotate_press_deposit_process;
    deposit_at_dbelt_process;
    rotate_deposit_press_process;
    deposit_at_press_process;
    rotate_press_table_process;
    proc_INC_NVAR
  END
END
```

The development structure of the production cell is shown in Figure 6.

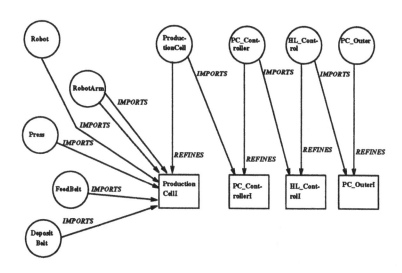

Fig. 6. Development Structure of Production Cell

5 Possible Extensions of B

A major deficiency of the current B language is that there is no capability for concurrent execution of operations. For example, in the robot and press interaction, we have to specify exactly how the press opening and closing actions synchronise with arm1 moving to and away from the table. It would be much more flexible to simply assert that the process consisting of steps

press_start_close; press_complete_close;
press_start_open; press_complete_open

executes in parallel with the extension and retraction process.

Likewise, the specification of the intended life histories of components, and of permission guards, can only be indirectly specified via preconditions. This leads to a lack of clarity, since preconditions are used for a number of other purposes: ensuring the feasibility of the operation substitution; ensuring that expressions within this substitution are well-defined, and ensuring that the machine invariant holds after any execution of the operation.

Liveness and timing constraints cannot be directly expressed, but only indirectly via the design-oriented mechanisms of variants and clocks.

Consequently, we suggest the following extensions of the language.

5.1 Trace Specifications

A *trace* specification describes in a concise manner the allowed histories of a B module. It consists of a list of trace set definitions and an overall history specification of the possible histories. It would be written as an additional clause of an abstract machine:

MACHINE **Name**

$\vdots$

TRACES
    GENERAL **TraceSetExpression;**
    ... **trace set definitions** ...
END

A trace set definition has the form

Identifier = TraceSetExpression

where a trace set expression is either an explicit definition (**trace, alphabet**) of a trace and an alphabet, or is built up from trace set identifiers and explicit trace set definitions using operators such as **w** (weave, or synchronised interleaving [23]).

An alphabet is a set of operation names from the machine and a trace is a regular expression in operation names, built using the operators ; (sequence), * (iteration zero or more times), **w** (synchronised interleaving) and | (binary choice). By default the **alphabet** is the set of operation names mentioned in the **trace**, and only needs to be explicitly written if this is not the case.

Thus the robot arm history could be expressed as follows:

ExtendSequence =
         (start_extend; continue_extend*; stop_extend)
RetractSequence =
         (start_retract; continue_retract*; stop_retract)
PickUpRelease = (pick_up; release)*

RobotArm =
      (ExtendSequence | RetractSequence | pick_up | release)* w
                         PickUpRelease

There are two possible interpretations of a trace as a contract:

1. "We are only guaranteed valid behaviour from **M** if its operations are called in an order allowed by the traces of **M**", or:

2. "No execution of operations of **M** is possible unless they conform to the specified traces".

The first implies that traces can become *weaker* (more traces allowed) in refinement, whilst the second implies that traces can become more restrictive in refinement. The latter is more appropriate in a concurrent environment. We will avoid this issue by only providing a TRACES clause in a MACHINE, and assuming that the trace sets of the refinements and implementations of this machine are unchanged.

An implementation **I** which invokes operations of a machine **M** must ensure that it calls **M**'s operations in an order which is some possible trace of **M**. For each operation or statement **S** of **I** we can define (approximately) the set $\underline{traces}_M(S)$ of possible sequences of calls that it may make on **M**. For example, if operation **op** is defined by a conditional IF **E** THEN S_1 ELSE S_2 END then

$$\underline{traces}_M(op) = \underline{traces}_M(S_1) \cup \underline{traces}_M(S_2)$$

and similarly for other statement constructs (in the program-like subset of B AMN).

Let $\underline{traces}(M)$ denote the set of sequences of operations which obey the trace specification of **M**. Then we require that given any trace **tr** that satisfies $\underline{traces}(I)$, subtraces $tr_1, \ldots, tr_n$ with $tr_1 \in \underline{traces}_M(op_1), \ldots, tr_n \in \underline{traces}_M(op_n)$, where these are the operations of **I**, we have

$$tr[tr_1/op_1, \ldots, tr_n/op_n] \in \underline{traces}(M)$$

Refinements or implementations cannot alter the traces of an abstract machine (in other words, the traces of a subsystem are defined by the trace specification of the abstract machine at the head of the refinement path of this subsystem). Thus $\underline{traces}(I)$ in the above conditions refers to $\underline{traces}(N)$ where **N** is the machine implemented by **I**.

5.2   Permission Guards

A permission guard G on an operation **op** is a condition which must be satisfied before execution of **op** can proceed. A machine clause for such constraints could have the form:

GUARDS
    op $\Rightarrow$ **Guard**;
    $\vdots$

where each **op** is an operation name (possibly with input parameters), and **Guard** is a condition involving the state variables, operation parameters, and counts **count(op′)** of the number of completed executions of operations **op′** of the machine.

Permission guards are used by *passive* server components to protect their internal integrity against inappropriate calls from multiple clients. For example, a machine encapsulating a shared buffer would guard its **get** action by a guard asserting that the contents of the buffer was non-empty. Clients requesting **get** would then be suspended on the call until another client had deposited elements in the buffer. As with traces, the MACHINE permission clauses define permission for the subsystem specified by this machine.

5.3   Threads

A *thread* specification could be included to provide the possibility of autonomous behaviour by 'active' modules, and to support asynchronous execution of operations. It would be a new clause of a refinement or implementation, written as

THREAD
    statement

where **statement** usually consists of an unbounded loop WHILE **true** DO **body** INVARIANT I END, and **body** is a select statement. We extend the syntax of select statements to include ANSWER clauses (akin to **accept** in Ada):

SELECT
        cond_1 ANSWER op_1 THEN stat_1
    WHEN
        cond_2 ANSWER op_2 THEN stat_2
    WHEN
        $\vdots$
END

If a guarded command within a select statement contains an ANSWER op_i clause, then it cannot proceed to execute the corresponding stat_i until an external caller attempts to call op_i, and cond_i must also be true at this point in time. The guarded command is executed by executing op_i and then executing stat_i (without interruption from other operations of the component). It is possible to mix guarded commands without ANSWER clauses with guarded commands with such clauses within the same select statement. If more than one guarded command can be executed, then the choice between them remains non-deterministic. This form of statement is essentially the same as that described in [6], where the labels of guarded commands are the names of actions.

Concurrency is introduced by a thread because the caller of an operation op_i is only blocked waiting for op_i to return whilst the ANSWER clause (and code of op_i in the OPERATIONS list) is executing – once the code of the operation is completed, the thread of the called component can then execute $stat_i$ concurrently with the caller.

Using threads we can simulate a parallel execution operator op1 ||| op2 by introducing intermediate components which define **async_op1** and **async_op2** as asynchronous operations invoking **op1** and **op2**, and define **complete_op1** and **complete_op2** operations which enable synchronisation of the termination of both parts of the ||| operator:

op1 ||| op2

in implementation M, where op1 is in machine N1, and op2 in N2, can be interpreted as the sequence

```
async_op1;  async_op2;
complete_op1;  complete_op2
```

where **async_op1** and **complete_op1** are operations of a machine **Async_N1**, with an implementation defined as:

```
IMPLEMENTATION Async_N1_imp
REFINES Async_N1
IMPORTS N1
OPERATIONS
    async_op1  =  skip;

    complete_op1  =  skip
THREAD
    WHILE true
    DO
        SELECT
            ANSWER async_op1
            THEN
                op1;
                ANSWER complete_op1
        END
    INVARIANT true
    END
END
```

and similarly for op2.

5.4 Histories

Finally *histories* are used to describe behaviours which concern liveness or fairness constraints which cannot be expressed using the other constructs. Here we use temporal logic operators ◊ (eventually), □ (henceforth) and ◯ (next) to describe temporal constraints in a more abstract manner than via regular expressions. For example, to assert that after any occurrence of **opa** there will eventually be an occurrence of **opb**, we could write:

MACHINE **Liveness**
OPERATIONS
    opa;
    opb
HISTORY
    □(**opa** ⇒ ◊**opb**)
END

A real-time logic such as RAL could be used to constrain the durations of operations and intervals between events [15]. The approach of Section 3 could then be used to implement such constraints.

5.5 Structuring

Contrary to conventional B, we allow shared write access to components via INCLUDES or IMPORTS, provided that the shared component does not possess a thread: it is a purely *passive* server object. If a component is shared for write access then its data can only be accessed via operation calls (ie., INCLUDES is full-hiding in this case). Similarly, because concurrency is introduced between a client and a server by threads in the server, clients of active servers must also be restricted to access its state via operation calls, so that INCLUDES and EXTENDS must be full-hiding in this case.

5.6 Proof Obligations

A consequence of allowing shared write access to a component is that every component that accesses this (by INCLUDES) must either:

1. Not refer to the variables of the shared component in its invariant; or

2. Have proof obligations for internal consistency which include that every operation of the *shared* component preserves such invariants.

This is not a major change in the B proof obligations, since these are the obligations for the EXTENDS construct in conventional B.

For the new constructs there are additional internal consistency obligations:

$$\exists(\mathbf{v}, \mathbf{x}).(\mathbf{Pre}_{op,M} \wedge \mathbf{Inv}_M \wedge \mathbf{G}_{op})$$

That is, there is at least one state $\mathbf{v}$ of the machine and input values $\mathbf{x}$ which satisfies both the guard $\mathbf{G}_{op}$ and the precondition, for each operation op. Usually we expect the precondition to imply the guard: the guard describes the set of *allowed* behaviours, whilst the precondition describes the set of *normal* behaviours.

For each operation **op** we also require that there exists at least one non-trivial history which satisfies the traces and guard conditions of the machine, and which contains **op**.

More generally, we could formulate the set of "normative histories", which are sequences of operation invocations where the post-state of one invocation is within the precondition of the successive invocation. We should expect that the trace sets defined for a machine have a non-empty intersection with this set of normative histories.

These additional obligations will detect errors such as initial deadlock (no non-trivial traces of the machine), and failing to provide any possibility of valid execution of a particular operation. They also help check the consistency of the permission guards with the trace specifications.

Animation techniques for B could be extended with checks that the history of operations up to the current interval is consistent with the traces, and that an operation guard is true when the operation is selected for execution.

5.7 Semantics

In [13] we present a semantics for the extended language by interpreting it in the Object Calculus [5]. This gives a more general definition of refinement as *theory extension*. An outstanding problem is the issue of refining synchronous abstract operations into asynchronous concrete operations (which may trigger a process that is intended to implement the abstract operation). Proving such refinement steps correct requires an analysis of the individual actions within each process, as discussed in [16].

6 Conclusions

We have presented some techniques for specifying reactive systems in the current B language, and some mechanisms by which B can be extended. The extended language still retains the desirable features of B in that a strict hierarchic design structure is used, and the semi and full hiding disciplines of the structuring mechanisms need only slight modification.

The approach of [3] to distributed system specification in B provides an alternative way to decompose a system specification on the basis of communication between processes modelled as B machines. The approach, based on CSP and Action Systems, allows synchronisation between an output action and multiple input actions. By using different channels (operation names) in the controller machine, several different machines can send messages to a single machine. However, issues of granularity of actions and recursive dependencies between machines are still to be developed in this approach.

The restrictions of B AMN mean that direct modelling of reactive systems is not possible, in particular, a two-way dependence between components **A** and

B, where **A** sends messages to **B** to signal the occurrence of an **A**-event, and vice-versa, must be decomposed into a hierarchical architecture whereby **B** (say) samples **A** to determine if **A** events have occurred. This implies that events may be missed by **B**, or the occurrence of events of different components may be detected in an order which is not that in which they actually occurred.

We have also not addressed the problems of *local* control of actuators for higher reliability and safety: it is preferable if few module boundaries need to be crossed in order to make a safety-critical decision, in order to reduce the possibilities of software errors causing hazards. This can be reconciled with our approach in part by having actuator operations check the state of other relevant components before carrying out a potentially safety-critical action. For example, the operation **c_open_gv** should check directly that the air valve is open before opening the gas valve.

References

1. J-R Abrial. *The B Book: Deriving Programs from Meaning*, Cambridge University Press, 1996.
2. J Bowen and V Stavridou. *Safety-critical systems, formal methods and standards*, Software Engineering Journal, July 1993, pages 189 – 209.
3. M Butler. Combining Action Systems and B AMN in the Design of Distributed Systems. Dept. of Electronics and Computer Science, University of Southampton, 1996.
4. E Durr and E Dusink. The role of VDM^{++} in the development of a real-time tracking and tracing system. In J Woodcock and P Larsen, editors, FME '93, Lecture Notes in Computer Science. Springer-Verlag, 1993.
5. J Fiadeiro and T Maibaum. *Temporal Theories as Modularisation Units for Concurrent System Specification*, Formal Aspects of Computing 4(3), pp. 239–272, 1992
6. J Fiadeiro and T Maibaum. *Interconnecting formalisms: supporting modularity, reuse and incrementality*, in Proc. 3rd Symposium on the Foundations of Software Engineering, ACM Press, 1996.
7. C Fidge. *Proof Obligations for Real-Time Refinement, Proceedings of 6th Refinement Workshop*, Springer-Verlag Workshops in Computing, 1994.
8. D Harel. *On Visual Formalisms*, Communications of the ACM 31(5), pp. 514–530, May 1988.
9. H Haughton and K Lano. *Testing and Safety analysis of AM specifications*, in Proceedings of the 6th Refinement Workshop, City University, London. Springer-Verlag Workshops in Computing, 1994.
10. I Hayes and M Utting. *Coercing Real-time Refinement: A Transmitter*, Northern Formal Methods Workshop, Ilkley, 1996. To appear in Springer-Verlag EWICS, 1997.
11. J Hoare. *The Use of B in CICS*, in **Applications of Formal Methods**, M G Hinchey and J P Bowen (Eds.), Prentice Hall, 1995.
12. K Lano. *The B Language and Method: A Guide to Practical Formal Development*, FACIT Series, Springer-Verlag, 1996.
13. K Lano, J Fiadeiro and J Dick. *Extending B AMN with Concurrency*, 3rd Theory and Formal Methods Workshop, Oxford. Imperial College Press, 1996.

14. K Lano, J Bicarregui and A Sanchez. *Using B to Design and Verify Controllers for Chemical Processing*, Conference on B, IRIN, Nantes, 1996.
15. K Lano, J Bicarregui and S Kent. *A Real-time Action Logic of Objects*, ECOOP 96 Workshop on Proof Theory of Object-oriented Systems, Linz, Austria, 1996.
16. K Lano and S Goldsack. *Refinement Rules for Concurrency and Real-time*, MEDI-CIS Workshop, April 1996.
17. K Lano and A Sanchez. *Specification of a Chemical Process Controller in VDM^{++} and B*, ROOS project document GR/K68783-11, Dept. of Computing, Imperial College, November 1996.
18. C Lewerentz and T Lindner (Eds.). *Formal development of reactive systems: case study production cell*, Berlin: Springer-Verlag, LNCS 891, 1995.
19. I Moon, G Powers, J R Burch and E M Clarke. *Automatic Verification of Sequential Control Systems using Temporal Logic*, AIChE Journal, 38(1):67–75, January 1992.
20. A Pnueli. *Linear and Branching Structures in the Semantics and Logics of Reactive Systems*, ICALP 85, Springer-Verlag LNCS Vol. 194, 1985.
21. A Sanchez and S Macchietto. *Design of Procedural Controllers for Chemical Processes*, ESCAPES Conference, June 1995, Slovenia.
22. A Sanchez, G E Rotstein, N Alsop and S Macchietto. *Procedural Control of Chemical Processes*, Chemical Process Control Conference, 1996.
23. J van de Snepscheunt. Trace Theory and VSLI Design, PhD thesis, Technische Logeschool Eindhoven, Eindhoven, The Netherlands, 1983.

An Improved Recipe for Specifying Reactive Systems in Z

Andy. S. Evans

Department of Computing,
University of Bradford, Bradford, UK.
email: a.s.evans@comp.brad.ac.uk
http://www.staff.comp.brad.ac.uk/~asevans1

1   Introduction

Z by itself is inadequate for specifying reactive systems

Leslie Lamport [Lam94]

How can a reactive system be specified in Z without having to use additional formalisms such as CSP or temporal logic? The conventional wisdom is that it cannot. Notations like Z and VDM traditionally describe a system as an abstract data type. Hence they concentrate on the 'static' system behaviour: that is why they define operations using state before and state after. However, it seems clear that in order to specify a reactive system, 'dynamic behaviour' must be described otherwise concurrent or real-time properties cannot be specified. It is this aspect which is entirely missing from conventional Z and VDM specifications.

During the late eighties, Duke et al [DHKR88, DS89] provided a partial solution to this problem. They showed how a conventional Z specification could be augmented with an additional specification describing its reactive behaviour. Their approach was to informally introduce a relation, OP, to represent all the possible before and after states of the operations of the system being specified. The behaviour of the system was then formalised as the history of state and operation executions resulting from the repeated application of OP (with concurrent operations arbitrarily ordered). Unfortunately, the specification approach they adopted was partially informal, and was oriented towards one specific example.

The aim of this paper is to show that this promising approach can be greatly improved and extended upon to the point where it can provide a practical method for specifying reactive systems in Z. It also adds to work originally presented in [Eva96b]. The four extensions made are:

Genericity - to introduce generic operations for specifying concurrent behaviour in Z.

Real-time - to extend these operations to allow the specification of real-time behaviour.

Modularity - to show how the notion of a module can be specified to encapsulate concurrent components of a Z specification.

276

Synchronised communication - to extend the definition of a module to allow CSP like communication between modules.

The work presented in this paper is based on the standard (ZRM [Spi92]) Z notation, thus it can be used with existing Z tools unchanged. The theoretical basis for the work is the fair-transition model of concurrency proposed by Manna and Pneuli [MP84].

2 The Basic Recipe

The basic recipe for specifying a reactive system in Z is the same as originally proposed in [DS89]: The state and operations of the system are specified in the conventional Z style (this is the *static* specification). This specification is then augmented with a specification of the system's allowable concurrent behaviour (this is the *dynamic* specification). It is the simplification of this second stage, via the use of generic operations, that is the focus of this paper.

To illustrate these ideas, a specification of part of a simple reactive system, a production cell, is presented[1]. The cell consists of two feedbelts, a robot and press. The *robot* is specified first, making a clear distinction between the conventional ('static') specification of its state and operations and the 'dynamic' specification which formalises its reactive behaviour:

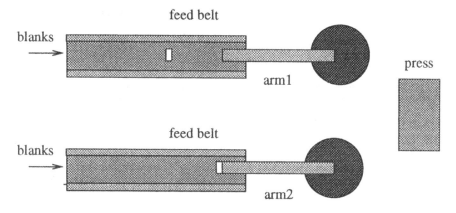

Figure 1: The Simple Production Cell

2.1 Static Specification

The robot's task is to transport uniquely identifiable blanks from the two feed belts and deposit them in the press. To do this, the robot has two independent arms. A generic arm is first specified:

---

[1] A full specification of the Karlsruhe production cell case study using the recipes can be found in [Eva96b] along with a comparison with a conventional Z specification of the case study developed by MacDonald and Carrington [MC95].

[BLANK_ID]

Arm_Position ::= at_belt | at_press
Blank_Loaded ::= loaded⟪BLANK_ID⟫ | unloaded

```
┌─ Arm ──────────────────────────────────────────────────
│ position : Arm_Position
│ status : Blank_Loaded
└──────────────────────────────────────────────────────
```

An arm is either positioned at the belt or at the press and is either loaded or unloaded. If loaded, a record is kept of the blank's identity.

Initially, the arm is at the belt and is unloaded:

```
┌─ InitArm ──────────────────────────────────────────────
│ Arm
│ ──────────
│ position = at_belt
│ status = unloaded
└──────────────────────────────────────────────────────
```

There are four operations of the generic arm. It may load and unload a blank:

```
┌─ Load ──────────────────────        ┌─ Unload ────────────────────
│ ΔArm                                │ ΔArm
│ blank? : BLANK_ID                   │ blank! : BLANK_ID
│ ──────────────────                  │ ──────────────────
│ position = at_belt                  │ position = at_press
│ status = unloaded                   │ status = loaded(blank!)
│ status' = loaded(blank?)            │ status' = unloaded
│ position' = position                │ position' = position
└─────────────────────────────        └─────────────────────────────
```

The pre-conditions for the generic arm to load are that it is at the belt and is unloaded. In order for the generic arm to unload it must be at the press and be loaded.

The generic arm can also move from the press and back again:

```
┌─ MoveToPress ──────────────          ┌─ MoveToBelt ────────────────
│ ΔArm                                 │ ΔArm
│ blank : BLANK_ID                     │ ──────────────────
│ ──────────────────                   │ position = at_press
│ position = at_belt                   │ status = unloaded
│ status = loaded(blank)               │ position' = at_belt
│ position' = at_press                 │ status' = status
│ status' = status                     │
└─────────────────────────────         └─────────────────────────────
```

The generic arm may move to the press provided that it is loaded and at the belt. It may move back again from the press once it is unloaded.

The robot can now be specified. It has two arms, arm1 and arm2 which are instances of Arm:

```
┌─ Robot ─────────────────────────────────────────────────────────
│  arm1, arm2 : Arm
│
└──────────────────────────────────────────────────────────────────
```

Two schemas are now specified which will be used to promote the operations of the generic arm to the robot state:

$$Arm1\,ToRobot \mathrel{\widehat{=}} [\Delta Robot;\ \Delta Arm\ |$$
$$arm2' = arm2 \land arm1 = \theta Arm \land arm1' = \theta Arm']$$
$$Arm2\,ToRobot \mathrel{\widehat{=}} [\Delta Robot;\ \Delta Arm\ |$$
$$arm1' = arm1 \land arm2 = \theta Arm \land arm2' = \theta Arm']$$

Each of the generic arm operations are then promoted to operations of the robot arms:

$$LoadArm1 \mathrel{\widehat{=}} Arm1\,ToRobot \land Load$$
$$LoadArm2 \mathrel{\widehat{=}} Arm2\,ToRobot \land Load$$
$$MoveArm1\,ToPress \mathrel{\widehat{=}} Arm1\,ToRobot \land Move\,ToPress$$
$$MoveArm2\,ToPress \mathrel{\widehat{=}} Arm2\,ToRobot \land Move\,ToPress$$
$$UnloadArm1 \mathrel{\widehat{=}} Arm1\,ToRobot \land Unload$$
$$UnloadArm2 \mathrel{\widehat{=}} Arm2\,ToRobot \land Unload$$
$$MoveArm1\,ToBelt \mathrel{\widehat{=}} Arm1\,ToRobot \land Move\,ToBelt$$
$$MoveArm2\,ToBelt \mathrel{\widehat{=}} Arm2\,ToRobot \land Move\,ToBelt$$

The promoted initial state schema for the robot is obtained similarly:

$$RobotInit \mathrel{\widehat{=}} Arm1\,ToRobot \land Arm2\,ToRobot \land InitArm$$

This completes the ('static') Z specification of the robot. However, the specification as it stands says nothing about the behaviour of the system over time. For example, it cannot express the intuitive property that the arms may operate concurrently. This is the purpose of the dynamic specification.

2.2 Dynamic Specification

Adopting Duke et al's approach, the dynamic behaviour of the robot is specified in terms of the allowable sequences of state changes (computations) that result from the execution of its operations. In this model, pre-conditions are viewed as determining when an operation can execute (a negated pre-condition implying that the operation is disabled). This contrasts with the conventional view of pre-conditions in Z, in which they are viewed as defining an operation's domain of definition [Str95] (outside these conditions the operation is viewed as undefined and may exhibit divergent behaviour). However, there is no contradiction here, as it is the purpose of the dynamic specification to formalise the reactive model. Another assumption made is that operations are *atomic*, i.e. they occur instantaneously. Events with duration are modelled by two operations which mark the beginning and end of an event's lifecycle. Concurrency is modelled by the non-deterministic interleaving of atomic operations.

In the approach used here a computation is always of infinite length, as opposed to a finite model in which termination is represented by a finite computation. The reason for choosing this particular model is that it saves the technical inconvenience of having to deal with both finite and infinite computations. Sequences are defined to be finite in Z. Thus, a new data type is introduced to model infinite sequences:

$$\text{comp } X == \mathbb{N}_1 \rightarrow X$$

Rather than informally introducing a next-state relation, OP, the next step is to construct a *next-state schema*. This is the disjunction of the file's operations. It represents the fact that an atomic step in the robot's behaviour may be caused by any one of its operations:

$RobotNS \; \widehat{=}$
  $LoadArm1 \vee LoadArm2 \vee$
  $MoveArm1\,ToPress \vee MoveArm2\,ToPress \vee$
  $UnloadArm1 \vee UnloadArm2 \vee$
  $MoveArm1\,ToBelt \vee MoveArm2\,ToBelt$

The set of before and after states of this schema correspond to the relation OP as introduced in Duke et al's work.

Once the next-state schema has been specified, it can be utilised within a specification to describe the behaviour of the robot. Although it would be possible to directly write a dynamic specification using $RobotNS$, it is at this point that the first of the generic operations is introduced to simplify the task.

The first operation validcomp describes what it means for a computation to be a valid behaviour a system in which I is the set of initial starting states, and R is its next-state relation:

$$
\begin{array}{l}
\rule{12cm}{0.4pt} \\
[STATE] \\
\rule[0.5ex]{12cm}{0.4pt} \\
\_\,\text{validcomp}\,\_ : \text{comp } STATE \leftrightarrow (\mathbb{P}\, STATE \times (STATE \leftrightarrow STATE)) \\
\rule{12cm}{0.4pt} \\
\forall \sigma : \text{comp } STATE;\; I : \mathbb{P}\, STATE;\; R : STATE \leftrightarrow STATE \bullet \\
\quad \sigma \;\text{validcomp}\; (I, R) \Leftrightarrow \\
\qquad \sigma(1) \in I \wedge \\
\qquad (\forall n : \mathbb{N}_1 \bullet \sigma(n)\; \underline{R}\; \sigma(n+1) \vee \sigma(n+1) = \sigma(n)) \\
\rule{12cm}{0.4pt}
\end{array}
$$

Formally, validcomp is true for all computations in which the first step in the computation belongs to the initial state of the system and each of the subsequent steps are related by the next-state relation, choosing non-deterministically between enabled state-changes, i.e. those belonging to the before states (domain) of R. Note the addition of an idling step $(\sigma(n+1) = \sigma(n))$ to guarantee infinite computations.

It is now possible to use validcomp to calculate the allowable (concurrent) behaviour of the robot. First, the robot's initial state set and next-state relation must be obtained from $RobotInit$ and $RobotNS$. To do this, schema binding is used to project out the relevant state components before direct substitution into validcomp:

$RobotBehaviour$
σ : comp $Robot$

σ validcomp $(\{RobotInit \bullet \theta Robot\}, \{RobotNS \bullet \theta Robot \mapsto \theta Robot'\})$

Any behaviour that satisfies σ will be a valid behaviour of the robot.

2.3 Adding Fairness

There is one further aspect that needs to be specified before the dynamic specification is complete: the robot's *progress* properties. As it stands, the specification allows behaviours that may never change $arm1$ or $arm2$ when required to do so. For example, it may choose to ignore the operations of one arm in favour of executing the operations of the other. To overcome this, *fairness* constraints must be added to the specification.

Rather than adding a predicate directly, a generic operation for 'weak fairness' is first defined which can be used more flexibly:

$[STATE]$
===

$\_ \, \text{wf} \, \_ :$ comp $STATE \leftrightarrow (STATE \leftrightarrow STATE)$

$\forall \sigma :$ comp $STATE$; $op : STATE \leftrightarrow STATE \bullet$
   σ wf $op \Leftrightarrow$
      $(\forall i : \mathbb{N}_1 \bullet (\exists j : \mathbb{N}_1 \mid j \geq i \bullet \sigma(j) \; \underline{op} \; \sigma(j+1))) \vee$
      $(\forall i : \mathbb{N}_1 \bullet (\exists j : \mathbb{N}_1 \mid j \geq i \bullet \sigma(j) \notin \text{dom} \, op))$

Weak fairness is asserted for any computation in which the set of state changes op (corresponding to an operation of the system) is always eventually executed, i.e. op eventually relates two consecutive states in the computation, or op is eventually disabled, i.e. there exists a state which does not belong to the set of before states (domain) of op.

An alternative type of fairness sometimes used is *strong* fairness. Strong fairness asserts that an operation must be executed if it is repeatedly enabled. Interpreting this to mean infinitely often, strong fairness asserts that either the operation eventually occurs, or its execution is eventually always impossible to execute. The first requirement is the same as that of weak fairness. To assert that op is eventually always impossible to execute is equivalent to asserting the existence of an index i in σ such that for each index $(j \geq i)$ in σ the operation is disabled (i.e. $\sigma(j) \notin \text{dom} \, op$):

$$
\begin{array}{l}
\text{=}[STATE]\text{=}\\
\_\,\text{sf}\,\_ : \text{comp}\ STATE \leftrightarrow (STATE \leftrightarrow STATE)\\
\hline
\forall \sigma : \text{comp}\ STATE;\ op : STATE \leftrightarrow STATE \bullet\\
\quad \sigma\ \text{sf}\ op \Leftrightarrow\\
\qquad (\forall i : \mathbb{N}_1 \bullet (\exists j : \mathbb{N}_1 \mid j \geq i \bullet \sigma(j)\ \underline{op}\ \sigma(j+1))) \vee\\
\qquad (\exists i : \mathbb{N}_1 \bullet (\forall j : \mathbb{N}_1 \mid j \geq i \bullet \sigma(j) \notin \text{dom}\ op)))
\end{array}
$$

2.4 Adding Fairness to the Dynamic Specification

To assert that an operation is weakly or strongly fair over a computation, the relevant constraint is simply conjoined to the dynamic specification. For example, to specify that all the *Move..* and *Unload..* operations of the robot are weakly fair the following is specified:

$$
\begin{array}{l}
\text{\_\_}FairRobotBehaviour\text{\_\_}\\
\sigma : \text{comp}\ Robot\\
\hline
\sigma\ \text{validcomp}\ (\{RobotInit \bullet \theta Robot\}, \{RobotNS \bullet \theta Robot \mapsto \theta Robot'\})\\
\sigma\ \text{wf}\ \{MoveArm1ToPress \bullet \theta Robot \mapsto \theta Robot'\}\\
\sigma\ \text{wf}\ \{MoveArm2ToPress \bullet \theta Robot \mapsto \theta Robot'\}\\
\sigma\ \text{wf}\ \{MoveArm1ToBelt \bullet \theta Robot \mapsto \theta Robot'\}\\
\sigma\ \text{wf}\ \{MoveArm2ToPress \bullet \theta Robot \mapsto \theta Robot'\}\\
\sigma\ \text{wf}\ \{UnloadArm1 \bullet \theta Robot \mapsto \theta Robot'\}\\
\sigma\ \text{wf}\ \{UnloadArm2 \bullet \theta Robot \mapsto \theta Robot'\}
\end{array}
$$

Notice that the operations *LoadArm*1 and *LoadArm*2 have been excluded from the list of weakly fair operations. This is because they represent 'spontaneous' events that may never occur (the robot has no control over the feed belt).

The final result is a intuitively simple and well structured specification of the robot's required behaviour.

3 A Soupçon of Time

A reactive system may be required to operate within strict time deadlines. For example, a realistic model of the robot might require that blanks must be loaded and unloaded within a certain time in order to guarantee safe and efficient operation.

This section considers how the basic recipe can be extended to *timed* specification. The approach adopted is to model the behaviour of a timed system using a *timed computation*. Formally, a timed computation is an infinite sequence of states and times:

```
┌─ TState[STATE]──────────────────────────────────────────────
│ time : N
│ state : STATE
└────────────────────────────────────────────────────────────
```

$\text{comp}_t\ X == N_1 \rightarrow TState[X]$

The relation validcomp$_t$ describes what it means for a timed computation to be a valid behaviour of a system. Its definition is analagous to validcomp. The only difference is that increases in time are arbitrarily interleaved within the computation. In other words, each step in the system's behaviour is either the result of executing an operation (chosen non-deterministically from the next-state relation) or an increase in time (the ticking of the clock). This model embodies a "coarse grain" interpretation of time which simplifies the modelling of real-time concurrency. For example, it is not necessary in this model to increase time whenever an operation occurs; again concurrency is modelled as the arbitrary interleaving of atomic operations. Events with duration must be modelled by dual operations.

```
═[STATE]══════════════════════════════════════════════════════
│ _ validcomp_t _ : comp_t STATE ↔ (P STATE × (STATE ↔ STATE))
├──────────────────────────────────────────────────────────────
│ ∀ σ_t : comp_t STATE;  I : P STATE;  R : STATE ↔ STATE •
│    σ_t  validcomp_t  (I, R) ⇔
│        (σ_t(1)).time = 0 ∧ (σ_t(1)).state ∈ I ∧
│        (∀ n : N_1 •
│            ((σ_t(n)).state R (σ_t(n + 1)).state ∧
│            (σ_t(n + 1)).time = (σ_t(n)).time) ∨
│            ((σ_t(n + 1)).state = (σ_t(n)).state ∧
│            (σ_t(n + 1)).time = (σ_t(n)).time + 1))
└──────────────────────────────────────────────────────────────
```

Formally, validcomp$_t$ is true for all computations in which the first step in the computation belongs to the initial state of the system at time zero and each of the subsequent steps are related by the next-state relation or an increment in time (but not both), choosing non-deterministically between enabled state-changes or time increments.

Not all behaviour allowed by validcomp$_t$ will represent actual behaviours of a timed system. Some subset will be required which will be determined by timing constraints on the operations of the system. These timing constraints will take the form of lower and upper time bounds.

Before considering how to specify these constraints it is important to understand their meaning. Suppose a particular operation becomes enabled at some moment in time. This moment in time is known as a choicepoint [Ost89]. A lower time bound l on the operation must prevent its occurrence until l ticks of the clock has occurred from any of the choicepoints of the operation. An upper time bound h must force the operation to occur within h ticks of the clock from any choicepoint, unless the operation is disabled.

The meaning of a choicepoint can be formalised as follows:

$$
\begin{array}{l}
\rule{6cm}{0.4pt}\,[STATE]\,\rule{6cm}{0.4pt} \\
\_\,\text{choicepoint}\,\_ : \mathbb{N}_1 \leftrightarrow \text{comp}_t\ STATE \times (STATE \leftrightarrow STATE) \\
\rule{10cm}{0.4pt} \\
\forall\, cp : \mathbb{N}_1;\ \sigma_t : \text{comp}_t\ STATE;\ op : STATE \leftrightarrow STATE \bullet \\
\quad cp\ \text{choicepoint}\ (\sigma_t, op) \Leftrightarrow \\
\qquad (\sigma_t(cp)).state \in \text{dom}\ op\ \wedge \\
\qquad ((\sigma_t(cp-1)).state\ \underline{op}\ (\sigma_t(cp)).state\ \vee \\
\qquad (\sigma_t(cp-1)).state \notin \text{dom}\ op)
\end{array}
$$

For any operation defined by the relation op and timed computation σ_t, cp will be a choicepoint of op iff: either op first becomes enabled at $\sigma_t(cp)$ after being disabled or op has occurred and remains enabled at $\sigma_t(cp)$.

A definition of a lower time bound constraint on an operation is now given:

$$
\begin{array}{l}
\rule{6cm}{0.4pt}\,[STATE]\,\rule{6cm}{0.4pt} \\
\_\,\text{lower}\,\_ : \text{comp}_t\ STATE \leftrightarrow (STATE \leftrightarrow STATE) \times \mathbb{N}_1 \\
\rule{10cm}{0.4pt} \\
\forall\, \sigma_t : \text{comp}_t\ STATE;\ op : STATE \leftrightarrow STATE;\ l : \mathbb{N} \bullet \\
\quad \sigma_t\ \text{lower}\ (op, l) \Leftrightarrow \\
\quad (\forall\, i : \mathbb{N}_1 \mid i\ \text{choicepoint}\ (\sigma_t, op) \bullet \\
\qquad \neg\, (\exists\, j : \mathbb{N}_1 \mid j \geq i \wedge (\sigma_t(j)).time \leq (\sigma_t(i)).time + l \bullet \\
\qquad (\sigma_t(j)).state\ \underline{op}\ (\sigma_t(j+1)).state))
\end{array}
$$

The occurrence of a state change belonging to op is prohibited before l seconds after any choicepoint.

$$
\begin{array}{l}
\rule{6cm}{0.4pt}\,[STATE]\,\rule{6cm}{0.4pt} \\
\_\,\text{upper}\,\_ : \text{comp}_t\ STATE \leftrightarrow (STATE \leftrightarrow STATE) \times \mathbb{N}_1 \\
\rule{10cm}{0.4pt} \\
\forall\, \sigma_t : \text{comp}_t\ STATE;\ op : STATE \leftrightarrow STATE;\ u : \mathbb{N} \bullet \\
\quad \sigma_t\ \text{upper}\ (op, u) \Leftrightarrow \\
\quad (\forall\, i : \mathbb{N}_1 \mid i\ \text{choicepoint}\ (\sigma_t, op) \bullet \\
\qquad (\exists\, j : \mathbb{N}_1 \mid j \geq i \wedge (\sigma_t(j)).time \leq (\sigma_t(i)).time + u \bullet \\
\qquad ((\sigma_t(j)).state \notin \text{dom}\ op\ \vee \\
\qquad (\sigma_t(j)).state\ \underline{op}\ (\sigma_t(j+1)).state)))
\end{array}
$$

The occurrence of a state change belonging to op must be guaranteed before h seconds after any choicepoint, unless the operation is disabled.

Rather than having to conjunct lower and upper for operations with both lower and upper bounds the relation bounds is introduced:

$$
\begin{array}{|l}
\hline \text{[STATE]}\\\hline
\_\text{ bounds }\_ : \text{comp}_t\ STATE \leftrightarrow (STATE \leftrightarrow STATE) \times \mathbb{P}\,\mathrm{N}\\\hline
\forall \sigma_t : \text{comp}_t\ STATE;\ op : STATE \leftrightarrow STATE;\ b : \mathbb{P}\,\mathrm{N} \bullet\\
\quad \sigma_t\ \text{bounds}\ (op, b) \Leftrightarrow\\
\quad\quad (\exists\, l, u : \mathrm{N} \mid b = l \mathinner{..} u \bullet\\
\quad\quad\quad \sigma_t\ \text{lower}\ (op, l) \wedge \sigma_t\ \text{upper}\ (op, u))\\\hline
\end{array}
$$

Note that upper and lower bounds are defined by the continuous set of natural numbers $l \mathinner{..} u$.

The above operations can now be used to add timing constraints to the allowable behaviour of the robot. The use of the subscripts I and R are used here for simplicity to denote the derivation of the initial state set and next-state relation from their parent schemas: $I_R \equiv \{I \bullet \theta State\}$ and $Op_R \equiv \{Op \bullet \theta State \mapsto \theta State'\}$:

$$
\begin{array}{|l}
\hline \text{TimedRobotBehaviour}\\\hline
\sigma_t : \text{comp}_t\ Robot\\\hline
\sigma_t\ \text{validcomp}_t\ (RobotInit_I, RobotNS_R)\\
\sigma_t\ \text{bounds}\ (MoveArm1ToPress_R, 5 \mathinner{..} 10)\\
\sigma_t\ \text{lower}\ (MoveArm2ToPress_R, 5)\\
\sigma_t\ \text{upper}\ (MoveArm2ToPress_R, 10)\\\hline
\end{array}
$$

Informally, this specification asserts that when first loaded both $arm1$ and $arm2$ must take at least 5 but no more than 10 seconds to move to the press.

4 A Twist of Modularity

In the real world, reactive systems are typically large and complex. Thus, any usable specification recipe must provide some sort of facility for modular development in order to help partition such systems into manageable chunks.

Unfortunately, Z does not provide any explicit facilities for modularizing specification other than the simple schema combinators of the schema calculus. Therefore, when specifying a modularized system in Z one is forced to adopt the following steps: (1) describe each of the components using separate Z specifications; (2) promote the component specifications to the global combined specification by conjoining their state schemas, defining a promotion schema for each of the components and using this to individually promote each operation. If components communicate with each other via synchronised inputs and outputs then the piping operator must be used to combine relevant operations. Obviously for large systems, this process can be both time-consuming and error-prone.

To help alleviate the above problem it is now shown how the idea of a *module* can be introduced into the specification techniques presented above. Generic operations for combining modules in parallel are also proposed.

4.1 Definition

Consider that a module has the following generic specification:

$$
\begin{array}{l}
\underline{\quad Module[State]\quad} \\
init : \mathbb{P}\ State \\
ns : State \leftrightarrow State \\
wf, sf : \mathbb{P}\ (State \leftrightarrow State) \\
\Sigma : \mathbb{P}\ (\text{comp}\ State) \\
\hline
\Sigma = \{\sigma : \text{comp}\ State \mid \\
\quad \sigma\ \textsf{validcomp}\ (init, ns) \wedge \\
\quad (\forall w : wf \bullet \sigma\ \textsf{wf}\ w) \wedge \\
\quad (\forall s : sf \bullet \sigma\ \textsf{sf}\ s)\}
\end{array}
$$

A module has a set of initial states, *init*, a next-state relation, *ns*, and a distinct set of weakly fair and strongly fair operation relations, *wf* and *sf*. Σ is the set valid behaviours that the module will generate if initialised in one of the initial states, arbitrarily executing the next state schema (modelled by *ns*) with weakly and strongly fair constraints specified by *wf* and *sf*.

It is now possible to specify a generic operation for calculating the *union* (parallel composition) of two modules:

$$
\begin{array}{l}
\underline{\underline{=[Z]=}} \\
\_\|\_ : Module[Z] \times Module[Z] \rightarrow Module[Z] \\
\hline
\forall m_1, m_2 : Module[Z] \bullet \\
\quad m_1 \parallel m_2 = (\mu\ m_3 : Module[Z] \mid \\
\quad\quad m_3.init = m_1.init \cap m_2.init \wedge \\
\quad\quad m_3.ns = m_1.ns \cup m_2.ns \wedge \\
\quad\quad m_3.wf = m_1.wf \cup m_2.wf \wedge \\
\quad\quad m_3.sf = m_1.sf \cup m_2.sf)
\end{array}
$$

The *union* of two modules is simply defined as the intersection of their initial states (corresponding to the conjunction of their initial state schemas) and the union of their next-state relations (corresponding to the disjunction of their next-state schemas). Weakly and strongly fair operations are appended together via set union,

Associativity and commutativity of the $\parallel$ operation is guaranteed by the by the definitions of $\cup$ and $\cap$, thus it can be used to compose any number of modules. However, as it stands, it may only be applied to modules which have the same state space. To overcome this limitation, a promotion function is given which takes a module and promotes it to the state of another:

$$
\begin{array}{l}
\underline{\qquad [X, Z] \qquad\qquad\qquad\qquad\qquad\qquad\qquad\qquad\qquad\qquad} \\
\uparrow : Module[X] \times ((X \times X) \to (Z \times Z)) \to Module[Z] \\
\hline
\forall\, mx : Module[X];\ fx : ((X \times X) \to (Z \times Z)) \bullet \\
\quad \uparrow (mx, fx) = (\mu\, mz : Module[Z]\ | \\
\qquad mz.init = \mathrm{dom}\ \{x : mx.init;\ y : X \bullet fx(x, y)\} \wedge \\
\qquad mz.ns = \{x : mx.ns \bullet fx(x)\} \wedge \\
\qquad mz.wf = \{w1 : mx.wf \bullet \{w2 : w1 \bullet fx(w2)\}\} \wedge \\
\qquad mz.sf = \{s1 : mx.sf \bullet \{s2 : s1 \bullet fx(s2)\}\})
\end{array}
$$

The function takes as its arguments the module that is to be promoted and a function which promotes the state changes of the module to the required state space. Promotion of the next-state relation and weakly and strongly fair operations is achieved by applying the promotion function to their before and after states. Promotion of the initial state set is obtained similarly with after states ignored.

The main advantage of the promotion function is that it can be applied to whole modules at a time, thus alleviating the need to individually promote each of the modules operations and state.

4.2 Example

A modular specification of the robot, which encapsulates its concurrent properties can be obtained like so:

$$
\begin{array}{l}
\underline{\quad RobotModule \qquad\qquad\qquad\qquad\qquad\qquad\qquad\qquad\qquad} \\
\quad robot : Module[Robot] \\
\hline
\quad robot.init = RobotInit_I \\
\quad robot.ns = RobotNS_R \\
\quad robot.wf = \\
\qquad \{MoveArm1ToPress_R, MoveArm2ToPress_R, MoveArm1ToBelt_R, \\
\qquad MoveArm2ToBelt_R, UnloadArm1_R, UnloadArm2_R\} \\
\quad robot.sf = \{\}
\end{array}
$$

Here, the module *robot* is instantiated from the *Module* class. Its initial, next-state and fairness sets are instantiated with the state and operations from the static specification of the robot (note the use of subscripts $_R$ and $_I$ to simplify the specification). The resultant behaviours, Σ is exactly those that would be obtained from the previous dynamic specification of the robot.

Next, the press component of the production cell is specified:

$$
|\ full : \mathbb{N}_1
$$

$$\boxed{\begin{array}{l} \underline{\textit{Press}}\\ \textit{Robot}\\ \textit{waiting} : \text{seq}\ \textit{BLANK\_ID}\\ \hline \#\textit{waiting} \leq \textit{full} \end{array}}$$

In this example, the press is required to have knowledge of the state of the robot in order to interface with it. The press can hold a number of blanks, the identity of which are recorded.

The operations of the press are as follows. Loading the press unloads the robot arm and appends the blank to those waiting:

$$\textit{Load\_Press} \cong [\Delta\textit{Press};\ \textit{UnloadArm}1 \mid \textit{waiting}' = \textit{waiting} \frown \langle \textit{blank!}\rangle]$$

Pressing the blanks and unloading them results in an empty press:

$$\textit{Press\_And\_Unload} \cong [\Delta\textit{Press};\ \Xi\textit{Robot} \mid \textit{waiting}' = \langle\rangle]$$

The press module can now be specified:

$$\textit{PressInit} \cong [\textit{Press} \mid \textit{waiting} = \langle\rangle]$$
$$\textit{PressNS} \cong \textit{Load\_Press} \vee \textit{Press\_And\_Unload}$$

$$\boxed{\begin{array}{l} \underline{\textit{PressModule}}\\ \textit{press} : \textit{Module}[\textit{Press}]\\ \hline \textit{press.init} = \textit{Press}_I\\ \textit{press.ns} = \textit{PressNS}_R\\ \textit{press.wf} =\\ \qquad \{\textit{Load\_Press}_R,\ \textit{Press\_And\_Unload}_R\}\\ \textit{press.sf} = \{\} \end{array}}$$

Once the press module has been specified, it can be composed with the robot. First, the combined state is calculated as the conjunction *Robot* and *Press*. Next, two functions *fr* and *fp* are specified which will serve to promote state changes of the robot and press to the combined state (ensuring that state variables not belonging to the promoted component stay the same):

$$\textit{Cell}1 \cong \textit{Robot} \wedge \textit{Press}$$
$$\textit{RobotP} \cong \textit{Cell}1 \setminus (\textit{Robot})$$
$$\textit{PressP} \cong \textit{Cell}1 \setminus (\textit{Press})$$
$$\textit{fr} == \{\Delta\textit{Cell}1 \mid \Xi\textit{RobotP} \bullet (\theta\textit{Robot}, \theta\textit{Robot}') \mapsto (\theta\textit{Cell}1 \mapsto \theta\textit{Cell}1')\}$$
$$\textit{fp} == \{\Delta\textit{Cell}1 \mid \Xi\textit{PressP} \bullet (\theta\textit{Press}, \theta\textit{Press}') \mapsto (\theta\textit{Cell}1 \mapsto \theta\textit{Cell}1')\}$$

The combined module can now be specified as the parallel composition of the two promoted modules:

```
┌─ Cell1Module ──────────────────────────────────────
│  RobotModule
│  PressModule
│  rp : Module[Cell1]
├────────────────────────────────────────────────────
│  rp = ↑ (robot, fr)  ‖  ↑ (press, fp)
└────────────────────────────────────────────────────
```

This partioning and combination of components results in a *shared variables* model of concurrent composition. The combined robot and press modules result in a module *rp* which is also concurrent. They communicate via the shared robot variables.

5 A Further Twist of Modularity

Shared variables are not the only type of communication mechanism that can be used between modules. In Z, one often specifies input and output variables using the '?' and '!' conventions. However, because these variables are not usually part of the state, they cannot be incorporated in the above approach. This limitation can however be overcome by adding *input* and *output* relations to the simple module definitions introduced above:

```
┌─ IOModule[STATE, PASSED] ──────────────────────────
│  Module[STATE]
│  in, out : ℙ(PASSED × (STATE × STATE))
├────────────────────────────────────────────────────
│  {r : (in ∪ out) • second r} ⊆ ns
└────────────────────────────────────────────────────
```

An input and output module (*IOModule*) is a simple module which has an additional pair of input and output relations. These relate an input or output of the module with its effect on the module state. It is required that the next-state relation also includes the state transitions belonging to *in* and *out*.

The parallel composition of two *IOModules* is defined as the intersection of their initial states and union of their next-state relations as above. In addition, the composition of the input and outputs sets is defined as their union with synchronised inputs and outputs removed. This is essentially equivalent to the hiding of shared input and output variables used by the piping operator in the schema calculus. The union of the weakly and strongly fair operations is then obtained by union as above:

$$\boxed{\begin{array}{l} =\!\![S,P]\!\!= \\ \underline{\ \|\ }: IOModule[S,P] \times IOModule[S,P] \to IOModule[S,P] \\ \hline \forall\, m_1 : IOModule[S,P];\ m_2 : IOModule[S,P]\ \bullet \\ \quad m_1 \parallel m_2 = (\mu\, m_3 : IOModule[S,P]\ | \\ \qquad m_3.init = m_1.init \cap m_2.init\ \wedge \\ \qquad m_3.ns = m_1.ns \cup m_2.ns\cup \\ \qquad\qquad (\{p:P;\ r:S\times S \mid (p,r) \in (m_1.out \cap m_2.in) \bullet r\})\ \wedge \\ \qquad m_3.in = (m_1.in \setminus m_2.out) \cup (m_2.in \setminus m_1.out)\ \wedge \\ \qquad m_3.out = (m_1.out \setminus m_2.in) \cup (m_2.out \setminus m_2.in)\ \wedge \\ \qquad m_3.wf = m_1.wf \cup m_2.wf\ \wedge \\ \qquad m_3.sf = m_1.sf \cup m_2.sf) \end{array}}$$

The promotion of *IOModules* is also the same as that for simple modules. In addition, each state change of an input or output is promoted to all the possible combinations of state changes allowed by the free variables of the combined state:

$$\boxed{\begin{array}{l} =\!\![X,Z,P]\!\!= \\ \uparrow: IOModule[X,P] \times ((X \times X) \to (Z \times Z)) \to IOModule[Z,P] \\ \hline \forall\, mx : IOModule[X,P];\ fx : ((X \times X) \to (Z \times Z))\ \bullet \\ \quad \uparrow (mx, fx) = (\mu\, mz : IOModule[Z,P]\ | \\ \qquad mz.init = \mathrm{dom}\ \{x : mx.init;\ y : X \bullet fx(x,y)\}\ \wedge \\ \qquad mz.ns = \{x : mx.ns \bullet fx(x)\}\ \wedge \\ \qquad mz.in = \{i : P;\ d,r : X;\ pd,pr : Z \mid (i,(d,r)) \in mx.in\ \wedge \\ \qquad\qquad pd \in \mathrm{dom}\{y : X \bullet fx(d,y)\}\ \wedge \\ \qquad\qquad pr \in \mathrm{dom}\{y : X \bullet fx(r,y)\} \bullet (i,(pd,pr))\}\ \wedge \\ \qquad mz.out = \{o : P;\ d,r : X;\ pd,pr : Z \mid (o,(d,r)) \in mx.out\ \wedge \\ \qquad\qquad pd \in \mathrm{dom}\{y : X \bullet fx(d,y)\}\ \wedge \\ \qquad\qquad pr \in \mathrm{dom}\{y : X \bullet fx(r,y)\} \bullet (o,(pd,pr))\}\ \wedge \\ \qquad mz.wf = \{w1 : mx.wf \bullet \{w2 : w1 \bullet fx(w2)\}\}\ \wedge \\ \qquad mz.sf = \{s1 : mx.sf \bullet \{s2 : s1 \bullet fx(s2)\}\}) \end{array}}$$

5.1 Example

As an example, consider the specification of the last component of the production cell - the feed belt:

$$\boxed{\begin{array}{l} \underline{\ FeedBelt\ } \\ belt\_state : Blank\_Loaded \end{array}}$$

The feed belt is either loaded or unloaded.

```
┌─ Load_Belt ─────────────────────────────────────────────
│ ΔFeedBelt
│ blank? : BLANK_ID
├──────────────────────────────────────────────────────────
│ belt_state = unloaded
│ belt_state = loaded(blank?)
└──────────────────────────────────────────────────────────
```

Provided that the belt is unloaded it may be loaded with a blank.

```
┌─ Unload_Belt ───────────────────────────────────────────
│ ΔFeedBelt
│ blank! : BLANK_ID
├──────────────────────────────────────────────────────────
│ belt_state = loaded(blank!)
│ belt_state' = unloaded
└──────────────────────────────────────────────────────────
```

Provided that the belt is loaded it may unload.

The *IOModule* for the feed belt and robot can be specified as follows:
First, the data passed between the two modules is specified as a free type:

$Passed ::=$
  $arm1top\langle\!\langle BLANK\_ID \rangle\!\rangle \mid btoarm1\langle\!\langle BLANK\_ID \rangle\!\rangle \mid$
  $arm2top\langle\!\langle BLANK\_ID \rangle\!\rangle \mid btoarm2\langle\!\langle BLANK\_ID \rangle\!\rangle \mid$
  $envtob\langle\!\langle BLANK\_ID \rangle\!\rangle$

Here, each choice belonging to the free type represents a communication channel and associated data type that will be assigned to the inputs and outputs of the modules.

The robot *IOmodule* can now be specified:

```
┌─ IORobotModule ─────────────────────────────────────────
│ robot : IOModule[Robot, Passed]
├──────────────────────────────────────────────────────────
│ robot.init = RobotInit_I
│ robot.ns = RobotNS_R
│ robot.wf =
│     {MoveArm1ToPress_R, MoveArm2ToPress_R, MoveArm1ToBelt_R,
│      MoveArm2ToBelt_R, UnloadArm1_R, UnloadArm2_R}
│ robot.sf = {}
│ robot.in = {LoadArm1 • (btoarm1(blank?), (θRobot, θRobot'))}∪
│     {LoadArm2 • (btoarm2(blank?), (θRobot, θRobot'))}
│ robot.out = {UnloadArm1 • (arm1top(blank!), (θRobot, θRobot'))}∪
│     {UnloadArm2 • (arm2top(blank!), (θRobot, θRobot'))}
└──────────────────────────────────────────────────────────
```

Note, how the inputs and outputs of the module are equated with the '?' and '!' variables of the schemas.

The feed belt *IOModule* can be specified:

$BeltInit \cong [FeedBelt \mid belt\_state = unloaded]$
$BeltNS \cong Load\_Belt \lor Unload\_Belt$

IOBeltModule

$belt : IOModule[FeedBelt, Passed]$

$belt.init = FeedBelt_I$
$belt.ns = FeedBeltNS_R$
$belt.wf = \{Unload\_Belt_R\}$
$belt.sf = \{\}$
$belt.in = \{Load\_Belt \bullet (envtob(blank?), (\theta FeedBelt, \theta FeedBelt'))\}$
$belt.out = \{Unload\_Belt \bullet (btoarm1(blank!), (\theta FeedBelt, \theta FeedBelt'))\}$

Promotion functions for the modules are specified:

$Cell2 \cong Robot \land FeedBelt$
$PRobot \cong Cell2 \setminus (Robot)$
$PBelt \cong Cell2 \setminus (FeedBelt)$
$rtoc == \{\Delta Cell2 \mid \Xi PRobot \bullet (\theta Robot, \theta Robot') \mapsto (\theta Cell2 \mapsto \theta Cell2')\}$
$btoc == \{\Delta Cell2 \mid \Xi PBelt \bullet (\theta FeedBelt, \theta FeedBelt') \mapsto (\theta Cell2 \mapsto \theta Cell2')\}$

Finally, the new *IOModule* resulting from the parallel composition of the feed belt and robot *IOModule* is obtained:

Cell2Module

IORobotModule
IOBeltModule
$rb : IOModule[Cell2, Passed]$

$rb = \uparrow (belt, btoc) \parallel \uparrow (robot, rtoc)$

The resultant module is fully concurrent and communicates synchronously via the shared channel *btoarm1*. All other channels not shared by the modules are included in the input and output channels of the new module and are available for communication with other modules as required.

The above definition of a module has one problem however: it requires prior knowledge of the channels and their associated data to be defined by *PASSED* before any of the modules can be written. This greatly hinders modular development due to the need to have prior knowledge of the data passed by the entire system. To get around this, one must also incorporate the promotion of input and output channels within the promotion function. This allows input and output data types to be declared locally to modules and then promoted to the system before composition. For example, the independent input and output data types for the robot and press would be:

$RobotIO ::=$
    $arm1top\langle\!\langle BLANK\_ID\rangle\!\rangle \mid btoarm1\langle\!\langle BLANK\_ID\rangle\!\rangle \mid$
    $arm2top\langle\!\langle BLANK\_ID\rangle\!\rangle \mid btoarm2\langle\!\langle BLANK\_ID\rangle\!\rangle$
$BeltIO ::=$
    $btoarm2\langle\!\langle BLANK\_ID\rangle\!\rangle \mid$
    $envtob\langle\!\langle BLANK\_ID\rangle\!\rangle$

The combined system IO would be:

$Shared ::=$
    $sarm1top\langle\!\langle BLANK\_ID\rangle\!\rangle \mid sbtoarm1\langle\!\langle BLANK\_ID\rangle\!\rangle \mid$
    $sarm2top\langle\!\langle BLANK\_ID\rangle\!\rangle \mid sbtoarm2\langle\!\langle BLANK\_ID\rangle\!\rangle \mid$
    $senvtob\langle\!\langle BLANK\_ID\rangle\!\rangle$

The IO promotion function for the feed belt would be defined as:

$$fbio : BeltIO \twoheadrightarrow Shared$$
$$\forall b : BLANK\_ID \bullet$$
$$fbio(envtob(b)) = senvtob(b) \wedge$$
$$fbio(btoarm2(b)) = sbtoatm2(b)$$

6 Conclusion

This paper has considered some new recipes for specifying reactive systems in Z. A number of generic techniques for specifying concurrent and real-time behaviour have been explored, whilst the use of generic modules has been proposed and illustrated as a means to structuring large specifications. An important part of the work is that it is based entirely on the standard Z (ZRM) notation, thus the specifications presented in this paper can be readily type checked using the $fuzz$ typechecker for example. Moreover, the paper has sought to build on the usability and expressibility of the Z notation, thereby providing tools and techniques to extend the Z toolkit to the specification of reactive systems.

How does the work presented in this paper compare with other state based concurrency notations such as TLA [Lam91] and Unity [CM88]? On the whole there is a favourable match. In particular, the recipes for specifying concurrent and real-time behaviour are as expressive and simple as those used in TLA, whilst the implicit operational semantics used by Unity could be explicitly captured by a dynamic Z specification. In general, any behavioural property that can be described by TLA and Unity could be captured using the techniques presented in this paper (they are all based on the same model of concurrency). Moreover, Z has the benefit of providing better abstraction facilities (i.e. schema types) for describing data than TLA and Unity, both of which eschew the use of data types.

Where there is a significant mismatch is in the specification of modules. In Unity and TLA, significant emphasis is placed on the importance of modules and their use in proof. Thus, composition in these notations is both simple and intuitive. In comparison, the modular approach proposed here is very primitive, requiring the use of promotion and other notational 'tricks'. These were needed because Z does not naturally support modular development. This is a serious limitation of Z, which is not just confined to reactive systems - virtually all systems are being developed using modular, component based approaches these days.

In terms of verification, a set of assertional proof rules now exist for verifying safety and liveness properties of Z specifications of concurrent systems based on $\mathcal{W}$, the emerging deductive system for Z. See [Eva96b, Eva96a] for further details. The issue of modularity will form a central part of further investigation into this aspect. In particular, it is hoped to develop *compositional* proof rules, which will allow the relevant properties of a a Z specification to be gained from the properties of its modules. Such a proof system will be essential if Z is to become a practical notation for the development of large systems.

One further aspect not considered in this paper is the role of the environment. A *closed* system model of the environment has been assumed in which both the system and its environment are treated as one. The specification of open systems will be examined elsewhere.

Finally, it is hoped that the recipes presented in this paper will go some way towards stimulating further debate on the application of Z to reactive systems and the importance of modularity in specification.

The generic operations presented in this paper are available in the form of LaTeX source from the author's home page.

7 Acknowlegments

I would like to thank the anonymous reviewers for their constructive comments, and the University of Bradford, Department of Computing for funding my present fellowship. Ian Hayes' insightful comments both on this paper and the author's PhD thesis are greatly appreciated.

References

[BH95]   J. P. Bowen and M. G. Hinchey, editors. *ZUM'95 – 9th International Conference of Z Users, Limerick 1995*, Lecture Notes in Computer Science. Springer-Verlag, 1995.

[CM88]   K. M. Chandy and J. Misra. *Parallel Program Design: A Foundation*. Addison Wesley, 1988.

[DHKR88] R. Duke, I. J. Hayes, P. King, and G. A. Rose. Protocol specification and verification using Z. In S. Aggarwal and K. Sabnani, editors, *Protocol Specification, Testing, and Verification VIII*, pages 33–46. Elsevier Science Publishers (North-Holland), 1988.

[DS89]   R. Duke and G. Smith. Temporal logic and Z specifications. *Australian Computer Journal*, 21(2):62–69, May 1989.

[Eva96a]   A. S. Evans. An assertional verification method for z specifications of concurrent systems. In *In Proceedings: ISCIS XI, Antalya, Turkey*, November 1996.

[Eva96b]   A. S. Evans. *Z for Concurrent Systems*. PhD thesis, Faculty of Information and Engineering Systems, Leeds Metropolitan University, 1996.

[Lam91]   L. Lamport. The temporal logic of actions. SCR Report 79, DEC, Palo Alto, December 1991.

[Lam94]   L. Lamport. TLZ. Workshops in Computing, pages 267–268. Springer-Verlag, 1994. Abstract.

[MC95]   A. MacDonald and D. Carrington.   Structuring Z specifications: Some choices. In Bowen and Hinchey [BH95], pages 203–223.

[MP84]   Z. Manna and A. Pneuli. Adequate proof principles for invariance and liveness properties of concurrent programs. *Science of Computer Programming*, 4:257–289, 1984.

[Ost89]   J. S. Ostroff. *Temporal Logic for Real-Time Systems*. Wiley, Chichester, 1989.

[Spi92]   J. M. Spivey. *The Z Notation: A Reference Manual*. Prentice Hall International Series in Computer Science, 2nd edition, 1992.

[Str95]   B. Strulo. How firing conditions help inheritance. In Bowen and Hinchey [BH95], pages 264–275.

A Z Specification of the Soft-Link Hypertext Model

Mark d'Inverno[1] and Michael Hu[2]

[1] School of Computer Science, University of Westminster, 115 New Cavendish Street, London, W1M 8JS, UK. Email: dinverm@westminster.ac.uk
[2] School of Electrical & Electronic Engineering, Nanyang Technological University, Singapore 639798, Email: emjhu@ntuvax.ntu.ac.sg

Abstract. This paper provides a formal specification in Z of a new intelligent hypertext model called the soft-link hypertext model (SLHM). This model has been implemented and extensively tested, and provides a new methodology for constructing the future generation of information retrieval systems. Its core is the adoption of a data structure called the conceptual index, which allows hypertext structure to be built automatically upon conventional Boolean systems. The functionality of resulting systems using this approach is then extended from Boolean search to more sophisticated information retrieval applications, including associative searches and information browsing. Compared with other similar projects, SLHM has the following three major advantages. First, the conceptual index is automatically formulated. Second, powerful neural learning mechanisms are applied to the conceptual index, thereby improving its efficiency and applicability. Third, machine intelligence installed on the conceptual index can be utilised for online assistance during navigating and information browsing. This specification has been developed by application of an existing formal framework for specifying hypertext systems. The specification presented here provides: a formal account of the state and operations of this new model; a sound basis for instantiations of the model to be built; a case study in the application of an existing formal framework; and an environment in which further refinements of new learning and hypertext strategies can be presented and evaluated.

1 Introduction

A hypertext system is a system which attempts to superimpose an external structure on data, in order that it can be accessed efficiently according to different users' needs. Essentially, generating a hypertext structure involves creating associations - typically manually - between various parts of different documents, resulting in a structure known as the information web. However, there are many difficulties in the generation, description, and presentation of this inter-connected hypertext structure. This leads to several well-documented problems [14], three of which are as follows.

1. Designing and constructing these information webs is expensive, laborious and tedious.

2. These manually built-up webs often confuse or mislead the users because there is no transparency nor any appropriate guidance facilities.
3. Once these inter-connected information webs are built, they usually exist as an unalterable structure. Consequently it is very difficult, if not impossible, to amend and improve these webs.

In response to these problems, a new design of the information web has been proposed [17]. It is called the conceptual index. Based on the conceptual index, a new information retrieval model called the soft-link hypertext model (SLHM) has been developed and implemented. The conceptual index extends the functionality of conventional Boolean search to more sophisticated applications including hypertext. Compared with other similar projects in building hypertext structures upon Boolean systems, *SLHM* provides the following three advantages.

1. The conceptual index is automatically formulated.
2. Powerful neural learning mechanisms are applied to the conceptual index. As the hypertext is used these mechanisms are invoked to train the hypertext thus improving its efficiency and applicability for user groups.
3. Statistical user data are collected in our model to formulate intelligent indicators. These indicators are utilised for online assistance during information search and browsing.

1.1 Background

Some solutions to the problems given above have been proposed [5, 7, 14]. In particular, different types of semantic networks have been considered and applied as a supplement for the conventional Boolean search [6, 21, 23]. However, these projects are only successful in providing alternative tools for refining the user's queries for information search and retrieval, but not tailored for information browsing from one document to another [23]. In our model, the associations between a pair of keywords are accommodated, so that the functionality of our model can be fully extended from the Boolean search to some more sophisticated information retrieval applications including hypertext. This is possible by using the conceptual index data structure and makes information search and browsing much more flexible. For a full review of this work see [17].

The algorithms used in SLHM are based on connectionist networks [1, 3, 13, 16]. Once the conceptual index is defined and represented as a connectionist network, various neural network learning algorithms can be applied to support its operation and applications [22]. Earlier attempts to apply the connectionist network methodology, include Mozer's first application to information retrieval [26], Cohen and Kjeldsen's constrained spreading activation in their semantic network for matching research funding agencies and research topics [4], Lelu's application of spreading activation in image databases [19] and finally, and most thoroughly, Belew's AIR project on spreading activation on query-index-abstract networks [2]. Nonetheless, these projects are only limited to small-scale bibliographic environments [16]. Compared with these projects, SLHM accommodates significantly more complicated information databases, and is potentially extendible to multi-media environments. Further, it differs from these earlier connectionist network projects in its collection of statistical data from previous applications and the use of this data as intelligent indicators for online assistance during information browsing [17].

1.2   An Overview of SLHM

It is the case that any information retrieval system can be treated as a self-inclusive Information Space. SLHM divides any such space into two layers: the Document Space and the Index Space. The Document Space is analogous to the Storage Layer of the Dexter Hypertext Model [15]; it contains all relevant documents in their original forms. The Index Space is analogous to the Link Layer of the Dexter Hypertext Model; it includes an abstract description of the Document Space, together with all the auxiliary mechanisms necessary to support the operations on, and applications of, the Information Space.

Traditionally, an index provides a collection of keywords (also called index terms) as an external description of the Document Space. In SLHM, these different index terms are further associated with one another to formulate a new data structure called the conceptual index. Consequentially, the conceptual index comprises two main elements: keywords and the associations between keywords, represented by concepts and links respectively. The mapping from concepts to related documents is used to support the Boolean search. Furthermore, the links between different keywords are used to support browsing in the Information Space from one concept to another, and from one document to another. In such a way, the conceptual index extends the functionality of the traditional index from the Boolean search to more sophisticated information retrieval applications including hypertext.

In most current systems the links are defined and assembled manually, and so liable to human prejudice and ignorance. In addition, two concepts are related by a link without an explicit motivation as to it was created. Lastly, once a link is made it remains unless it is manually removed. We refer to this group of links as hard links.

In SLHM, however, general rules are used to identify the links automatically, ensuring that all links are systematically generated. Further, the motivation for the creation of each link can be understood since these rules are made explicit and available for the users once the conceptual index is formulated. Furthermore, powerful learning mechanisms are installed to record intelligent statistical information and to train the conceptual index as it is used in information retrieval. This information can be utilised for online assistance during an information retrieval session. We call such links soft links and argue that they do not suffer from the limitations of hard links. They enhance the users' mobility in an Information Space, and make the application of the Index Space more flexible and powerful.

1.3   The Specification

In this paper we provide a formal specification of SLHM which represents a formal, concise and readable definition of the *SLHM* and its applications. The specification can then be used as the basis for implementation, as well as a framework which can be further refined to develop and evaluate new hypertext and learning strategies in information retrieval. We choose the language Z [25] to formalise our model because of experience of the using it in a number of fields including interactive conferencing systems [8], multi-agent systems systems [11] and design methodologies [9]. In addition, we have previously considered the requirements for the structures or *formal frameworks* that are necessary to provide a rigorous approach to any discipline [20]. In particular we have presented a formal framework for hypertext systems [12], which provides an explicit formal environment for the presentation, evaluation and comparison of hypertext systems. The specification given in this paper has been derived from applying this framework to SLHM. This approach of deriving formal specifications from formal frameworks has also been considered in multi-agent systems [10, 11].

1.4   Overview of the Paper

In Section 2 we give a specification of the structure of the conceptual index and define the operations of indexing and hyperization on the conceptual index. This represents the high level description of SLHM. Section 3 presents the specification of the intelligent conceptual index, Section 4 shows how the SLHM can be applied in information retrieval and Section 5 shows how SLHM can be updated whilst maintaining the validity of machine-learnt information. Lastly, Section 6 provides a summary of the work presented in this paper.

2   The Conceptual Index

The conceptual index is a data structure which presents an abstract view of a Document Space. It defines all the concepts and the links between these concepts. We envisage the Universe consisting of a set of concepts.

$[CONCEPT]$

Any Information Space of concern is some subset of this Universe. Further, in our Universe, a pair of concepts can be related to each other by an association called a link. A link is therefore characterised by the two concepts it connects.

$LINK == CONCEPT \times CONCEPT$

The structure, or the state space, of the conceptual index is represented as a collection of concepts and links where links can only exist between a pair of concepts in the space. Concepts may be linked to themselves and may also be totally isolated. Initially, the state space of any Information Space contains no concepts.

```
┌─ Concept ──────────────────────────────────────────────────
│ Concepts : P CONCEPT
│
└────────────────────────────────────────────────────────────
```

```
┌─ Link ─────────────────────────────────────────────────────
│ Links : P LINK
│
└────────────────────────────────────────────────────────────
```

```
┌─ ConceptualIndex ──────────────────────────────────────────
│ Concept
│ Link
│─────────────────────────────────────────────────────────────
│ (dom Links ∪ ran Links) ⊆ Concepts
└────────────────────────────────────────────────────────────
```

```
┌─ InitConceptualIndex ──────────────────────────────────────
│ ConceptualIndex
│─────────────────────────────────────────────────────────────
│ Concepts = { }
└────────────────────────────────────────────────────────────
```

We next define the following terms which enable us to make our specification more readable.

1. The *LeavingLinks* of a concept are those links which connect from that concept.
2. The *ArrivingLinks* of a concept are those links which connect to that concept.
3. The *Children* of a concept are all those concepts which are possible destinations reachable using the leaving links of that concept.
4. The *Parent* of a link is the concept from which a link leaves.
5. The *Child* of a link is the concept to which a link arrives.

```
┌─ ReadabilityFunctions ─────────────────────────────────────
│ ConceptualIndex
│
│ LeavingLinks, ArrivingLinks : CONCEPT ⇻ P LINK
│ Children : CONCEPT ⇻ P CONCEPT
│ Child, Parent : LINK ⇻ CONCEPT
│─────────────────────────────────────────────────────────────
│ ∀ c : Concepts; l : Links •
│         LeavingLinks c = {c} ◁ Links ∧
│         ArrivingLinks c = Links ▷ {c} ∧
│         Children c = second (|{c} ◁ Links|) ∧
│         Parent l = first l ∧
│         Child l = second l
└────────────────────────────────────────────────────────────
```

The above schema introduces five functions which are used later in presenting a readable specification. Whenever we wish to use these functions the above schema is included.

2.1 Updating the Conceptual Index

The construction of the conceptual index includes two groups of operation known as indexing and hyperization. Indexing, concerns isolating, or extracting, all concepts in the Information Space and is typified by the conventional indexing process used in most current Boolean systems. Hyperization, concerns defining the relationship between a particular pair of concepts typified by the construction of traditional hard hypertext links.

In our model, concepts and links may be introduced, as long as they are not already part of the state space. However, it should be noted that no link can be added unless both of the concepts it connects currently exist in the state space.

$$
\begin{array}{l}
\_\_AddConcept_____ \\
\Delta ConceptualIndex \\
\Xi Link \\
c? : CONCEPT \\
\hline
c? \notin Concepts \\
Concepts' = Concepts \cup \{c?\}
\end{array}
$$

$$
\begin{array}{l}
\_\_AddLink_____ \\
\Delta ConceptualIndex \\
\Xi Concept \\
c_1?, c_2? : CONCEPT \\
\hline
(c_1?, c_2?) \notin Links \\
\{c_1?, c_2?\} \subseteq Concepts \\
Links' = Links \cup \{(c_1?, c_2?)\}
\end{array}
$$

Associated with each adding operation is a corresponding removing operation. A link can be removed from the state space without affecting other links and concepts. However, no concept can be removed unless all leaving and arriving links to the concept have been removed first.

$$
\begin{array}{l}
\_\_RemoveLink_____ \\
\Delta ConceptualIndex \\
\Xi Concept \\
c_1?, c_2? : CONCEPT \\
\hline
(c_1?, c_2?) \in Links \\
Links' = Links \setminus \{(c_1?, c_2?)\}
\end{array}
$$

```
┌─ RemoveConcept ──────────────────────────────────────
│ Δ ConceptualIndex
│ Ξ Link
│ c? : CONCEPT
├──────────────────────────────────────────────────────
│ c? ∈ Concepts
│ c? ∉ (dom Links ∪ ran Links)
│ Concepts' = Concepts \ {c?}
└──────────────────────────────────────────────────────
```

The specification presented in this section represents a high-level description of SLHM and is equivalent to the higher-level specification provided by our formal framework [12]. We now refine this specification in the next three sections to provide a lower-level description of the intelligent mechanisms of SLHM.

3   The Intelligent Conceptual Index

In order to introduce some intelligent mechanisms into our specification and to apply the conceptual index more efficiently in information retrieval, two read states are defined to record statistical information. These are the system read state, which records the statistical information regarding the applications of the conceptual index, and the user read state, which records general user information such as the user browsing history.

3.1   System Read State

The system read state consists of the statistical information recorded as a collection of indicators. An indicator is either accumulative, in which case it is a measure of the combined use of the system, or non-accumulative, sometimes called current, in which case it is a measure dependent on only the current user. We first consider and discuss the indicators associated with concepts.

Concepts Whenever a concept is visited as a hypertext source anchor, we record that concept. This indicator is accumulative.

```
┌─ ReadConceptsVisited ────────────────────────────────
│ ConceptsVisited : bag CONCEPT
└──────────────────────────────────────────────────────
```

For each concept, we also record a measure of the likelihood that it is relevant to the current user's information needs, known as the activation. The activation is defined as a rational in the interval $[0, 1]$. This indicator is non-accumulative which means that is is only valid and meaningful for the current information retrieval session. If the activation of a concept is equal to 0, it suggests that the concept is totally irrelevant to the user's needs, and if equal to 1, it suggests that the concept is exactly appropriate. We define $[RAT_0^1]$ as the rationals between 0 and 1, assuming basic arithmetic operations are applicable [27].

```
┌─ ReadConceptActivation ─────────────────────────────────────
│  ConceptActivation : CONCEPT ⇸ RAT_0^1
└─────────────────────────────────────────────────────────────
```

We define *ReadConcepts* as the schema which includes all the concepts of the systems along with this other information.

```
┌─ ReadConcepts ──────────────────────────────────────────────
│  Concept
│  ReadConceptsVisited
│  ReadConceptActivation
├─────────────────────────────────────────────────────────────
│  dom ConceptsVisited ⊆ Concepts
│  dom ConceptActivation = Concepts
└─────────────────────────────────────────────────────────────
```

Links Whenever a link is used during an information retrieval session we record that link in the accumulative indicator *LinksVisited*. For each link, we also record a measure of its popularity known as the weight. The accumulative indicator *LinkWeight* is defined as a rational number in the interval $(0, 1]$ so that the weight of a link cannot take the value 0. This indicator is totally dependent on other indicators. If the weight of a link is 1, it suggests that the link is the only link leaving some concept whereas as the weight of a link approaches 0, it suggests that the link is so unpopular it may never be used.

```
┌─ ReadLinksVisited ──────────────────────────────────────────
│  LinksVisited : bag LINK
└─────────────────────────────────────────────────────────────
```

```
┌─ ReadLinkWeight ────────────────────────────────────────────
│  LinkWeight : LINK ⇸ RAT_0^1
└─────────────────────────────────────────────────────────────
```

```
┌─ ReadLinks ─────────────────────────────────────────────────
│  Link
│  ReadLinksVisited
│  ReadLinkWeight
├─────────────────────────────────────────────────────────────
│  dom LinksVisited ⊆ Links
│  dom LinkWeight = Links
└─────────────────────────────────────────────────────────────
```

Note, that the variables *ConceptsVisited* records a concept every time it is used as a hypertext source anchor in a session and consequently its domain is a subset of all system concepts. This is in contrast to *ConceptActivation*, which maps *every* system concept to a rational between 0 and 1. Analogously the domain of *LinksVisited* is a subset of all system links whilst *LinkWeight* maps every system link to a rational. In response, we define the frequency of a system link or concept to be the number of times it has been visited.

```
┌─ FrequencyFunctions ──────────────────────────────────────────────
│  ReadConceptsVisited
│  ReadLinksVisited
│  ConceptualIndex
│
│  ConceptFrequency : CONCEPT ⇸ N
│  LinkFrequency : LINK ⇸ N
├───────────────────────────────────────────────────────────────────
│  ConceptFrequency = (λ x : Concepts • 0) ⊕ ConceptsVisited
│  LinkFrequency = (λ x : Links • 0) ⊕ LinksVisited
└───────────────────────────────────────────────────────────────────
```

The functions *ConceptFrequency* and *LinkFrequency* in the schema above can be defined by overriding the function which maps all system links and concepts to zero with the respective bag representations. In this way, these functions are defined for all system links and concepts, and return their frequency. In future, we may treat these data structures as either bags, or functions (by including this schema), depending on our purpose.

The System Read State Combining the records about concepts and links defines the system read state of the conceptual index. The indicators in the system read state are related to each other as follows:

- The weight of any link is given by the frequency of the link plus 1 divided by the sum of the number of leaving links from the parent concept and the frequency of the parent concept.
- The number of times a concept has been visited as the hypertext source anchor is equal to the sum of the number of times each of its leaving links have been visited.
- As long as there are links leaving a concept, the sum of the weights of all the links leaving a concept is equal to 1.

```
┌─ ReadSystemState ─────────────────────────────────────────────────
│  ReadConcepts
│  ReadLinks
│  ConceptualIndex
│
│  ReadabilityFunctions
│  FrequencyFunctions
├───────────────────────────────────────────────────────────────────
│  ∀ c₁, c₂ : CONCEPTS | (c₁, c₂) ∈ Links •
```

$$\forall\, c_1, c_2 : CONCEPTS \mid (c_1, c_2) \in Links \bullet$$
$$LinkWeight\,(c_1, c_2) = \frac{1 + (LinkFrequency\,(c_1, c_2))}{\#(LeavingLinks\ c_1) + (ConceptFrequency\ c_1)}$$

$$\forall\, c : Concepts \bullet ConceptFrequency\ c =$$
$$\text{sumseq}\,(\text{mapsettoseq}\ LinkFrequency\,(LeavingLinks\ c))$$

$$\forall\, c : \text{dom}\ Links \bullet$$
$$\text{sumseq}\,(\text{mapsettoseq}\ LinkWeight\,(LeavingLinks\ c)) = 1$$

The generic function mapsettoseq takes a function, and applies it to every member of a set creating a sequence. This is to protect against the partial function not being injective. The function sumseq simply sums the elements of a sequence of rationals.

$$
\begin{array}{l}
=[X, Y]= \\
\hline
\text{mapsettoseq}: (X \nrightarrow Y) \rightarrow \mathbb{P}\, X \nrightarrow \text{seq}\, Y \\
\hline
\forall f : X \nrightarrow Y;\ xs : \mathbb{P}\, X;\ x : X \mid (xs \cup \{x\}) \subseteq \operatorname{dom} f \bullet \\
\qquad \text{mapsettoseq}\, f\,(\{x\} \cup xs) = \langle f(x)\rangle \frown \text{mapsettoseq}\, f\, xs \land \\
\qquad \text{mapsettoseq}\, f\,\{\} = \langle\rangle
\end{array}
$$

$$
\begin{array}{l}
\text{sumseq}: \text{seq}\ RAT_0^1 \nrightarrow RAT_0^1 \\
\hline
\forall n : RAT_0^1\ ;\ ms, ns : \text{seq}\ RAT_0^1 \bullet \\
\qquad \text{sumseq}\ \langle n\rangle = n \land \\
\qquad \text{sumseq}\ (ms \frown ns) = \text{sumseq}\ ms + \text{sumseq}\ ns
\end{array}
$$

The initial state of the read state is where there is no statistical information. No links or concepts have been visited, the activation of all system concepts is 0, and the weight of any link is the reciprocal of the number of leaving links of the parent concept of that link, and so initially, all the leaving links of any concept have equal weights.

$$
\begin{array}{l}
\_InitReadSystemState_____ \\
ReadSystemState \\
\hline
ConceptsVisited = [\,] \\
LinksVisited = [\,] \\
ConceptActivation = \{c : Concepts \bullet (c, 0)\} \\
LinkWeight = \{c_1, c_2 : CONCEPT \mid \\
\qquad (c_1, c_2) \in links \bullet ((c_1, c_2), \frac{1}{\#(LeavingLinks\ c_1)})\}
\end{array}
$$

3.2 User Read State

In SLHM, the accumulative history records all past users' visited concepts, and the current history records the current user's set of visited concepts (known as their browsing history). The use of the current history is standard and enables, for example, the user to re-visit any concepts within the current session. The history of a session refers to both the accumulative and current histories, and is used by the system to infer the current user's information needs. For example, by comparing the current history with the accumulative history, the system may obtain a more accurate user model of the current information user, and so, reason about their next move. This feature of the model is not specified in this paper. For more details, the interested reader is asked to consult [17, 18].

```
┌─ AccumulateHistory ─────────────────────────────────────────
│  AccHistory : F(F CONCEPT)
│
└──────────────────────────────────────────────────────────────
```

```
┌─ CurrentHistory ────────────────────────────────────────────
│  CurrHistory : F CONCEPT
│
└──────────────────────────────────────────────────────────────
```

```
┌─ History ───────────────────────────────────────────────────
│  AccumulateHistory
│  CurrentHistory
└──────────────────────────────────────────────────────────────
```

In addition, the user may have a current position in the conceptual index known as the current concept and a destination concept known as the next concept. These indicators do not contribute to the intelligence of SLHM but are included in the user read state for the dynamic description of the user's movement. Note too, that the next concept which the user visits can only be defined if the current concept has been defined. (In other words, the user needs a position in order to choose a link).

```
┌─ Position ──────────────────────────────────────────────────
│  CurrentConcept : optional [CONCEPT]
│  NextConcept : optional [CONCEPT]
├──────────────────────────────────────────────────────────────
│  undefined CurrentConcept ⇒undefined NextConcept
└──────────────────────────────────────────────────────────────
```

We have found it useful in this and other specifications — as in the previous schema — to be able to assert that an element is optional. The following definitions provide for a new type optional T for any existing type T. In addition, we define predicates defined and undefined to test whether an element of optional T is defined or not, and an operation the to extract the T element from a defined member of optional T, which are used later in this paper.

$$\text{optional } [X] == \{ xs : \mathsf{P}\, X \mid \#\, xs \leq 1 \}$$

```
╔═[X]═══════════════════════════════════════════════════════════
║  defined _ : P( optional [X])
║  undefined _ : P( optional [X])
║  the : optional [X] ↠ X
╟───────────────────────────────────────────────────────────────
║  ∀ xs : optional [X] • defined xs ⇔ # xs = 1 ∧
║                        undefined xs ⇔ # xs = 0 ∧
║                        (defined xs) ⇒the  xs = (μ x : X | x ∈ xs)
╚═══════════════════════════════════════════════════════════════
```

The user read state comprises the history and the position.

```
┌─ ReadUserState ─────────────────────────────────────────────
│  History
│  Position
└─────────────────────────────────────────────────────────────
```

Initially, there is no statistical information.

```
┌─ InitReadUserState ─────────────────────────────────────────
│  ReadUserState
│ ─────────────────────────────────────────
│  AccHistory = {}
│  CurrHistory = {}
└─────────────────────────────────────────────────────────────
```

3.3 The Soft-Link Hypertext Model

The combination of system read state and user read state defines the read state of SLHM. We refer to this as a **session**. The read state constitutes the intelligence in the model; it is applied as an auxiliary navigation tool for the more efficient information retrieval operations. In the read state, the accumulative indicators are the frequencies of the links and concepts, the weight of the links and the accumulative history, the current indicators are concept activations, the current history, the current (if known) and next (if known) position.

The combined history of the past and present users is the set of all concepts which have been visited. Furthermore, although any concept which has been visited during the current session should have an activation value of 1, the converse is not necessarily true: there may exist some circumstances when the activation of a concept may reach 1 without ever being visited. Finally, the position of a user must be within the Information Space.

```
┌─ SoftLinkHypertext ─────────────────────────────────────────
│  ReadSystemState
│  ReadUserState
│ ─────────────────────────────────────────
│  (⋃ AccHistory) ∪ CurrHistory = dom ConceptsVisited
│  CurrHistory ⊆ {c : Concepts | ConceptActivation c = 1 • c}
│  CurrentConcept ⊆ Concepts
│  NextConcept ⊆ Concepts
└─────────────────────────────────────────────────────────────
```

Before the first user starts, all session indicators are set at their initial value. From this moment, the accumulative indicators start collecting statistical information about the operations of the conceptual index. The life-span of these indicators is the same as that of the system, whereas the life-span of non-accumulative indicators is that of one information retrieval session.

```
┌─ InitSoftLinkHypertext ─────────────────────────────────────
│  SoftLinkHypertext
│  InitReadSystemState
│  InitReadUserState
└─────────────────────────────────────────────────────────────
```

4 Applications of the Soft-Link Hypertext Model

In this section we detail the application of SLHM in information retrieval. This is concerned with the user's movement in the Information Space, and how the statistical indicators of the conceptual index are updated. Any such read application does not affect the structure of the conceptual index.

```
┌─ ΔSoftLinkHypertext ──────────────────────────────
│  SoftLinkHypertext
│  SoftLinkHypertext'
├────────────────────────────────────────────────
│  Ξ ConceptualIndex
└────────────────────────────────────────────────
```

4.1 Starting an Information Retrieval Session

When an information retrieval session starts, all non-accumulative indicators are reset, whilst accumulative indicators remain unchanged. The user does not have a position in the Information Space.

```
┌─ Start ────────────────────────────────────────
│  ΔSoftLinkHypertext
│  Ξ ReadLinks
│  Ξ ReadConceptsVisited
│  Ξ AccumulateHistory
├────────────────────────────────────────────────
│  ConceptActivation' = { c : Concepts • (c, 0) }
│  CurrHistory' = {}
│  undefined CurrentConcept
└────────────────────────────────────────────────
```

4.2 Starting a New Trail

Once a session is started, a concept in the Index Space may be chosen from which to start the information retrieval session. This corresponds to a conventional Boolean search operation: the user makes an information request explicitly and the system responds by highlighting the requested concept in the Index Space; and the user retrieves a document in the Document Space in which the information requested is included. It should be noted that this operation does not affect any statistical information of the read state, it is only the position which changes. We call any such move starting a new trail.

```
┌─ StartNewTrail ────────────────────────────────
│  ΔSoftLinkHypertext
│  Ξ ReadSystemState
│  Ξ History
│  NewStartingConcept? : CONCEPT
├────────────────────────────────────────────────
│  NewStartingConcept? ∈ Concepts
│  the CurrentConcept' = NewStartingConcept?
│  undefined NextConcept'
└────────────────────────────────────────────────
```

From this stage on, the user has essentially two types of movement available, each representing a different information retrieval methodology:

1. The user can repeat the above operation and start another new trail. In this case, another Boolean search can be initialised, or the user may just want to re-visit a concept of the current history.
2. The user can treat the current concept as a hypertext source anchor and move to a hypertext destination concept by following one of the soft links leaving the current concept. Here, the user utilises the soft links provided by the conceptual index, and moves around in the Information Space by following these links. In this case, some indicators of the read state are updated. This process is known as **browsing**.

4.3 Browsing

Browsing is defined as a process of moving to a specific concept in the Information Space by following a soft link from a concept. In this situation, the user treats the current concept as a hypertext source anchor, specifies a soft link leaving the current concept and moves to the new concept to which the link connects. To move in such a manner, the user must choose a link. There are three possible scenarios.

1. The user has not started a trail, and so does not have a position in the Information Space. In other words, no concept can be currently used as a hypertext source anchor. In this case, there is no change to any of the session indicators, and a report is given.

$Report ::= No\_Such\_Leaving\_Link \mid Must\_Start\_Trail\_Before\_Browsing$

```
┌─ UserBrowseMoveError1 ──────────────────────────────────
│ link? : LINK
│ Ξ SoftLinkHypertext
│ report! : Report
├─────────────────────────────────────────────────────────
│ undefined CurrentConcept
│ report! = Must_Start_Trail_Before_Browsing
└─────────────────────────────────────────────────────────
```

2. The user has a position at a concept, but the link specified is not one of the leaving links of that concept. There is no change to any of the session indicators, and a report is given.

```
┌─ UserBrowseMoveError2 ──────────────────────────────────
│ link? : LINK
│ Ξ SoftLinkHypertext
│ report! : Report
│
│ ReadabilityFunctions
├─────────────────────────────────────────────────────────
│ defined CurrentConcept
│ link? ∉ LeavingLinks (the CurrentConcept)
│ report! = No_Such_Leaving_Link
└─────────────────────────────────────────────────────────
```

3. The user has a current position at a concept and the specified link is a leaving link of the current concept. In this case the user successfully makes a legitimate browsing move. The next concept becomes defined; it is the child of the user-specified link.

```
┌─ UserBrowseMove ──────────────────────────────────
│ ΔSoftLinkHypertext
│ link? : LINK
│
│ ReadabilityFunctions
├───────────────────────────────────────────────────
│ defined CurrentConcept
│ link? ∈ LeavingLinks (the CurrentConcept)
│ the NextConcept' = Child link?
│ CurrentConcept' = CurrentConcept
└───────────────────────────────────────────────────
```

We now describe how the statistical information is updated as a consequence of a browsing mode. The frequency of the hypertext source anchor (current concept), is incremented along with the frequency of the specified leaving link. The activation of the current concept is set to 1. If there is more than one leaving link from the current concept, the weights of all these links are automatically updated so maintaining the state invariant definition of weight.

```
┌─ BackwardLearning ────────────────────────────────
│ ΔSoftLinkHypertext
│ Ξ ReadUserState
│
│ ReadabilityFunctions
├───────────────────────────────────────────────────
│ ConceptsVisited' = ConceptsVisited ⊎ [the CurrentConcept]
│ LinksVisited' = LinksVisited ⊎
│                      [(the CurrentConcept, the NextConcept)]
│ ConceptActivation' = ConceptActivation ⊕
│                      {(the CurrentConcept, 1)}
└───────────────────────────────────────────────────
```

Then the new activation of the current node spreads to the children of the current concept as described by the following algorithm:

For the child of each leaving link of the activated concept, add the activation of that child to the sum of the input activation of each of the arriving links to that child concept, where the input activation of any link is defined as the product of its weight and the concept activation of its parent concept. If this value is greater than 1 then set the activation to 1.

```
┌─ ForwardActivationSpreading ──────────────────────────────
│ ΔSoftLinkHypertext
│ Ξ ReadLinks
│ Ξ ReadConceptsVisited
│ Ξ ReadUserState
│
│ ReadabilityFunctions
│
│ InputActivation : LINK ⇸ RAT_0^1
├──────────────────────────────────────────────────────────
│ ConceptActivation' = ConceptActivation ⊕
│   {c : Concepts | c ∈ Children (the CurrentConcept) •
│   (c, min (1, (ConceptActivation c)+
│     sumseq (mapsettoseq InputActivation (ArrivingLinks c))))}
│
│ ∀ c_1, c_2 : CONCEPT | (c_1, c_2) ∈ Links •
│     InputActivation (c_1, c_2) =
│         LinkWeight (c_1, c_2) * ConceptActivation c_1
└──────────────────────────────────────────────────────────
```

The schema above makes use of the intermediate concept of the Input Activation of a link, as defined above, in order to make the definition of the forward activation algorithm spreading more readable.

At the end of each browsing move, the system outputs a sequence of concepts, listed in the order of their activations. This is used as the online navigation guide for the current user to search and retrieve information more efficiently.

```
┌─ OutputConceptOrder ──────────────────────────────────────
│ Ξ SoftLinkHypertext
│ reply! : seq CONCEPT
├──────────────────────────────────────────────────────────
│ reply! = sort Concepts ConceptActivation
└──────────────────────────────────────────────────────────
```

The function sort simply takes a set of elements and sorts them in decreasing order according to an ordering function.

```
┌═[X]═══════════════════════════════════════════════════════
│ sort : (P X) → (X ⇸ RAT_0^1) ⇸ seq X
├──────────────────────────────────────────────────────────
│ ∀ f : (X ⇸ RAT_0^1); x_1, x_2 : X; xs : P X | xs ⊆ dom f •
│             ran (sort xs f) = xs ∧
│             #(sort xs f) = #xs ∧
│             ⟨x_1, x_2⟩ insert xs f ⇒ f x_1 ≥ f x_2
└──────────────────────────────────────────────────────────
```

Finally, the current concept is added to the current history, the user is transferred to the child of the specified link, which becomes the current concept, whilst the next concept is now undefined.

```
┌─ UpdateUserHistory ──────────────────────────────────
│ △SoftLinkHypertext
│ Ξ ReadSystemState
│ Ξ AccumulateHistory
├──────────────────────────────────────────────────────
│ CurrHistory' = CurrHistory ∪ CurrentConcept
│ CurrentConcept' = NextConcept
│ undefined NextConcept'
└──────────────────────────────────────────────────────
```

The total move operation is then given by either of the above three scenarios.

$$Browse \;\hat{=}\; (UserBrowseMove \;\wedge$$
$$(BackwardLearning \;;\; ForwardActivationSpreading \;;$$
$$OutputConceptOrder \;;\; UpdateUserHistory))$$
$$\vee$$
$$UserBrowseMoveError1$$
$$\vee$$
$$UserBrowseMoveError2$$

4.4 Quitting

A user may finish an information retrieval session at any time, in which case the local history is added to the accumulative history.

```
┌─ Quit ───────────────────────────────────────────────
│ △SoftLinkHypertext
│ Ξ ReadSystemState
│ Ξ CurrentHistory
├──────────────────────────────────────────────────────
│ AccHistory' = AccHistory ∪ {CurrHistory}
└──────────────────────────────────────────────────────
```

5 Authoring the Soft-Link Hypertext Model

Any changes made to the state space of the conceptual index through indexing or hyperization should not invalidate the statistical information collected by any accumulative indicators. We consider each of the four authoring cases defined in section 2.1 in turn, refining the previously defined schemas. First, a new concept is given an activation of 0.

```
┌─ UpdateAddConcept ───────────────────────────────────
│ △ReadSystemState
│ Ξ ReadLinks
│ Ξ ReadConceptsVisited
│ AddConcept
├──────────────────────────────────────────────────────
│ ConceptActivation' = ConceptActivation ∪ {(c?, 0)}
└──────────────────────────────────────────────────────
```

When a new link is introduced, its frequency is set to 0 and its weight given the appropriate value automatically.

```
┌─ UpdateAddLink ─────────────────────────────────────────────
│ Δ ReadSystemState
│ Ξ ReadConcepts
│ Ξ ReadLinksVisited
│ AddLink
│
│ ReadabilityFunctions
└──────────────────────────────────────────────────────────────
```

When a link is removed the frequency of that link is subtracted from the frequency of its parent concept.

```
┌─ UpdateRemoveLink ──────────────────────────────────────────
│ Δ ReadSystemState
│ Ξ ReadConceptActivation
│ RemoveLink
│
│ ReadabilityFunctions
│ FrequencyFunctions
├──────────────────────────────────────────────────────────────
│ LinksVisited' = {(c_1?, c_2?)} ⊲ LinksVisited
│ ConceptsVisited' = ConceptsVisited ⊎
│                              {(c_1?, LinkFrequency(c_1?, c_2?))}
└──────────────────────────────────────────────────────────────
```

4. When we remove a concept, we must delete any occurrence of it from the accumulative history in the system read state.

```
┌─ UpdateRemoveConcept ───────────────────────────────────────
│ Δ ReadSystemState
│ Ξ ReadLinks
│ RemoveConcept
│ Δ AccumulateHistory
├──────────────────────────────────────────────────────────────
│ ConceptActivation' = {c?} ⊲ ConceptActivation
│ ConceptsVisited' = {c?} ⊲ ConceptsVisited
│ AccHistory' = {cs : AccHistory • cs \ {c?}}
└──────────────────────────────────────────────────────────────
```

6 Conclusions

This paper has presented a Z specification of a new information retrieval model called the soft-link hypertext model (SLHM). SLHM differs from other current information retrieval models by its conceptual index data structure, which allows information organisation and retrieval to be achieved more efficiently and effectively. The conceptual index is constructed automatically according to well-defined rules. Subsequently, it can improve itself by collecting statistical data and also by applying automatic learning algorithms to this data in order to produce a more detailed map of the hypertext structure. These new features solve many of the problems of current information retrieval systems; consequently this new model can become the paradigm for a new generation of hypertext systems.

Our specification has provided a formal, precise and unambiguous account of SLHM, which can be used as the basis from which sound implementations can be designed and built. This specification was derived from applying the formal framework described in [12]. We claim the benefits of doing this are twofold:

- We have gained confidence in the validity and utility of the framework.
- If we use this framework to specify other existing large-scale and reconfigurable information systems then we can more readily incorporate our conceptual index and its related learning algorithms into them.

Finally, we believe that the SLHM specification provides an environment for research on further hypertext and learning strategies. In particular, our own future work will concentrate on installation of intelligent mechanisms on other aspects of the model, for example to extend the Document Space to multimedia applications.

Acknowledgements

Particular thanks to Colin Myers and also to Claire Cohen, Jennifer Goodwin, Mike Hinchey, Paul Howells, Michael Luck and Mark Priestley for comments on earlier versions of this paper. Thanks too to the anonymous referees who made valuable suggestions and isolated several errors in the original specification. The specification in this paper was checked using the *fuzz* package [24].

References

1. M. Arbib. *The Metaphorical Brain 2: Neural Networks and Beyond.* Wiley, 1989.
2. R. Belew. Adaptive Information Retrieval. In *Proceedings of the 12th Annual International ACM SIGIR Conference on Research Development in Information Retrieval*, 1989.
3. A. Carling. *Introduction to Neural Networks.* Sigma Press, 1992.
4. P. Cohen and R. Kjeldsen. Information Retrieval by Constrained Spreading Activation in Semantic Networks. *Information Processing & Management*, 23(4):255 – 268, 1987.
5. J. Conklin. Hypertext: An Introduction and Survey. *Computer*, 1987.
6. W. B. Croft and H. Turtle. A Retrieval Model for Incorporating Hypertext Links. In *Hypertext'89*, November 1989.
7. S. J. DeRose. Expanding the Notion of Links. In *Hypertext'89*, November 1989.
8. M. d'Inverno and J. Crowcroft. Design, specification and implementation of an interactive conferencing system. In *Proceedings of IEEE Infocom, Miami, USA. Published IEEE*, 1991.
9. M. d'Inverno, G. R. Justo, and P. Howells. A formal framework for specifying design methodologies. In *29th Annual Hawaii International Conference on System Sciences*, pages 741–750. IEEE Computer Society Press, 1996.

10. M. d'Inverno and M. Luck. A formal view of social dependence networks. In *Distributed Artificial Intelligence: Architecture and Modelling, First Australian Workshop on DAI, Lecture Notes in Artificial Intelligence, 1087*, pages 115–129. Springer Verlag, 1996.

11. M. d'Inverno and M. Luck. Formalising the contract net as a goal directed system. In W. Van de Velde and J. W. Perram, editors, *Agents Breaking Away: Proceedings of the Seventh European Workshop on Modelling Autonomous Agents in a Multi-Agent World, LNAI 1038*, pages 72–85. Springer-Verlag, 1996.

12. M. d'Inverno and M. Priestley. Structuring a Z specification to provide a unifying framework for hypertext systems. In J. P. Bowen and M. G. Hinchey, editors, *ZUM'95: 9th International Conference of Z Users, LNCS 967*, pages 83–102, Heidelberg, 1995. Springer-Verlag.

13. S. Grossberg. *Studies of Mind and Brain: Neural Principles of Learning, Perception, Development, Cognition, and Motor Control*. D.Reidel Publishing Company, 1992.

14. F. G. Halasz. Reflections on NoteCards: Seven Issues for the Next Generation of Hypermedia Systems. *Communications of the ACM*, 31(7), July 1988.

15. F. G. Halasz and M. Schwartz. The Dexter Hypertext. *Communications of the ACM*, 37(2):30–39, 1994.

16. G. E. Hinton. Connectionist learning procedures. *Artificial Intelligence*, 40(1 – 3):185 – 234, 1989.

17. M. J. Hu. *An Intelligent Information System*. PhD thesis, Department of Computer Science, Gower Street, University College London, 1994.

18. M. J. Hu and P. Kirstein. An Intelligent Hypertext System. In *The First International Workshop on Intelligent Hypertext (CIKM-93)*, Washington, 1993.

19. A. Lelu. Browsing through image databases via data analysis and neural networks. In *User-oriented Content-based Text and Image Handling: Proceedings of RIAO'88,*, pages 1034 – 1043, 1993.

20. M. Luck and M. d'Inverno. A formal framework for agency and autonomy. In *Proceedings of the First International Conference on Multi-Agent Systems*, pages 254–260. AAAI Press / MIT Press, 1995.

21. R. M. Fung et. al. An Architecture for Probabilistic Concept-based Information Retrieval. In *Proceedings of the ACM SIGIR'90 conference, Brussels, Belgium*, pages 455 – 467, 1990.

22. D. E. Rose and R. K. Belew. A connectionist and symbolic hybrid for improving legal research. *International Journal of Man-Machine Studies*, 35(1):1–31, 1991.

23. J. Savoy. Bayesian Inference Networks and Spreading Activation in Hypertext Systems. *Information Processing and Management*, 28(3):389 – 406, 1992.

24. J. M. Spivey. *The fUZZ Manual*. Computing Science Consultancy, 2 Willow Close, Garsington, Oxford OX9 9AN, UK, 2nd edition, 1992.

25. J. M. Spivey. *The Z Notation*. Prentice Hall, 2nd edition, 1992.

26. J. Reggia T. E. Doszkocs and X. Lin. Connectionist models and information retrieval. *Annual Review of Information Science and Technology*, 25, 1990.

27. S. H. Valentine. Putting numbers into the mathematical toolkit. In J. P. Bowen and J. E. Nicholls, editors, *Z User Workshop, London 1992*, Workshops in Computing, pages 9–36. Springer-Verlag, 1993.

Experience with Z
Developing a Control Program for a
Radiation Therapy Machine

Jonathan Jacky [*], Jonathan Unger, Michael Patrick,
David Reid and Ruedi Risler

Radiation Oncology, Box 356043
University of Washington
Seattle, WA 98195-6043

Abstract. We are developing a control program for a unique radiation therapy machine. The program is safety-critical, executes several concurrent tasks, and must meet real-time deadlines. Development employs both formal and traditional methods: we produce an informal specification in prose (supplemented by tables, diagrams and a few formulas) and a formal description in Z. The Z description includes an abstract level that expresses overall safety requirements and a concrete level that serves as a detailed design, where Z paragraphs correspond to data structures, functions and procedures in the code. We validate the Z texts against the prose specification by inspection. We derive most of the code from the Z texts by intuition and verify it by inspection but a small amount of code is derived and verified more formally. We have produced about 250 pages of informal specification and design description, about 1200 lines of Z and about 6000 lines of code. Experiences developing a large Z specification and writing the program are reported, and some errors we discovered and corrected are described.

1 Introduction

This paper reports on the development of the control program for the therapy operator's console for a unique radiation therapy machine, including our use of formal methods with the Z notation. This is not a pilot study or demonstration project; we are developing the program we will use to administer neutron therapy at our clinic. The program is safety-critical; it could contribute to delivering a treatment that differs from the prescribed one, irradiating the wrong volume within the patient or delivering the wrong dose.

The program we are developing is a replacement for a program that was developed by the therapy machine manufacturer and has been in use since the machine was installed in 1984 [18]. There is no reuse of code, design or specifications from the earlier program.

---

[*] email jon@radonc.washington.edu

2 System description

The purpose of the therapy console program is to help ensure that patients are treated correctly, as directed by their prescriptions. The treatment console computer stores a database of prescriptions for many patients. Each patient's prescription usually includes several different beam configurations called fields. Each field is defined by about fifty machine settings (positions, dose etc.) that must be set properly to deliver the prescribed treatment. The console program enables the therapist to choose fields from the prescription database. The program sets some settings automatically, but others (external motions that present collision hazards) must be set manually by the therapist. The program checks all settings against the prescription and ensures that the radiation beam can only turn on when the correct settings for the chosen field have been achieved (subject to override by the therapist for some settings in some circumstances). The therapist can turn on the therapy beam (by a separate nonprogrammable mechanism) after the program indicates that the machine is ready.

The program provides a user interface (so the therapist can select prescriptions and view machine status) and controls devices. Low-level device control (such as turning the beam on and off and guiding machine motions) is performed by other nonprogrammable mechanisms, programmable logic controllers (PLC's), and simple embedded computers. The therapy console program provides some of these low-level controllers with endpoints (such as positions and doses), and it enables (or disables) motions and activation of the beam. The delegation of functions among the software and hardware components was a prerequisite to the work discussed here and is described elsewhere [12, 13].

The therapy console program is coded in ANSI C; concurrency and device control is provided through EPICS [3], a library originally developed for controlling research accelerators, which is built on a commercially available real-time embedded software development product[2]. The user interface (display and keyboard) is handled by the X window system, programmed using Xlib only.

3 Project description

3.1 Personnel

The project was performed by the authors, who are full-time technical staff members in the Radiation Oncology department, a clinical department at the University of Washington Medical Center. The software was developed by a team of two people (Jacky throughout, working with Unger, then Patrick, then Reid). Risler is chief engineer of the facility and and is responsible for the overall system. We have full technical responsibility for the project; there is no separate quality assurance group or certifying body. Degrees are: Jacky, Ph.D, physiology; Unger, M.S., computer science; Patrick, M.S., control engineering; Reid: B.S., physics; Risler, Ph.D., physics.

---

[2] Our early work [14] was coded in Pascal, using a different product.

All had years of experience with radiation therapy, accelerator engineering or software development before joining the project. None had experience with formal methods before this project. Before beginning work on the therapy console, Jacky wrote some smaller Z descriptions of portions of the accelerator controls [5, 6, 7] (106, 166 and 178 lines of Z, respectively); these have not yet been implemented. Jacky wrote most of the formal description; Unger and Patrick also contributed. All software developers wrote code from Z specifications.

3.2 Products

We wrote about 250 pages of pertinent informal specification and design description, about 1200 lines of Z and about 6000 lines of code[3]. Table 1 reports the size of each product.

PRODUCT	SIZE
Informal specification: overview, entire facility [12]	106 pages
Informal specification: user interface, entire facility [13]	235 pages
Informal specification: user interface, therapy only (in [13])	45 pages
Informal specification: hardware and files, therapy only [10]	131 pages
Formal description (Z texts) [11]	77 pages (1137 lines)
Implementation Guide [15]	42 pages
Program code	4786 lines (41 files)
Test scripts	35 pages

Table 1. Development products (documents and code)

The informal specification [12, 13, 10] describes the system in prose, diagrams, tables and a few formulas. Part I [12] is an overview of the entire facility, part II [13] specifies the user interface for every console (including accelerator operators' consoles) and part III [10] describes hardware interfaces and external file formats for the therapy console program only. The formal description [11] is expressed in the Z notation [20]; its contents are discussed in Section 4 below. Software developers also wrote an implementation guide in prose [15] to explain our use of the platform's system software and other design information not expressed in the Z texts.

Software developers derived most of the code directly from the Z texts by intuition and verified it by inspection, without any intermediate formal refinement steps. We wrote down *post hoc* correctness arguments in a few cases where the intuitive derivation was not obvious to all (for example see section 28.9 in [8]). In one instance we refined part of the specification and formally derived about one page of code from the refined specification [14]. The code can be classified

---

[3] Our estimates for the completed project. Table 1 reports the sizes at this writing.

into categories defined in the implementation guide. Table 2 shows the lines of Z description, code[4], and any derivation or proof in each category. The table reveals that we modelled different parts of the program at very different levels of detail, according to our judgment about the novelty and difficulty of each portion. Some portions of the program have no formal description at all, while the formal description of some portions is as large as the code itself.

3.3 Assurance

Our primary assurance method is to review the documents, and review the code against the documents. An essential feature of the method is our willingness to rewrite or discard documents and code that seem unclear or overly complicated, whether or not we discover any errors. We spent more time reviewing, revising and rewriting each product than we spent writing the initial version. The program that we are preparing for clinical use is in its third major version.

The authors and reviewers of the informal specification include the physicist who defined the physical and clinical requirements for the original machine, engineers who installed and maintain the machine, and clinical users of the machine.

The formal description was validated by inspection: we read the Z texts closely, confirming to our (subjective) satisfaction that they expressed the intent of the informal description. The formal description was written to facilitate this inspection: each Z paragraph is accompanied by cross references to pertinent section, page numbers (and often paragraph and line numbers) in the informal specification. The report also contains a glossary that identifies Z variables with items described in the informal specification.

CATEGORY	Z	PROOF	CODE
Process and event handling	40	40	433
Pervasive constants and types	73	—	200
File handling and persistent data	52	—	672
Operations and volatile data	764	20	747
Graphics utilities	—	—	589
Graphics displays	—	—	1233
Low-level device control	208	—	912
Total	1137	60	4786

Table 2. Lines of formal description, proof, and code

All software developers participated in the reviews and all discovered and corrected errors where the Z formulas did not express the intended behaviors.

---

[4] Noncomment, nonblank lines of code

Some errors in the Z were not discovered until the coding stage; writing code necessarily involves another intensive review. Two errors in the Z were not detected until testing. See section 4.5 below.

We did not do any formal proofs or automated analyses to confirm that the Z texts express the intended behaviors. We did use a type checker [19] and we corrected the numerous but trivial errors that it found (misspellings, confusion about variable names, using a set element instead of a singleton set, etc.). We also tried a well-formedness checker [17]. This checker found a few expressions where functions were applied to arguments that could not be proved to lie within their domains; these were all deliberate elisions, not oversights.

Recently we translated a small portion of the formal description to the input language of the SMV temporal logic model checker; this checker detected some seeded errors we had discovered during our inspections, but detected no errors in our lastest (corrected) version [9].

We review all code before execution, and discover many coding errors in these reviews. Developers may execute code, discover errors, and make corrections *ad lib*. This *ad-hoc* testing activity is not scripted or recorded.

Finally we follow written test plans called *scripts* that describe exactly to how the execute the program and what the results should be (example scripts from another project appear in [16]). *Functional tests* attempt to cover the specification in some systematic way. Most functional tests can be performed on incomplete versions of the program; all will be performed on the completed version as well. Functional tests are derived from the formal description; they are designed to execute code that implements particular Z operation schemas.

Acceptance tests are simulated treatment sessions, using the completed program exactly as a therapist would use it, that rehearse both typical and unusual clinical situations (but with no patient present). Acceptance tests are derived from treatment scenarios described in the informal specification.

3.4   Effort

We estimate that we have devoted about six person-years to developing the therapy console program. This includes learning the platform and its programming environment, learning Z, and producing prototypes, documents and code that we later discarded. The effort was about equally divided between learning the requirements and writing the informal specification, creating the design including writing the Z texts, and implementation including coding and testing.

3.5   Status

The program is not yet complete and we have not begun acceptance testing. Tables 1 and 2 report the work accomplished at this writing (December 1996).

4 Z description

The formal description comprises 1137 lines of Z (207 paragraphs, including 131 schema definitions), presented in a 77 page report [11] (most of this report is prose)[5]. We found it quite difficult to produce a formal description that faithfully expresses the informal requirements and also serves as a useful basis for developing code. There are problems of scale and organization and not much guidance from the small examples in the literature. After many revisions we obtained a satisfactory solution by applying these strategies:

- Split the Z description into two levels: an abstract level that expresses overall safety requirements (section 4.1 below) and a concrete level that serves as the detailed design (section 4.2). The requirements level expresses properties that we might wish to check or prove; the design level is the system to be checked.
- Partition the detailed design. Much of the apparent complexity of the therapy machine arises from the interaction of several subsystems which, by themselves, are simpler. We partition the system into subsystems and describe one state schema and several (or many) operation schemas on each. For each operation on the system as a whole, we define a separate operation on each affected subsystem. The complex behaviors of the whole system emerge when we compose these simpler operations together. Identifying the best partition requires much exploration; we eventually discovered that the obvious partition corresponding to the hardware subsystems that the users see is not the best one.
- Decide what *not* model in Z. We left out the details of file organization and access (because they are obvious), and the appearance of the displays (because they are adequately described in the informal specification).
- Model what remains in sufficient detail so the translation to code is obvious.

4.1 Safety requirements

The central idea of the therapy control program is this safety requirement: the beam can only turn on when the actual state or *setup* of the machine is physically safe, and matches a *prescription* that the operator has selected and approved. We must only deliver setups that are physically consistent and reasonable or *safe*. The control program helps ensure that we can only treat a patient when the *measured* machine setup *matches* a *prescribed* setup.

[*SETTING, VALUE, FIELD*]

$SETUP == SETTING \rightarrow VALUE$

---

[5] The line count is the number of nonblank lines output by running the Fuzz tool -v option [19] on the report. This output is similar to the LaTeX source for the Z formulas.

$$safe\_ : \mathbf{P}\ SETUP$$
$$match\_ : SETUP \leftrightarrow SETUP$$
$$prescription : FIELD \nrightarrow SETUP$$

SafeTreatment
_____
measured, prescribed : SETUP
_____
safe(measured)
match(measured, prescribed)
prescribed ∈ ran prescription

The whole design arises from elaborating this simple model. The entire purpose of the control program is to establish and confirm the *SafeTreatment* condition. The *prescribed* setup must be selected; the *measured* setup must be achieved; the *safe* and *match* conditions must be tested.

Essential safety requirements can be expressed at this level. For example, the condition of the machine when the radiation beam is on could be expressed by a state schema, *BeamOn*. The safety requirement that the beam can only be on when the machine is in a safe condition can be expressed

$$BeamOn \Rightarrow SafeTreatment$$

Checking or proving this property would be a significant achievement. It would provide additional evidence (besides inspection and testing) for the soundess of our detailed design. We have not yet attempted this, but it is a long term goal of our work. Soon it may be feasible using model checking [1].

4.2 Detailed design

Most of our Z texts express a detailed design. We used Z to discover the design, not just to document an already existing design. There are few descriptions of other therapy control systems and these do not provide enough detail to serve as examples. We had to create our own. We used no other design notation (except prose).

The design is very detailed: each X window system event (including every keystroke) and transmission or receipt of every message to or from a controller is modelled by a Z operation schema. The translation to code is almost obvious: Z basic types and free types become enumerations. Z state variables become program variables and data structures. Z operation schemas become functions and procedures. A Z state schema and the operation schemas on that state become a module (in C, an .h file and a .c file that define a collection of constants, types and variables and the functions that use them). Z schema inclusion becomes module inclusion (C #include directives). Some examples appear in [8].

4.3 Concurrency and real time

In our detailed design concurrency is handled by the interleaving model recommended by Evans [4]: the state is shared by all processes, but operations might be invoked by different processes; when an operation's preconditions are satisfied, that operation occurs. The real time clock (needed for timeouts) is modelled in Z as an ordinary state variable.

4.4 Reviewing large Z specifications

We did not do any formal proofs or automated analyses to confirm that the Z texts express the intended behaviors, but we devoted much effort to reviews (inspections). Partitioning the design makes review feasible: each review session considers one state schema and all the operations that use it.

It is not sufficient to review Z operations one by one, it is necessary to consider how they work together. This can be difficult because related operations are usually spread across many pages. We find it helpful to collect all the pertinent formulas together in tables. For example, Table 3 summarizes all of the operation schemas that model interactions with a single embedded controller (for more details see [9]). It is similar to the mode transition tables of the SCR notation [2].

There is one entry in the table for each Z operation schema. The four columns in the table show the operation name, the conjuncts of the precondition that only involve the state variables, the conjuncts of the precondition that only involve input variables, and conjuncts from the postcondition that indicate a change of state (postconditions of the form $x' = x$ indicating that x does not change are not shown). In addition to the thirteen top-level operations, there are entries for three building-block operations used to define others. Each group of top-level operations is headed by the building-block operation whose predicates are conjoined with theirs. For example, *Wait* appears above *StartTimer* so the full state precondition of *StartTimer* is *processing = run ∧ WaitCmd*.

Many properties of our design can be confirmed by inspecting Table 3:

Invariance: Every operation has either *WaitCmd* or *WaitReceive* in its precondition. If the controller process entered a state where neither predicate were true, no operations would be enabled; the process would deadlock. Therefore *WaitCmd ∨ WaitReceive* must be an invariant. The negation of the invariant is the forbidden state where an unsent command is pending but no reply is expected.

Completeness: The design is complete if the response to all events is defined in all states permitted by the invariant. The table shows this is true: the disjunction of all the preconditions forms a tautology. For example in the second column, there are two large sections for *processing = run* and *processing = wait*, together these cover all values of *processing*. Within each section there are subsections for *WaitCmd* and *WaitReceive*, together these cover the invariant *WaitCmd ∨ WaitReceive*, etc.

Determinism: The design is deterministic if only one operation is enabled for each input in each state. The table shows that all the preconditions are

325

Z operation name	State precondition	Input precondition	Progress postcondition
Wait	processing = run	true	processing' = wait
StartTimer	WaitCmd	true	timer' = clock + period
RestartTimer	WaitReceive	true	timer' = clock + deadline
HandleEvent	processing = wait	e? ∈ events	processing' = run events' = events \ {e?}
NewCommand (HandleCmd ∧ Send)	WaitCmd ∧ p	e? ∈ cmd	WaitReceive' expected' = responses c pending' = tail cs
CommandError	WaitCmd ∧ ¬ p	e? ∈ cmd	status' = error
ReceiveUnsolicited	WaitCmd	e? = receive	status' = error
ReceiveAck	WaitReceive tail expected ≠ ⟨⟩	e? = receive ∧ q	status' = ok expected' = tail expected
NextCommand (ReceivePending ∧ Send)	WaitReceive tail expected = ⟨⟩ pending ≠ ⟨⟩	e? = receive ∧ q	status' = ok expected' = responses c pending' = tail pending
ReceiveComplete	WaitReceive tail expected = ⟨⟩ pending = ⟨⟩	e? = receive ∧ q	WaitCmd' ∧ status' = ok expected' = tail expected
ReceiveError	WaitReceive	e? = receive ∧ ¬ q	WaitCmd' ∧ status' = error
ReceiveTimeout	WaitReceive	e? = timeout	WaitCmd' ∧ status' = error
Signal	true	true	events' = events ∪ {e?}
SignalEvent	true	e? ≠ timeout	
SignalTimeout	clock > timer	true	timer' = stopped
Tick	true	true	clock' > clock

$WaitReceive \Leftrightarrow expected \neq \langle \rangle$
$WaitCmd \Leftrightarrow pending = \langle \rangle \wedge expected = \langle \rangle$
$WaitCmd \vee WaitReceive$ is invariant; $pending \neq \langle \rangle \wedge expected = \langle \rangle$ is forbidden.

$c = head\ pending \wedge cs = tail(commands\ e?)$
$p \Leftrightarrow status = ok \vee e? \in restart$; $\langle cmd, \{receive\} \rangle$ partition $EVENT$
$q \Leftrightarrow message? \in dom\ response \wedge response\ message? = head\ expected$

Table 3. Controller operations, preconditions, and progress postconditions

disjoint. For example in the second column we find the disjoint pairs *processing* = *run* and *processing* = *wait*, *WaitCmd* and *WaitReceive*, etc.

Progress (liveness): To make progress, it is necessary to make state transitions from one table entry to another. For example unhandled command events can accumulate in the *WaitReceive* state. However all the operations that begin in *WaitReceive* either end in *WaitCmd* or make progress toward *WaitCmd* by establishing *expected'* = *tail expected* or *pending'* = *tail pending*. Therefore the system will eventually reach *WaitCmd* and pending command events will be handled.

4.5 Error discovery

A large Z specification will contain many errors at first. We discover many trivial errors by inspection (such as $S = \{x, y\}$ where $S \subseteq \{x, y\}$ is intended, etc.). Simple pencil-and-paper analyses such as preparing Table 3 reveals deeper errors involving the interactions of two or more schemas. Here are some examples:

Omitted case: Sometimes occurs because the author was thinking of a poorly chosen (overly restricted) example. Common variations are **Missing operation** and **Precondition too strong**.

Inconsistent includes: Schema C includes A and B, but A is inconsistent with B. Usually occurs because A was revised after C was written.

Displaced precondition: Operations B and C both include operation A. Precondition P should apply only to B but appears in A, so the precondition of C is too strong. Usually occurs because A was revised by splitting into a new A, plus B and C.

Overbooked state variable: Operation A stores a value in state variable x, intending it to be found by operation B. However operation C can store a different value in x before B is invoked.

Implicit invariant: The implicit invariant formed by the disjunction of the preconditions of all the operations is stronger than the explicit state invariant. However the postcondition of one operation does not satisfy this implicit invariant. After that operation executes, no operations are enabled and the system deadlocks.

Two errors in the detailed Z design were not discovered until *ad hoc* testing. They were instances of **Omitted case** (see [14] p. 329) and **Overbooked state variable**.

5 Conclusions

We find the formal notation useful for discovering a design, and then documenting the detailed design so that it can be validated against the prose specification, and serve as a guide for writing, inspecting and testing the code.

Not many conclusions can be made until we complete the program and gain experience in acceptance testing and clinical use. Meanwhile we can offer some observations based on our experience so far:

- Formal methods can help create novel designs and develop original code. They are not just for documenting existing designs and analyzing code that has already been written.

- A detailed and explicit informal specification that has been reviewed by the systems' designers and users (not just software developers) is an indispensable prerequisite to any use of formal methods. Only this can serve as the standard for validation. It is a major portion of the whole project effort, not just a preliminary.

- A useful formal description is not just a paraphrase of the informal specification into mathematical notation. Creating the formal description requires design judgment in addition to understanding the requirements and the formal notation.
- All documents and code require much revision for clarity and organization, not just content and correctness.
- Software developers who have the education and experience needed to work on this kind of project can learn to read, review, and implement Z and even write small amounts of it fairly quickly. Writing a useful formal description of a complex system is much more difficult and requires much experience on progressively harder problems.
- Simply having a good formal description does not guarantee that a good implementation will come easily. Diligent ongoing review is required to ensure that the implementation is simple and clear enough to review against the formal description. This is a prerequisite to checking that the implementation is correct.
- Inspection and simple paper-and-pencil analyses can detect most, but not all, errors in a large Z specification if it is sufficiently well partitioned to be reviewed in small sections. Some other assurance technique is needed to discover the remaining errors (we used testing).

Acknowledgments

The authors thank Mark Saaltink for running the well-formedness checks and thank Ira Kalet for assistance with the project.

References

1. Richard J. Anderson, Paul Beame, Steve Burns, William Chan, Francesmary Modugno, David Notkin, and Jon Reese. Model checking large software specifications. In David Garlan, editor, *SIGSOFT '96: Proceedings of the Fourth ACM SIGSOFT Symposium on the Foundations of Software Engineering*, pages 156–166, 1996. (also published as *ACM Software Engineering Notes* 21(6), Nov. 1996).
2. Joanne M. Atlee and John Gannon. State-based model checking of event-driven system requirements. *IEEE Transactions on Software Engineering*, 19(1):24–40, January 1993.
3. L. R. Dalesio, M. R. Kraimer, and A. J. Kozubal. EPICS architecture. In C. O. Pak, S. Kurokawa, and T. Katoh, editors, *Proceedings of the International Conference on Accelerator and Large Experimental Physics Control Systems*, pages 278–282, 1991. ICALEPCS, KEK, Tsukuba, Japan.
4. Andy S. Evans. Specifying and verifying concurrent systems using Z. In Maurice Naftalin, Tim Denvir, and Miquel Bertran, editors, *FME '94: Industrial Benefit of Formal Methods*, pages 366–380. Springer-Verlag, 1994. (Lecture Notes in Computer Science number 873).
5. Jonathan Jacky. Formal specifications for a clinical cyclotron control system. In Mark Moriconi, editor, *Proceedings of the ACM SIGSOFT International Workshop on Formal Methods in Software Development*, pages 45 – 54, Napa, California, USA, May 9 – 11 1990. (Also in *ACM Software Engineering Notes*, 15(4), Sept. 1990).

6. Jonathan Jacky. Formal specification and development of control system input/output. In J. P. Bowen and J. E. Nicholls, editors, *Z User Workshop, London 1992*, pages 95 – 108. Proceedings of the Seventh Annual Z User Meeting, Springer-Verlag, Workshops in Computing Series, 1993.

7. Jonathan Jacky. Specifying a safety-critical control system in Z. *IEEE Transactions on Software Engineering*, 21(2):99–106, 1995.

8. Jonathan Jacky. *The Way of Z: Practical Programming with Formal Methods*. Cambridge University Press, 1997.

9. Jonathan Jacky and Michael Patrick. Modelling, checking, and implementing a control program for a radiation therapy machine. In Daniel Jackson, editor, *AAS '97: First ACM SIGPLAN Workshop on Automated Analysis of Software*, 1997. (in press).

10. Jonathan Jacky, Michael Patrick, and Ruedi Risler. Clinical neutron therapy system, control system specification, Part III: Therapy console internals. Technical Report 95-08-03, Radiation Oncology Department, University of Washington, Seattle, WA, August 1995.

11. Jonathan Jacky, Michael Patrick, and Jonathan Unger. Formal specification of control software for a radiation therapy machine. Technical Report 95-12-01, Radiation Oncology Department, University of Washington, Seattle, WA, December 1995.

12. Jonathan Jacky, Ruedi Risler, Ira Kalet, and Peter Wootton. Clinical neutron therapy system, control system specification, Part I: System overview and hardware organization. Technical Report 90-12-01, Radiation Oncology Department, University of Washington, Seattle, WA, December 1990.

13. Jonathan Jacky, Ruedi Risler, Ira Kalet, Peter Wootton, and Stan Brossard. Clinical neutron therapy system, control system specification, Part II: User operations. Technical Report 92-05-01, Radiation Oncology Department, University of Washington, Seattle, WA, May 1992.

14. Jonathan Jacky and Jonathan Unger. From Z to code: A graphical user interface for a radiation therapy machine. In J. P. Bowen and M. G. Hinchey, editors, *ZUM '95: The Z Formal Specification Notation*, pages 315 – 333. Ninth International Conference of Z Users, Springer-Verlag, 1995. Lecture Notes in Computer Science 967.

15. Jonathan Jacky, Jonathan Unger, and Michael Patrick. CNTS implementation. Technical Report 96-04-01, Department of Radiation Oncology, University of Washington, Box 356043, Seattle, Washington 98195-6043, USA, April 1996.

16. Jonathan Jacky and Cheryl P. White. Testing a 3-D radiation therapy planning program. *International Journal of Radiation Oncology, Biology and Physics*, 18:253–261, January 1990.

17. Irwin Meisels and Mark Saaltink. The Z/EVES reference manual. Technical Report TR-95-5493-03, ORA Canada, 267 Richmond Road, Suite 100, Ottawa, Ontario K1Z 6X3 Canada, December 1995.

18. Ruedi Risler, Jüri Eenmaa, Jonathan P. Jacky, Ira J. Kalet, Peter Wootton, and S. Lindbaeck. Installation of the cyclotron based clinical neutron therapy system in Seattle. In *Proceedings of the Tenth International Conference on Cyclotrons and their Applications*, pages 428–430, East Lansing, Michigan, May 1984. IEEE.

19. J. M. Spivey. *The fUZZ Manual*. J. M. Spivey Computing Science Consultancy, Oxford, second edition, July 1992.

20. J. M. Spivey. *The Z Notation: A Reference Manual*. Prentice-Hall, New York, second edition, 1992.

Preliminary Evaluation of a Formal Approach to User Interface Specification

John C. Knight[1] and Susan S. Brilliant[2]

[1] University of Virginia, Charlottesville, VA 22903, USA
[2] Virginia Commonwealth University, Richmond, VA 23284, USA

Abstract. In this paper we report on a research project in which the user interface for a research nuclear reactor was specified using a combination of formal notations. The goal of the project was to evaluate the use of a combination of techniques and to assess their utility in specifying a user interface for a non-trivial safety-critical application. We conclude that the techniques worked well and scale up easily to the size of the application studied.

1 Introduction

User interfaces can be very complex. The user interface in a modern nuclear power plant in the U.S.A., for example, might include hundreds of illuminated indicators, digital and analog gauges, strip-chart paper displays, switches, rotary controls, and so on. Such systems usually require several human operators to maintain proper control.

In many applications, computer-based control systems are replacing systems based on older technology to increase flexibility, speed up operation, and enhance functionality. Modern user interfaces in these new systems offer significant benefits over their electromechanical counterparts. These benefits include the use of new devices such as voice input, the provision of new display concepts such as three-dimensional visualization, and new analysis capabilities such as filtered and time-history presentation of signals [12].

These new user-interface capabilities bring with them a variety of challenges including proper ergonomic design and reliable software implementation. These challenges have to be addressed in order to develop safe systems [16] because the user interface is a safety-critical element of a control system. If a display contains an error, such as incorrect data or mislabeling of an object, an accident might occur when the operator acts on this incorrect information. Also, if the interface does not respond to user actions as the operator expects, the result might also be unsafe. Examples of such problems have already occurred [17].

An incident reported recently that occurred in a nuclear power plant highlights the importance of correct and understandable specifications [22]. The overhead annunciator system in the control room (procured from a major vendor) stopped processing and sounding messages without any indication of failure. The system failed when the primary process that sorted the alarm input buffer into a time sequenced output buffer aborted. Software that displayed the data and actuated the overhead annunciator windows continued to operate and so the system appeared to be operating. Root-cause investigation revealed that the vendor thought the operating system would force a restart if any process aborted (it does not), and that a watchdog timer would protect the system (it does not).

In general, the development of user interfaces is a difficult and important problem. Because many errors in computer systems can be traced to defects in their specifications, a critical aspect of user-interface development is the specification of exactly what the user interface is to do. As safety-critical applications rely more heavily on increasingly complex user interfaces, the need to specify the user interface precisely and correctly in such systems becomes essential.

Despite the importance of specification, in practice user interfaces are rarely specified with any degree of care. Most user interfaces evolve from an initial prototype implementation derived from an informal and incomplete description provided by application experts to a final implementation for which there is no specification. This evolution takes place as a series of iterations in which the implementation is demonstrated, evaluated by the application experts, and then enhanced.

This is not an ideal situation and we hypothesize two reasons why it has arisen. The first reason is that the formal specification of a user interface requires a massive amount of detail if it is to be done properly. For example, to specify precisely the details of a pull-down menu on a computer screen requires detail such as colors, shape and appearance, and location as well as all the semantics associated with selecting a menu item.

The second reason that we hypothesize to explain why user interfaces are rarely specified carefully is the variety of material that has to be specified. No single formal specification language has the facilities to describe all that is needed, and very few have any kind of animation mechanism that would permit application engineers to check that what is described is what they require.

In two previous papers [3, 7] we have described an early version of an approach to formal specification of user interfaces that we have developed. In this paper, we present a preliminary evaluation of the approach based on the development of formal specifications for the user interfaces of two safety-critical systems: a medical robot and a nuclear reactor [6]. This latter system is the subject of a case study in which we are developing a prototype experimental (non-operational) control system for the research nuclear reactor at the University of Virginia. Examples from this project are used for illustration in this paper.

The reactor specification was developed informally because its creation drove the refinement of the techniques described here. The requirements for the user interface were determined by examination of existing documentation, observation of the current system in operation, and extensive discussion with reactor operators and staff. Several reviews of different aspects of the specification have been held, and the user interface has been connected to a high-fidelity reactor simulator for user evaluation. All of these activities were informal and in no way validate the specification. Evaluation is ongoing.

The specification approach that we are using employs three existing formal notations in an integrated framework—no new notations are involved. Our goals with the project that we describe are twofold. The first is to determine the capabilities that can be achieved and determine the difficulties raised by integrating more than one formal notations. The second goal is to evaluate the utility of the notations themselves in spec-ifying the user interface for a non-trivial safety-critical application. In other words, how well do these techniques scale-up?

We begin with an overview of the reactor application, and we follow this with a summary of the current version of the approach to specification that we have used. We then present examples of the specification taken from the reactor case study. These examples are followed by details of our evaluation criteria and our current evaluation results. Finally, we present our conclusions.

2   University Of Virginia Reactor

The case study application providing most of the information for the evaluation of the specification technique is the *University of Virginia Reactor* (UVAR). This is a research reactor that is used for the training of nuclear engineering students, service work in the areas of neutron activation analysis and radioisotope generation, neutron radiography, radiation damage studies, and other research [23]. As part of a research program in software engineering, a digital control system is being developed for the UVAR and is currently in the specification stage.[1]

The UVAR is a "swimming pool" reactor, i.e., the reactor core is submerged in a very large tank of water. The water is used for cooling, shielding, and neutron modera-tion. The core uses Low Enriched Uranium (LEU) fuel elements and is located under approximately 22 feet of water on an 8x8 grid-plate that is suspended from the top of the reactor pool. The reactor core is made up of a variable number of fuel elements and in-core experiments, and always includes four control rod elements. Three of these control rods provide gross control and safety. They are coupled magnetically to their drive mechanisms, and they drop into the core by gravity if power fails or a safety shutdown signal (known as a "scram") is generated either by the operator or the reactor protection system. The fourth rod is a regulating rod that is fixed to a drive mechanism and is therefore non-scramable. The regulating rod is moved automatically by the drive mechanism to maintain fine control of the power level to compensate for small changes in reactivity associated with normal operations [23].

The heat capacity of the pool is sufficient for steady-state operation at 200 kW with natural convection cooling. When the reactor is operated above 200 kW, the water in the pool is drawn down through the core by a pump via a header located beneath the grid-plate to a heat exchanger that transfers the heat generated in the water to a second-ary system. A cooling tower located on the roof of the facility exhausts the heat and the cooled primary water is returned to the pool. The overall organization of the system is shown in Fig. 1.

The existing reactor control system, shown in Fig. 2, is comprised primarily of analog instrumentation that is used by the reactor operators to monitor and regulate operating parameters over all ranges of operation, from start-up to full power. A first-generation of the digital control system will replicate the functionality of the existing control console. The majority of that functionality is the display of process variables

---

1. At present there is no intention of putting the digital control system into operation.

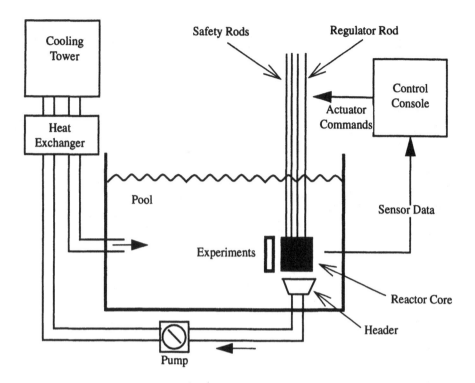

Fig. 1. The University of Virginia reactor system.

including gross output, neutron flux and period, temperature difference between water entering the core and water leaving the core, control and regulating rod positions, primary system flow, and pool water level. The control console also provides facilities for operator input to the reactor system, including control of the regulating and control rods, a means to test instrumentation, and responses to unsafe conditions.

3 Specification Approach

A user interface is a complex entity; specifying such an entity is correspondingly complex. The interface is far more than the graphics, the operator commands, or even these two combined. Informally, the items that have to be defined if a specification is to be in any sense complete include everything that is presented to the operator, everything that the operator can do to the interface, everything that it would be erroneous for the operator to do together with the actions that are required in each case, and the exact meaning of each input that the operator can enter. In a comprehensive approach to user-interface specification, all of the these aspects need to be addressed, and the specification technique(s) used must deal with each aspect completely and consistently.

Formal specification of user interfaces is not new. Various texts [5, 10] and surveys [2] have been prepared, and many research contributions published. Some of the

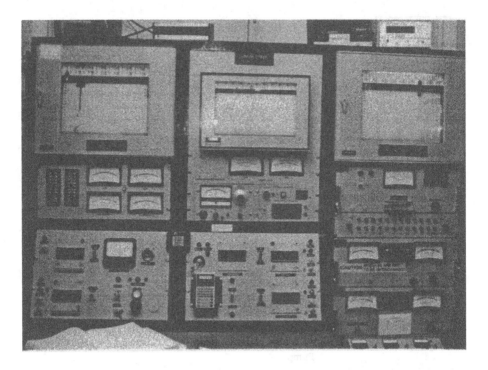

Fig. 2. Primary part of the existing main control console of University of Virginia reactor.

existing work has been focused on problems such as specifying the graphical element of a user interface [1] or the details of the valid interactions [15, 20]. Other work has been concerned more with the tools needed to develop interfaces rapidly and accurately [18]. Such tools can be thought of as application generators controlled by formal notations. An advantage of the use of such tools is that they permit exploration of ergonomic issues associated with user interfaces.

Our concern with the work described in this paper is with the precise specification of all aspects of the user interface for safety-critical systems. We are seeking specification with the greatest possible accuracy rather than rapid or flexible generation, and we are interested in a rather specialized area of application rather than general applicability. We also assume that ergonomic analysis is performed separately. In other words, we expect that ergonomic analysis will be undertaken at appropriate points by appropriate experts and that our task is to specify and analyse the resulting requirements.

It could be argued that errors in the user interface of a safety-critical system, such as a nuclear reactor, are not themselves especially important because a protection systems is present (in principle) in the control system. A defective or omitted command should not lead to a hazard because the protection system will intervene. Although this will often be the case, protection systems usually only guard against the most catastrophic of situations and plenty of damage can be done within the range of operations accepted by the protection system—and protection systems do not always work. The position taken in the work described here is that the user interface should be viewed as a critical element of the system and not impose an additional safety burden on other elements of the system.

The specification approach we use builds on a view of the user interface introduced by Foley in 1974 [9]. This view models the interface as a dialog between the operator and the computer system carried out in a fixed interaction language. The specification problem is to define this language completely and accurately. In practice, the dialog cannot be defined just in terms of the inputs to be received from the operator since responses from the computer system change the state of the user interface and thereby change the operator inputs that are valid.

Defining the interaction language using just *one* of the available formal specification notations is certainly possible but extremely difficult. It requires, for example, that a graphic screen be modelled using a mechanism such as a sequence of pixels if another notation, such as a picture is not to be used. Our approach, therefore, uses several different formal languages—each one suited to the part of the specification for which it is used.

3.1 Structure of the Specification

In practice, the dialog between the operator and the computer system takes place in a *set* of languages—in effect, the operator is engaged in a set of sub-dialogs going on in parallel. In the alarm system for the UVAR, for example, an alarm can be signalled and possibly dealt with by the operator whilst he or she is adjusting control rod heights for a separate purpose. There are in effect two separate sub-dialogs going on in that case which take place in two separate if somewhat simple interaction languages. These languages are not entirely separate, however. In many circumstances, actions taken in one sub-dialog can affect the possible actions in another. In the UVAR alarm system, for example, a scram alarm forces the reactor to be shut down thereby limiting the possible actions in most other sub-dialogs.

In our specification approach, each member of the set of languages required for the sub-dialogs is defined separately. The overall structure of each specification uses the three levels employed traditionally to describe formal languages: *lexical*, *syntactic*, and *semantic* levels.

Each of these three levels is specified separately. Since different notations are used, a remaining issue is how the communication between these levels is defined. To provide this communication, we adopt abstract interfaces between the three specification levels, and also between the user interface specification and the application software specification. The role of the user interface is to accept and check user commands, and to update the material presented to the operator. The actual system functionality is implemented in the application software and that is assumed to be specified separately. However, it is essential that the interaction between the user interface and the application be defined. The overall structure of the specification, including the communication paths between specification levels and communication with the application itself, is shown in Fig. 3.

The interface between the lexical and the syntactic levels is the same as that used in the formal definition of other languages—the lexical level defines a set of tokens, possibly including parameters, that are input to the syntactic level. Communication

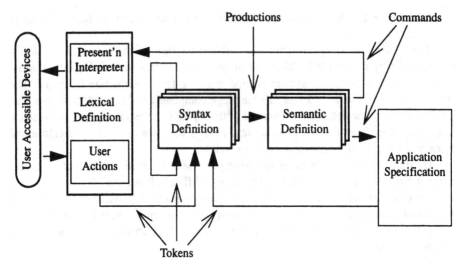

Fig. 3. User interface specification structure.

from the application to the user interface is also accomplished using tokens thereby enabling state changes in the application to effect the necessary changes to the valid input sequences of the sub-dialogs.

The application also generates a token to communicate a change in state that may need to be reflected in the user interface. Any changes that are required in the material presented to the user are recognized by the syntax level and invoked by the semantic level. These changes are specified by a set of messages that are generated by the semantic level and received and effected by the presentation interpreter (see Fig. 3).

The interface between the syntactic level and the semantic level is borrowed from syntax-directed compiler technology. We specify the context-free syntax using BNF and each production in the specification is associated with a (possibly null) semantic action specified in Z so that for each production an appropriate action is specified. The semantic level communicates with the application by messages that are transmitted to the application after the recognition of valid sequences of user actions that require a response from the application.

We summarize the details of our use of formal notations and how the inter-level communication is defined in the following sections.

3.2 Lexical Level

The lexical level defines exactly how the user interface effects its dialogue, i.e., what instruments and controls the user sees and employs to control the system [8]. Part of the lexical level corresponds to the graphical items seen by the user on a computer screen. In our reactor example, precisely how the presence of an alarm is signalled to the operator—colored light, flashing light, bell, horn, voice, or a combination of these—is a lexical issue.

A number of formal notations for specifying the lexical level of the user interface have been proposed [1, 11, 14]. No matter what the notation, however, the specification of the lexical level is extremely complex. The source of the complexity lies in the sophistication of the available input and output devices. For example, in specifying a graphic user interface, all of the objects that are to appear on the screen must be described in complete detail. Some idea of the difficulty that this entails can be seen by considering the semantics of something as simple as a screen menu item. It is essential that the actions of the menu item be completely defined for all possible user actions including the following: depressing any mouse button over the item; releasing any mouse button over the item; releasing the mouse button over the item having depressed it elsewhere; typing text with the focus over a menu item; and so on.

Rather than define an entire lexical specification using one of the existing notations for the graphic element of our specification approach, we have chosen to adopt a complete existing lexical specification framework and tailor it to our needs. The framework we employ is Borland's *Object Windows Library* (OWL). This framework is defined using C++ classes and provides predefined specifications for all common graphic elements including command buttons, menus, bit-mapped graphics, and so on. OWL allows easy description of a graphic interface by inheriting from classes of graphic entities.

Of great importance is the fact that the OWL documentation defines the base classes used in the specification precisely and hence provides the exact meaning of actions that manipulate those graphical objects (such as those involving the effect of pressing mouse buttons over a menu item mentioned above). In essence what we are doing is reusing an existing set of specifications and accepting whatever definitions they include. This is a very satisfactory trade-off, and there are a number of similar frameworks that could be used with similar results, e.g. TurboVision, Delphi, or Visual Age. The graphic component of a user interface specified with our approach consists of a simple C++ program that utilizes the library classes to define graphic objects. However, this relatively simple specification is accompanied by the library documentation and hence provides a complete specification of a complicated aspect of the interface. Any doubt that remains after the documentation is consulted can be resolved by regarding the C++ program as an operational definition.

3.3 Syntactic Level

The syntax of the human-computer dialogue defines valid sequences of user input and computer output. The "dialogue" as we have noted is actually several sub-dialogues that proceed concurrently and that are interleaved arbitrarily. The syntactic level in our specification approach is documented with a set of context-free grammars with one grammar for each of the concurrent, asynchronous dialogues that might be taking place.

The tokens that are input to each context-free grammar can be generated by three sources:

* *The lexical level.*
  The user interface generates tokens in response to user manipulation of the interface.

- *The application program.*
  Tokens arriving from the application program are the mechanism by which the application communicates with the user interface.
- *One of the other grammars in the set.*
  This is the mechanism by which communication is achieved between the different languages that effect the sub-dialogs.

The concept of a multi-party grammar described by Shneiderman [21] is appropriate for representing grammars in which tokens are generated by more than one source. However, we have elected to use a conventional context-free grammar representation together with a naming convention to distinguish sources of tokens. The primary advantage of using conventional grammars is the ready availability of tools that provide automatic generation of syntax analyzers from the grammars. If one of these tools is used, the implementation is guaranteed to implement the specification if the tool is working correctly, and so verification of the translation is simplified.

3.4 Semantic Level

In our specifications, we use Z [4, 13, 19] to define the user-interface context-sensitive syntax (i.e., the rules of syntax that are dependent on context) and the semantics. Z provides all the necessary mechanisms to define the various operations that are effected by the operator.

It is important to keep in mind that the semantics to which we refer are just the semantics of the set of interaction languages. Once a command is deemed valid, its "meaning" in almost all cases is merely to send a message requesting some operation either to the application software itself or to the presentation interpreter. Thus important functional semantics, the reactor safety rules for example, are defined in the application specification.

4 An Example

Many aspects of the reactor control system user interface are straightforward. We focus here on one of the more complex subsystems in order to illustrate the division of labor and communication among the layers of the specification of the computerized control system. The subsystem we use is the safety control rod system.

Recall that the three safety control rods (also called *shim* rods) provide gross control of reactivity and are one element of the safety system. For each of these control rods, the present operator's console includes a set of four lights:

- *Engaged.*
  This light indicates whether the rod is magnetically coupled to its drive mechanism.
- *Up.*
  This light indicates whether the rod is fully withdrawn.
- *Down.*
  This light indicates whether the drive mechanism is at its lowest level.

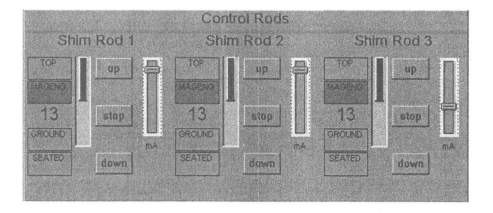

Fig. 4. Rod controls provided by new digital system. This is just a small part of the user interface.

- *Seated.*
  This light indicates whether the rod is fully inserted into the core, regardless of the position of the drive mechanism.

For each control rod, there are also displays indicating the rod height and the magnetic current to the rod attachment mechanism. To allow operator control of each rod, there are controls allowing the operator to set the magnet current and to raise and lower the rod. The operator can either move the rod by a small increment or move the rod continuously at a predetermined speed. The standard operating procedures in the current manual system allow the operator (a) to move the rods only when the neutron flux measurement exceeds a prescribed threshold (thereby indicating that the instrument is operating), and (b) to move two control rods simultaneously only if all three rods are raised less than 10 inches from their seated positions. These restrictions will be enforced by the software control system.

Fig. 4 shows the small portion of the screen that results from executing the OWL lexical specification of the interface for the control rods, and Fig. 5 shows part of that specification. At this level the screen objects associated with the display of control-rod-related data are specified. For example, the rod heights are displayed in a digital format and are also shown graphically in an analog-style display for operator convenience. Either one of these could be used alone. An "up" and a "down" button are supplied for controlling the movements of each rod. Pressing and quickly releasing a button provides the incremental move action; holding the button down moves the rod continuously until the button is released. Another approach might have used a dialog box in which the direction of movement could be chosen, and then separate buttons might be provided to move incrementally, or start and stop the movement.

Recall that the communication from the lexical level to the syntactic level uses tokens. For example, the action of depressing the left mouse key over the "up" button for a particular safety rod generates the token LEXStartRod with parameters indicating which safety rod button has been pressed and that the "up" action has been selected.

```
ControlRod::ControlRod
                (TWindow* parent,      int Id,   float current, int X,
                 int Y,                int W,    int H)
               : TWindow(parent, 0, 0)

{
    up_button     = new ControlButton
                    (this,                     IDC_BUTTON_UP,
                     "up",                     BUTTON_LEFT,
                     UP_BUTTON_TOP,            BUTTON_WIDTH,
                     BUTTON_HEIGHT);

    top_light     = new Light
                    (LIGHT_LEFT                TOP_LIGHT_TOP,
                     TOP_LIGHT_RIGHT,          TOP_LIGHT_BOTTOM,
                     "TOP",                    TFont("Arial", FONT_SIZE),
                     TColor::LtGray,           TColor::LtBlue);

    CurrentControl = new TVSlider
                    (this,                     IDC_SLIDER_CURRENT,
                     CURRENT_CONTROL_LEFT,     CURRENT_CONTROL_TOP,
                     CURRENT_CONTROL_WIDTH,    CURRENT_CONTROL_HEIGHT);}
```

Fig. 5. Lexical specification fragment.

The prefix LEX on the token is used to indicate that the lexical level is the source for this particular token. The parts of the control rod syntactic and semantic specification related to this action are shown in Fig. 6.

Production 11 immediately preceding the Z schema *StartRod* will be applied when the LEXStartRod token is generated in response to the user action described above. The *StartRod* schema gives the semantic action associated with the application of this production. The precondition:

$$(WhichWay? = Up \wedge Which? \notin RodUp) \vee (WhichWay? = Down \wedge Which? \notin RodDown)$$

reflects the fact that the control rod can move up only if it is not already all the way up, and down only if it is not already down. If the precondition is satisfied, the output *Msg-ToAP!*, a message to the application, will send the operation code *APStartRod* with parameters to indicate that *Which?* is the control rod to be moved in the direction *WhichWay?*.

The syntactic level imposes the required temporal ordering on user actions and system responses. For example, the context in which the LEXStartRod token can be recognized within the grammar enforces an ordering on the user actions in the usual sense associated with a context free grammar.

In response to the message with the *APStartRod* operation code, the application will cause the indicated control rod to begin moving in the indicated direction. As the control rod moves, its new position will be communicated periodically to the user interface through the generation of a token by the application. This token is read by a

```
(1)    <control_rod_seq>    ::=    <rod_enable> <rod_move>

(2)    <rod_enable>         ::=    <rod_engage> <instrument_check>  |
(3)                                <instrument_check> <rod_engage>

(4)    <rod_engage>         ::=    RodG2RodsEngaged

(5)    <instrument_check>   ::=    ReactorG1InstrumentCheck

(6)    <rod_move>           ::=    <move_1_rod_or_2>+

(7)    <move_1_rod_or_2>    ::=    <move_1_rod>  |
(8)  .                             <move_2_rods>

(9)    <move_1_rod>         ::=    <incremental_move>  |
(10)                               <start_rod_move> <stop_rod_move>

(11)   <start_rod_move>     ::=    LEXStartRod (Which?, WhichWay?)
```

___ StartRod _____

$\Xi RodControl$

Which?	:	Rod
WhichWay?	:	Direction
MsgToAP!	:	APRodMsg

___

$(WhichWay? = Up \wedge Which? \notin RodUp) \vee (WhichWay? = Down \wedge Which? \notin RodDown)$
$MsgToAP! = (APStartRod, Which?, WhichWay?)$

Fig. 6. Part of the specification associated with moving a control rod.

separate, concurrent grammar. The relevant part of this grammar and the associated semantic specification are shown in Fig. 7. The fact that this grammar fragment is part of a separate grammar from that used for control-rod movement allows the application to update the rod-height display any number of times between the user activities corresponding to the <start_rod_move> and <stop_rod_move> nonterminals that must occur contiguously in the grammar in Fig. 6.

The response to the recognition of the APRodHeight token is the generation of a message to the presentation interpreter. This message directs that the appropriate rod height be displayed. The rod heights are also stored in a state variable, *Heights*, so that the standard operating procedure restriction on raising two rods simultaneously can be enforced. The specification of the allowable movement of two rods is shown in Fig. 8. The precondition in the first schema specifies that the action specified in the schema can be taken only when all of the rods are below the height threshold. The second schema defines an attempt to violate the standard operating procedure as an operator error.

```
(1)    <display_rod_data>    ::=    <display_rod_item>+

(2)    <display_rod_item>    ::=    <display_mag_engaged>     |
(3)                                 <display_rod_up>          |
(4)                                 <display_rod_down>        |
(5)                                 <display_rod_seated>      |
(6)                                 <display_not_engaged>     |
(7)                                 <display_not_up>          |
(8)                                 <display_not_down>        |
(9)                                 <display_not_seated>      |
(10)                                <display_rod_height>

(11)   <display_rod_height>::= APRodHeight (Which?, Height?)
```

```
┌──── DisplayRodHeight ───────────────────────────────────────────
│ Δ RodControl
│ Which?                          :    Rod
│ Height?                         :    R
│ MsgToPI!                        :    PIRodMsg
│ ────────────────────────────────────────────────
│ Heights' (Which?)      = Height?
│ MsgToPI!               = (PIDisplayRodHeight, Which?, Height?)
│
└──────────────────────────────────────────────────────────────────
```

Fig. 7. Part of the specification associated with displaying control rod information.

5   Evaluation of the Approach

Our goal with this work is to develop an approach to user interface specification that
has all the well-known benefits of formal specification and can be applied effectively
to realistic safety-critical systems. To evaluate the utility of the approach in as thor-
ough a manner as possible, we have used a simple set of evaluation criteria.

Our evaluation framework involves assessment in the following areas: expressiv-
ity, usability, changeability, implementability, analyzability, verifiability, and accuracy.
Naturally, good performance in these areas of concern were objectives when the specification approach was originally formulated and during its subsequent development—
that is in part why formal notations are used.

Of particular concern to us in this evaluation was to assess the impact of using
three quite dissimilar formal mechanisms in one specification technique. From the out-
set it became clear that using three separate notations to specify the three major aspects
of the interface provides complete separation of concerns between the levels, i.e., portions of the user interface that address different concerns are specified in distinct
pieces. This very rigid separation of concerns has yielded numerous advantages that
are discussed below.

Now that we have some experience with the approach, we are in a position to see

(1)   <start_rod_moves> ::= LEXStartRod(Which1?, Which2?, WhichWay?)

```
┌─── StartRods ─────────────────────────────────────────────────
│
│ Ξ RodControl
│ Which1?, Which2?              :    Rod
│ WhichWay?                     :    Direction
│ MsgToAP1!, MsgToAP2           :    APRodMsg
├───────────────────────────────────────────────────────────────
│
│ ∀ r : Rod | Heights(r) < HeightThreshold
│ (WhichWay? = Up ∧ Which1? ∉ RodUp) ∨
│         (WhichWay? = Down ∧ Which1? ∉ RodDown) ⇒
│         MsgToAP1! = (APStartRod, Which1?, WhichWay?)
│ (WhichWay? = Up ∧ Which2? ∉ RodUp) ∨
│         (WhichWay? = Down ∧ Which2? ∉ RodDown) ⇒
│         MsgToAP2! = (APStartRod, Which2?, WhichWay?)
│
└───────────────────────────────────────────────────────────────
```

```
┌─── StartRodsError ────────────────────────────────────────────
│
│ Ξ RodControl
│ Which1?, Which2?              :    Rod
│ WhichWay?                     :    Direction
│ MsgToAP!                      :    APErrMsg
├───────────────────────────────────────────────────────────────
│
│ ¬(∀ r : Rod | Heights(r) < HeightThreshold)
│ MsgToAP! = APStartRodError
│
└───────────────────────────────────────────────────────────────
```

Fig. 8. Part of the specification of multiple rod movement.

how well the objectives are being met, and in this section, we examine each of these areas in turn. Our evaluation is mostly subjective and limited to our experience with two systems.

5.1 Expressivity

We have found the three notations we used to be entirely adequate to express the various user interface requirements. The notations matched the needs of the various levels very well and were quite simple to use. Once all three levels had been specified we initially had some difficulty documenting the interfaces between the levels. For example, the tokens defined by the lexical level appear in the syntactic specification but not in a focused manner making understanding the interface a little difficult. This has been resolved by tabulating the token list and using cross references to the two levels.

5.2 Usability

The separation of concerns mentioned above has proved to be a major benefit in the area of usability. It facilitates validation of the interface because it allows the use of the most appropriate specification vehicle for each level and because each piece of the specification can be examined separately. Nuclear engineers have been able to follow semantic specifications in Z quite easily once the notation was explained.

The use of a formal notation that can be executed for the lexical level made part of the validation fairly simple. A screen mock-up of the graphical part of the user interface was produced by compiling and executing the specification. The executable mock-up then served as a vehicle for communication with application experts because it acts as a prototype for the presentation. The lexical level of the reactor control system has been reviewed carefully by reactor technicians twice and revised considerably each time to suit their needs.

5.3 Changeability

The separation of concerns between the specification levels provides flexibility by allowing multiple specification solutions for a particular level to be developed and prototyped without affecting the remainder of the user interface specification. This flexibility is especially useful in permitting changes in the lexical level, allowing the most natural and error-resistant presentation to be developed through prototyping and usability testing, without requiring changes in the rest of the interface specification. In the reactor specification, for example, the safety rod control system illustrates how the interfaces between levels of the specification maintain isolation thereby permitting change. We have changed the lexical specification associated with the controls to move the control rods without changing the syntactic specification about what rod movements are legal nor the semantic specification about what rod movement means. Also, if the requirement to check that the neutron flux registers at or above a threshold level is dropped, only the syntactic specification must be changed. Finally, if the restriction on the simultaneous movement of rods is dropped or changed, only the semantic specification is affected.

Needless to say, the current specification that we are using has evolved and changed considerably during development and will change in the future. These changes have been accommodated easily by the specification notations.

5.4 Implementability

The structure of the specifications yielded by this approach maps easily into an implementation structure that is easy to work with but quite different from that typically found in user-interface implementations. Most implementations are event-driven and the majority of the functionality of the user interface (including error checking and semantic interpretation) is included in the call-back functions associated with the various events. This is undesirable since it precludes the adoption of modern design techniques.

The implementation structure associated with this specification approach is, of course, also event driven but the call-back functions do no more than generate tokens. The syntax analysis is done by separate syntax analyzers and these are constructed with the typical syntax-directed-translation structure in which semantic actions are tied to specific productions. This implementation structure matches the specification structure exactly.

The lexical specification is immediately executable since it is written in C++. The syntactic specification can be implemented automatically using a parser generator and so requires no human implementation effort. The semantic specification is written in Z and maps fairly easily into an implementation in a procedural language like C++.

A complete implementation of the user interface and the associated control system have been developed although they are viewed as prototypes only at this point. The entire system operates in conjunction with the reactor simulator mentioned earlier and provides a valuable environment for discussion with users.

5.5 Analyzability

Many analyses of a specification using the approach we describe have been performed and others are possible. Clearly, various type rules and other rules of syntax are checked for the lexical specification by a C++ compiler. The use of a parser generator for implementing the syntactic specification also ensures a variety of checks on the grammars that are in the specification. Finally, all the tools that are available for analyzing Z could be applied to the semantic specification although this has not yet been done.

As well as the above, several analyses have been performed on the specification as a whole to increase confidence in the completeness and consistency of the specification. As an example, consider the token set defined by the lexical specification—the following checks have been performed by hand (but they could easily be automated): each token is generated by only one operator action; each operator action generates a token; and all tokens defined in the lexical specification are referenced in the grammars included in the syntactic specification.

The use of synthesized syntax analyzers enables other analyses to be performed. The generated syntax analyzers can determine from the grammars which tokens are valid at any given point, and can utilize the information to disable dynamically elements of the interface that are not valid at any point in time. This process of disabling parts of the interface is often referred to as "graying out" because it is usually done by showing text on menus and buttons in grey rather than black. With the necessary information obtained automatically from the syntactic specification and implemented directly from that information, this again provides a part of the implementation that is guaranteed, thereby eliminating the need for verification. As with the analyses mentioned above, at this point this analysis and synthesis has been performed by hand but could easily be automated.

Finally, we note that rigorous human inspection of the specification is facilitated by its structure and by the choice of notations used, and such inspections have been carried out.

5.6 Verifiability

The verification of an implementation derived from a user interface specification using this approach is simplified considerably by the use of an executable specification for the lexical level and by the use of a notation from which an implementation can be synthesized for the syntactic level. In both cases, verification is immediate provided the parser generators and compilers involved can be trusted (this is a separate issue that we will not discuss here).

The remaining verification issue is the verification of the semantic specification. Since this is written in Z, at least there is hope that a formal or at least a rigorous approach to verification can be undertaken. In particular, all of the tools and techniques developed for Z can be applied.

5.7 Accuracy

A significant area in which we have no results at present is the accuracy of the specification that we are building. If the formal notations and the specification structure are to be of real value, they have to contribute to a reduced rate of specification errors. Informally, we have observed that the approach we are following has yielded many questions about the interface requirements and thereby has almost certainly improved their quality.

6 Conclusions

Our experience to-date with the specification approach that we have described is very positive. The development of all three layers of the specification has been relatively straightforward, and many benefits have accrued from the formalism.

As we have noted, no new notations or novel application of existing notations are being reported here. The issues of interest were the integration of techniques and their scalability. We are confident as a result of this work that the three notations that we have used can be integrated effectively and the resulting structure can be applied to a non-trivial safety-critical system.

7 Acknowledgments

Matt Elder was heavily involved in the initial stages of this research and his contribution is gratefully acknowledged. It is a pleasure to acknowledge many helpful discussions about the user interface requirements for UVAR with a variety of our colleagues including Tom Doyle, Bo Hosticka, Don Krause, and Bob Mulder. We are also very grateful to Charles Odell and Meng Yin for their assistance in developing the lexical specification. This work was supported in part by the National Science Foundation under grant number CCR-9213427, and in part by NASA under grant number NAG1-1123-FDP.

References

1. Abowd, G., Dix, A.: Integrating status and event phenomena in formal specifications of interactive systems, Proc. FSE 2: Second ACM Sigsoft Symposium on Foundations of Software Engineering, New Orleans, LA (1994).
2. Abowd, G., et al, User interface languages: a survey of existing methods, Technical Report PRG-TR-5-89, Oxford University Computing Laboratory (1989).

3. Brilliant, S., Knight, J., Elder, M.: Formal specification of a user interface, American Nuclear Society International Topical Meeting on Nuclear Plant Instrumentation, Control, and Human Machine Interface Technologies, University Park, PA (1996)

4. Diller, A.: *Z: An Introduction to Formal Methods*, John Wiley and Sons, Inc., New York (1990).

5. Dix, A.: *Formal Methods for Interactive Systems*, Academic Press (1991).

6. Elder, M.: *Specification of User Interfaces for Safety-Critical Systems*, Technical report CS-95-30, Department of Computer Science, University of Virginia (1995).

7. Elder, M., Knight, J.: Specifying user interfaces for safety-critical medical systems, Proceedings, MRCAS '95, 1995 International Symposium on Medical Robotics and Computer Assisted Surgery, Baltimore, MD (1995).

8. Foley, J., Van Dam, A.: *Fundamentals of Interactive Computer Graphics*, pp. 217-242, Addison-Wesley Inc., New York (1982).

9. Foley, J., Wallace, V.: The art of natural graphic man-machine conversation, *Proceedings of the IEEE*, 62, 4, pp. 462-471 (1974).

10. Harrison, M., Thimbleby, H.: *Formal Methods in Human-Computer Interaction*, Cambridge University Press (1990).

11. Hartson, H., Siochi, A., Hix, A.: The UAN: A user-oriented representation for direct manipulation interface designs, *ACM Transactions on Information Systems,* 8, 3, pp. 181-203 (1990).

12. Hix, D., Hartson R: *Developing User Interfaces: Ensuring Usability Through Product and Process*, John Wiley and Sons, Inc., New York (1993).

13. Ince, D.: *An Introduction to Discrete Mathematics and Formal System Specification*, Clarendon Press (1988).

14. Jacob, R.: A specification language for direct-manipulation user interfaces, *ACM Transactions on Graphics,* 5, 4, pp. 283-317 (1986).

15. Jacob, R.: Using formal specifications in the design of a human-computer interface, *CACM* 26, 4, pp. 259-264 (1983).

16. Leveson, N.: Software safety: why, what, and how, *Computing Surveys*, 18, 2, pp. 125-163 (1986).

17. Leveson, N., Turner, C.: An investigation of the Therac 25 accidents, *IEEE Computer,* 26, 7, pp. 18-41 (1993).

18. Myers, B. et al, Garnet: Comprehensive support for graphical, highly interactive user interfaces, *IEEE Computer*, 23, 11, pp. 71-85 (1990).

19. Potter, B., et al.: *An Introduction to Formal Specification and Z*, Prentice Hall, Inc., New Jersey (1991).

20. Reisner, P.: Formal grammar and human factors design of an interactive graphics system, *IEEE Trans. on Software Engineering*, SE-7, 2, pp. 229-240 (1981).

21. Shneiderman, B.: Multiparty grammars and related features for defining interactive systems," *IEEE Transactions on Systems, Man, and Cybernetics,* 12, 2, pp. 148-154 (1982).

22. Waite, C.: electronic mail posted to safety-critical newsgroup (1996).

23. University of Virginia Reactor, *The University of Virginia Nuclear Reactor Facility Tour Information Booklet,* http://minerva.acc.virginia.edu/~reactor.

Analyzing and Refining an Architectural Style

Paolo Ciancarini and Cecilia Mascolo

Dipartimento di Scienze dell'Informazione,
Università di Bologna,
e-mail: {cianca,mascolo}@cs.unibo.it

Abstract. Architectural styles have been introduced in [1] in order to classify and analyze software architectures. In that paper, Z was used as a notation to specify and study architectural styles, however some problems remained open concerning specification and analysis of their behavioral properties. We use a new operational semantics to describe and analyze an architectural style of distributed systems. We introduce three refinements of a "Message Router" style, useful to describe distributed applications like e-mail or news systems; we also formalize and prove some properties of the style and, henceforth, of derived software architectures.

1 Introduction

Software architectures are structures that can help in designing and understanding complex software systems. Notations and tools for specifying and analyzing software architectures are currently widely investigated [17, 14, 18].

Abstractly, software architectures can be classified using the concept of "architectural style", introduced in [17] and formalized using the Z notation [1]. Architectural styles are abstractions including entities like components and connectors, and some composition rules. An architectural style is a sort of skeleton which can help in understanding and analyzing concrete software architectures which are instances of such a style. An architectural style is less constrained and less complete than a specific architecture [17]. For instance, the architectural style of *pipes and filters* can be used to study the architecture of a compiler organized as a pipeline of analysis and translation tools.

There are at least three reasons why it is important and useful to systematically study architectural styles:

1. to build a library of styles to be made available for designers, so that they can choose the most appropriate one;
2. to help designers to recognize the need to apply a specific style in a given design situation;
3. to offer designers analysis methods and tools suitable to deal with concrete software architectures, reasoning on and understanding their properties.

However, using the approach described in [1] most behavioral analyses on formal software architectures are difficult, because Z has no operational semantics.

For instance, to compare a sequential pipeline to a parallel pipeline is difficult, because in Z there is no concept of control, either sequential or parallel.

Instead, operational formalisms like the Chemical Abstract Machine [3] have been successfully used in the specification of software architectures to describe and study their behavioral aspects [16].

We show here how an approach which maps the Z notation to a formal operational semantics based on the CHAM [8] allows us to both specify and analyze dynamic properties of an architectural style using tools for both notations.

The example we consider is a style describing a Message Router, a key component of modern distributed systems. A formal specification and design of this kind of systems has to take into consideration their architectural structure. In fact, the Message Router can be abstracted by a communication style that can be formalized and analyzed as in [1].

We present three refinements of the style and study their properties at different levels of detail: note that, in this context, refinement is thought as incremental definition of a specification, as in [10]. We show that the Z formalization helps to give a modular structure to the architectural style, whereas the operational semantics supports the analysis of the behavioral aspects.

The paper has the following structure: Sec.2 contains a short description of the semantics we associate to Z and the logic we use in reasoning on our formal specifications. Sec.3 defines the architectural approach we consider; Sec.4 contains three refined versions of the Message Router style specified and analyzed. Finally, in Sec.5 we describe some conclusions and future work.

2 A chemical semantics for Z and its logic

The elementary components of any Z document are State schemas and Operation schemas. This means that we can see a Z specification as a pair: $< S, O >$, where S is the set of the State schemas and O the set of the Operation schemas.

Semantically, a State schema s in S can be seen as the set of all its possible instantiations: the standard Z semantics [19, 6, 5] is declarative and does not offer any formalization for any behavior, be it concurrent or distributed. For this reason, we introduce an operational semantics based on a chemical model.

In the Chemical Abstract Machine (CHAM) model [3] *Molecules*, *Solutions*, and *Rules* are the fundamental elements. A Chemical Abstract Machine is a triple (G, C, R), where G is a grammar, C is a set of configurations (the language generated by the grammar) or molecules, and R is a set of the rules of the form $condition(C) \times$ bag $C \times$ bag C.

A solution is a multiset of molecules, namely bag C. Two rules can fire concurrently if they do not need the same molecules to react on; hence, several rules can progress simultaneously on a solution. If two rules conflict, in the sense that they "consume" the same molecules, one of them is chosen non deterministically to react.

We give a CHAM interpretation of Z specifications which allows us to deal with concurrent dynamics. Intuitively, an instance of a state schema is associated to a solution where, in some way, each variable is a subsolution (in many cases a single molecule). Instead, an operation schema corresponds to a chemical rule where premises and consequences are solutions composed of pre and post conditions of the operation. We now describe the formalization of such an interpretation.

A molecule is a tuple of a name, a type and a value:

$$MOLECULE == NAME \times TYPE \times VALUE$$

A solution is a bag of molecules:

$$SOLUTION == \text{bag } MOLECULE$$

and a rule is composed of a conditional part that define the applicability of the rule and of two solutions that indicate molecules to delete and add to the state solution:

$$RULE == CONDITION \times SOLUTION \times SOLUTION$$

We will call the first $SOLUTION$ "pre-solution" and the second "postsolution" in order to avoid ambiguity.

A rule is applicable to a solution if the solution contains molecules that satisfy the conditional part ($CONDITION$) of the rule and molecules that match the presolutions of the rule.

The function $FSem$ associates a schema_instance to a solution:

$$Fsem : SCHEMA\_INSTANCE \rightarrow SOLUTION$$

Every identifier of the schema instance is associated to a subsolution (not necessarily a single molecule): we remark that Z sets and bags are decomposed by this function in several molecules so as to increase potential concurrency of the model.

$Fsem\_op$ associates a rule to an operation schema:

$$Fsem\_op : SCHEMA\_OP \rightarrow RULE$$

$Fsem\_Op$ associate to pre and postcondition of a schema different part of the rule:

- Every schema postcondition requiring the removal of an element from a set or bag is mapped to a presolution of the rule , namely it becomes a molecule to be deleted.
- Every postcondition that specifies the insertion of an element in a set or bag is mapped to a postsolution of the rule (molecules to be added).
- Every schema precondition that defines a membership predicate ($\in$, in) is mapped to a presolution (a removal) and to a postsolution (reinsertion) if the postcondition does not contain an indication of removal of that element.
- Postconditions containing mathematical operators ($+$, $-$,...) on naturals are encoded deleting one molecule and adding the molecule updated.

- Preconditions containing relational operators are encoded as conditions, whereas the molecule corresponding to the variable is deleted and added again, as already described. This conforms to the chemical semantics where conditions can only be stated on local molecules involved in a rule [4].

It is possible to define many other functions to describe, for instance, when rules, namely operation schemas, can fire concurrently: it usually depends on their postconditions.

2.1 The logic

We define now an *execution model*, that is a way of abstractly executing a Z specification document, and a Unity-like logic [7] to reason on properties exhibited by such a execution model.

The execution model represents the unfolding of the application of the rules of the operational semantics.

For every state schema s an abstract execution tree can be constructed: every node corresponds to a particular instance of the state schema and from each node several different applicable operation sets can exist, (chosen among all the enabled operations on that node), thus introducing non-determinism in the choice of the operations being in conflict. Each branch corresponds to the application of a group of enabled operations which could be applied without conflicts, as dictated by the Cham model.

It is now possible to formalize a logic to reason on the dynamics of a Z specification.

In order to be able to reason on dynamic properties, like deadlock and starvation, we introduce our logic borrowing a few constructs from the Unity logic. Properties are expressed as predicates related by Unity logic operators; predicates now have chemical semantics and are interpreted as chemical solutions. We can state a predicate p is valid on a state solution s if all the molecules in the chemical interpretation of p are also in s.

- p **unless** q says that whenever p is true during the execution, surely either q will become true or p continues to hold. In particular, on the tree: if p is true on some nodes then on their children q is true or p still holds.
- **Stable** is an alias for p **unless** false, that is when p becomes true it will hold forever. On the tree: if p is true on a node it will remain true for the whole subtree of that node.
- **Invariant** p says that p is true forever. That is, for every node of the execution tree p is valid.
- p **ensures** q means that when p becomes true then eventually q will hold and before that moment p is still valid. That is, if p is true on a node N, then in each branch through N there is a node M below N where q holds and on nodes between nodes N and M in the path, p holds.
- p **leads_to** q has quite the same meaning as **ensures** except that it does not ensure that p is valid until q becomes true. On the tree: if p is true on a

node N, then in each branch through N there is a node M below N where q holds.

A formalization of this interpretation can be found in [8]. Having introduced the meaning of these logic predicates on the model, we can reason and define new dynamic properties on the abstract concurrent interpretation of Z specifications.

3   The message router as an architectural style

Architectural properties are especially important in the design of a complex software system. This means that great care must be paid to behavioral aspects of the system being designed. The study of the dynamics of a software architecture is a very important phase in the development process. Other approaches to the specification of software architectures put emphasis on the analysis of system dynamics: see for instance [16] for an example of direct application of CHAM for such a task. However, we believe that a simple CHAM specification *per se* is not sufficient to offer a complete view of the style of a system architecture.

We combine the best features of two approaches: in fact we use Z for studying the static, modular architecture, whereas we use CHAM for analyzing its dynamics. In this way we obtain a coherent methodology useful to analyze all aspects of software architectures.

Let us first shortly recall the architectural ontology described in [1], which considers an abstract syntax useful for classifying architectural styles.

Components are active computational entities of an architecture; they accomplish tasks through internal computation and external communication with the rest of the system (the interaction points are the *ports*) [1].

A *component* is

```
┌─ Component ──────────────────────────────────────
│  ports : P PORT
│  descr : COMPDESC
└──────────────────────────────────────────────────
```

Connectors define the communication between components. Each connector provides a way for a collection of ports to come into contact.

A *connector* is defined as:

```
┌─ Connector ──────────────────────────────────────
│  roles : P ROLE
│  descr : CONNDESC
└──────────────────────────────────────────────────
```

Finally, a *configuration* attaches connectors to components:

```
┌─ Configuration ──────────────────────────────────────────
│ components : COMPNAME ⇸ Component
│ connectors : CONNNAME ⇸ Connector
│ attachment : RoleInst ⇸ PortInst
├──────────────────────────────────────────────────────────
│ ∀ cn : CONNNAME; r : ROLE | (cn, r) ∈ dom attachment •
│     cn ∈ dom connectors ∧ r ∈ (connectors(cn)).roles
│ ∀ cn : COMPNAME; p : PORT | (cn, p) ∈ ran attachment •
│     cn ∈ dom components ∧ p ∈ (components(cn)).ports
└──────────────────────────────────────────────────────────
```

This is a meta-description useful for defining and classifying architectural styles: for instance, it has been used to describe a pipe/filter style [1].

We apply it to the Message Router. Consider a communication network that connecting N senders to M receivers via a message router. Each sender is connected to one of the input ports of the router, whereas each receiver is connected to one output port.

The specifications of the message router given in [11, 10] use pictures to illustrate the architecture of the system and help in the understanding of its properties. In [10] the specification of an abstract router is especially detailed and six refinements are provided and studied using the Unity language.

In the first version we give here each message consists of a single packet and the router as a black box delivering messages. Messages are actually sequences composed of a sender, a receiver, a number, and a content; the number identifies different messages exchanged between the same pair sender/receiver.

$$MESSAGE == NAME \times NAME \times \mathbb{N} \times DATA$$

These are the structures of the senders and the receivers:

```
┌─ Sender_Struct ──────────────
│ Sending : P MESSAGE
│ PermMsg : P MESSAGE
│ Sender : P NAME
└──────────────────────────────
```

$PermMsg$ permanently records sent messages.

```
┌─ Receiver_Struct ────────────
│ Receiving : P MESSAGE
│ Receiver : P NAME
└──────────────────────────────
```

The Router is composed of a table that records the numbers of the last sent messages from each sender to each receiver $NumMR$; $Router$ stores routed messages.

```
┌─ Router_Struct ──────────────────────
│ NumMR : P(NAME × NAME × ℕ)
│ Router : P MESSAGE
└──────────────────────────────────────
```

The *Generic_Router* is composed of:
Sender_Struct, *Receiver_Struct*, and *Router_Struct*.

Generic_Router _____
Sender_Struct
Receiver_Struct
Router_Struct

For this style components are *Sender_Struct* and *Receiver_Struct*, whereas the connector is *Router_Struct*. Component ports are *Sending* and *Receiving* variables that record the messages to be sent and the received ones for each sender and receiver; roles are associated with *Router* variables where the routed messages for each pair of sender and receiver are stored. The configuration is associated to the *Generic_Router* schema: ports and roles are attached by messages sent for each pair of sender and receiver.

The meaning function helps in clearly define the structure of the architecture and the points of interaction.

Following [1], we give the formalization of the mapping from the abstract syntax to our specification:

$\mathcal{M}_{Comp} : Component \rightarrow Struct$
$\mathcal{M}_{port} : PortInst \rightarrow MESSAGE$

$\forall c : Component;\ n : COMPNAME;\ st : Struct;$
$\quad s : Sender\_Struct \mid c \in \operatorname{dom}\mathcal{M}_{Comp} \wedge$
$\quad\quad S(s) = st \bullet s.Sending = \mathcal{M}_{port}(\!|\ \{n\} \times c.ports\ |\!)$
$\forall c : Component;\ n : COMPNAME;\ st : Struct;$
$\quad r : Receiver\_Struct \mid c \in \operatorname{dom}\mathcal{M}_{Comp} \wedge$
$\quad\quad R(r) = st \bullet r.Receiving = \mathcal{M}_{port}(\!|\ \{n\} \times c.ports\ |\!)$
---

where
$$PortInst == COMPNAME \times PORT$$

$$Struct ::= S\langle\!\langle Sender\_Struct \rangle\!\rangle$$
$$\mid\ R\langle\!\langle Receiver\_Struct \rangle\!\rangle$$

The mapping function $\mathcal{M}$ associates the variables *Sending* and *Receiving* to component ports.

$\mathcal{M}_{Conn} : Connector \rightarrow Router\_Struct$
$\mathcal{M}_{role} : RoleInst \rightarrow MESSAGE$

$\forall c : Connector;\ n : CONNNAME;\ r : Router\_Struct \mid$
$\quad c \in \operatorname{dom}\mathcal{M}_{Conn} \wedge r = \mathcal{M}_{Conn}(c)$
$\quad \bullet\ r.Router = \mathcal{M}_{role}(\!|\ \{n\} \times c.roles\ |\!)$
---

where

$$RoleInst == CONNNAME \times ROLE$$

The following function associates variable *messages* to a connector role.

$$\mathcal{M}_{Conf} : Configuration \rightarrow Generic\_Router$$

$\forall\, cfg : \text{dom}\,\mathcal{M}_{Conf} \bullet (\mathcal{M}_{Conf}.Sender\_Struct =$
$\quad \{n : COMPNAME;\ c : Component;\ s : Sender\_Struct \mid$
$\quad (n, c) \in cfg.components \wedge s \in \mathcal{M}_{Comp}(c)$
$\quad \wedge\, s.Sending = \mathcal{M}_{port}(\!|\,\{n\} \times c.ports\,|\!) \bullet s\})$
$\quad \wedge\, Same\_definition\_for\_Receiver\_Struct$
$\quad \wedge\, (\mathcal{M}_{Conf}.Router\_Struct =$
$\quad \{n : CONNNAME;\ c : Connector;\ s : Router\_Struct \mid$
$\quad (n, c) \in cfg.connectors \wedge s \in \mathcal{M}_{Conn}(c)$
$\quad \wedge\, s.Router = \mathcal{M}_{roles}(\!|\,\{n\} \times c.roles\,|\!) \bullet s\})$

Send_Message is the schema describing the operation to transmit a message from a sender to a receiver, provided that such a message is not the first one from that sender to that receiver (a special operation is defined for the first message).

Send_Message

$\Delta\, Generic\_Router$

$\exists\, s, r : NAME;\ n : \mathbb{N};\ dat : DATA \mid (s, r, n, dat) \in Sending\ \wedge$
$\quad s \in Sender \wedge (s, r, n - 1) \in NumMR \bullet$
$\quad Router' = Router \cup \{(s, r, n, dat)\}$
$\quad Sending' = Sending \setminus \{(s, r, n, dat)\}$

This is the chemical rule which denotes the semantics of such an operation schema.

This is the pre-solution:

$(Sending, \mathbb{P}\, MESSAGE, (s, r, n, dat)),$
$(Sender, \mathbb{P}\, NAME, s),$
$(NumMR, \mathbb{P}(NAME \times NAME \times \mathbb{N}), (s, r, n - 1))$

This is the post-solution:

$(Router, \mathbb{P}\, MESSAGE, (s, r, n, dat)),$
$(Sender, \mathbb{P}\, NAME, s),$
$(NumMR, \mathbb{P}(NAME \times NAME \times \mathbb{N}), (s, r, n - 1))$

The following operation delivers the message:

Deliver_Message

$\Delta\, Generic\_Router$

$\exists\, s, r : NAME;\ n : \mathbb{N};\ dat : DATA \mid (s, r, n, dat) \in Router\ \wedge$
$\quad r \in Receiver \wedge (s, r, n - 1) \in NumMR \bullet$
$\quad Router' = Router \setminus \{(s, r, n, dat)\} \wedge$
$\quad Receiving' = Receiving \cup \{(s, r, n, dat)\}$
$\quad \wedge\, NumMR' = NumMR \setminus \{(s, r, n - 1)\} \cup \{(s, r, n)\}$

This is the chemical rule which denotes the semantics of such an operation schema. This is the pre-solution:

$(Router, \mathsf{P}\ MESSAGE, (s, r, n, dat)),$
$(Receiver, \mathsf{P}\ NAME, r),$
$(NumMR, \mathsf{P}(NAME \times NAME \times \mathsf{N}), (s, r, n - 1))$

This is the post-solution:

$(Receiving, \mathsf{P}\ MESSAGE, (s, r, n, dat)),$
$(Receiver, \mathsf{P}\ NAME, r),$
$(NumMR, \mathsf{P}(NAME \times NAME \times \mathsf{N}), (s, r, n))$

We omit the formalization of the initialization operation introducing the senders, the receivers, and the messages to be sent.

3.1 Analysis

We consider some classes of properties already introduced in [11]. However, our analysis is quite different; in fact, in [11] properties themselves define the specification of the system, while here they are introduced only to analyze behavioral features of a Z specification document.

We will study the same properties for all refinements, showing how the granularity of detail used for describing the software architecture influences its analysis.

The classes of properties we consider are the following:

1. **Eventual Delivery**: every sent message will eventually be delivered.
2. **Order Preserving**: the order of the messages sent by the same sender to the same receiver will be preserved.
3. **Prefix Invariant**: delivered messages have been messages to be sent.

This is the analysis of the first refinement:

- **Class Eventual Delivery**

Theorem 1. $(s, r, n, dat) \in Sending$ **ensures** $(s, r, n, dat) \in Router$

That is: a message to be sent will eventually be routed.

Theorem 2. $(s, r, n, dat) \in Router$ **ensures** $(s, r, n, dat) \in Receiving$

That is: a message routed will eventually be delivered.

Theorem 3. $(s, r, n, dat) \in Sending$ **leads_to** $(s, r, n, dat) \in Receiving$

That is: a message to be sent will eventually be delivered; this theorem in some sense summarizes the previous two theorems.

Proof of Theorem 1:

To prove p **ensures** q

(where p is $(s, r, n, dat) \in Sending$ and q is $(s, r, n, dat) \in Router$) we must prove p **unless** q and that a set of operations exists, which when it is applied to a state where $p \wedge \neg q$ is valid leads to a state where q holds.

We now prove p **unless** q; for every operations set enabled on the solution containing the molecule $(Sending, \mathbf{P} \ MESSAGE, (s, r, n, dat))$ the application has to generate a state where the molecule is not present and $(Router, \mathbf{P} \ MESSAGE, (s, r, n, dat))$ is present or the previous molecule is still in the solution (in fact, this is the semantics of **unless**).

Considering our case study, we remark that applying operations on a different sender or receiver we reach a state where p still holds, while applying operation $Send\_Message$ with sender s and receiver r, we obtain a state where q holds. Then p **unless** q holds.

We now prove the second part of our theorem: given a state where $p \wedge \neg q$ holds, there exists an enabled set of applicable operations, which generates a state where q holds. Such a set is composed of operation $Send\_Message$ acting on the molecule

$(Sending, \mathbf{P} \ MESSAGE, (s, r, n, dat))$

and other operations not acting on that molecule. Since our CHAM model is *fair*, we can state that the set will eventually be applied. This completes the proof.

Proofs of Theorems 2 and 3 are similar: we omit them.

- **Class Order Preserving**

Theorem 4. $(s, r, n, dat) \in Sending$ **unless** $(s, r, n - 1, dat1) \in Receiving$

That is: a message will be delivered only after the message previously sent from that sender to the same receiver has been delivered.

Proof of Theorem 4:

Let p be the predicate $(s, r, n, dat) \in Sending$ and q be $(s, r, n - 1, dat) \in Receiving$.

We have to prove that, given a state where p holds, every set of applicable operations generates a state where p is still valid or q holds. The premise of the operation $Send\_Message$ forbids the sending of a message if the previous message from that sender to that receiver has not been sent (the check is on variable $NumMR$). Then, no operations set can act on message (s, r, n, dat) until the message $(s, r, n - 1, dat1)$ has been sent.

Predicate p continues to hold for operations in the set which do not act on messages from same sender to same receiver. In the other cases, acting on message $(s, r, n - 1, dat')$, will make true q.

- **Class Prefix Invariant**

Theorem 5. Invariant *Receiving* ⊆ *PermSMsg*

That is: messages received had previously been in the set of messages to be sent. Variable *PermSMsg* records messages to be sent but it differs from *SendMsg* because messages are not removed from it: it plays the role of a log file.

Proof of Theorem 5:

Let us name p the invariant. The initialization operation sets the variable *Receiving* to void. Thus, after the initialization the property holds. Then we have to prove that the property is stable, that is, when p holds on a state the application of every set of operations leads to a state where p still holds. This is true because the operation that increases the cardinality of *Receiving*(*Deliver_Message*) checks that the message has been in *Router* set. The operation that increases the cardinality of *Router* (*Send_Message*) also checks that the message has been in *Sending*. Finally, the operation updating *Sending* (not formally described here) also updates *PermSMsg*. Hence, the theorem follows.

4 Two refinements

In the second version the messages are no more monolithic packets: they are composed of a header packet, a set of data packets, and a trailer packet.

In the third refined version the router itself is refined and seen as a grid on which packets flow and switch to reach their destinations.

4.1 Second version: token-level routing

In this version the router is still a black box delivering messages, but messages are refined: they are decomposed in packets. Thus, each message consists of a header packet, a sequence of data packets, and a trailer packet.

We study the architectural style of this refinement and notice that components still correspond to *Sender_Struct* and *Receiver_Struct*, and connectors to *Router_Struct*. However, there are some changes in the ports and roles structures. In fact we introduce different ports and roles for different packet types.

Components ports are variables *HeaderSend*, *ValueSend*, *TrailerSend*, *HeaderRec*, *ValueRec*, *TrailerRec* and connector roles are *HeaderRouter*, *ValueRouter*, and *TrailerRouter*. We omit the formalization of the mapping function.

$HEADER == NAME \times NAME \times \mathsf{N}$
$VALUE == NAME \times NAME \times \mathsf{N} \times DATA \times \mathsf{N}$
$TRAILER == NAME \times NAME \times \mathsf{N}$

Sender records the number of packets sent for each type and sender.

```
__ Sender_Struct _____
  HeaderSend : P HEADER
  ValueSend : P VALUE
  TrailerSend : P TRAILER
  PermSH : P HEADER
  PermSV : P VALUE
  PermST : P TRAILER
  Sender : P NAME × N × N × N
```

"*PermXX*" variables record packets sent.

```
__ Receiver_Struct _____
  HeaderRec : P HEADER
  ValueRec : P VALUE
  TrailerRec : P TRAILER
  Receiver : P NAME × N × N × N
```

The router keeps trace of the last packets sent from any sender to any receiver.

```
__ Router_Struct _____
  HeaderRouter : P HEADER
  ValueRouter : P VALUE
  TrailerRouter : P TRAILER
```

The transmission operation for the header packet is described in the following schema.

```
__ Transmit_Header _____
  ΔGeneric_Router
 _____
  ∃ s, r : NAME;  n : N | (s, r, n) ∈ HeaderSend ∧ s ∈ Sender
     ∧ (s, r, n − 1) ∈ TrailerRouter •
  HeaderSend′ = HeaderSend \ {(s, r, n)}
     ∧ HeaderRouter′ = HeaderRouter ∪ {(s, r, n)} \ {(s, r, n − 1)}
```

This is the corresponding rule with pre-solution:

$(HeaderSend, \mathbf{P}\ HEADER, (s, r, n))$,
$(Sender, \mathbf{P}\ NAME, s)$,
$(TrailerRouter, \mathbf{P}\ TRAILER, (s, r, n − 1))$,
$(HeaderRouter, \mathbf{P}\ HEADER, (s, r, n − 1))$

and post-solutions:

$(Sender, \mathbf{P}\ NAME, s)$,
$(TrailerRouter, \mathbf{P}\ TRAILER, (s, r, n − 1))$,
$(HeaderRouter, \mathbf{P}\ HEADER, (s, r, n))$

The following schema specifies the delivery of a header packet:

```
┌─ Deliver_Header ──────────────────────────────────────────
│ ΔGeneric_Router
├────────────────────────────────────────────────────────────
│ ∃ s, r : NAME;  n : N | (s, r, n) ∈ HeaderRouter ∧ r ∈ Receiver •
│     HeaderRec' = HeaderRec ∪ {(s, r, n)}
└────────────────────────────────────────────────────────────
```

This is the corresponding CHAM rule with pre-solution:

$(HeaderRouter, \mathbf{P}\ HEADER, (s, r, n))$,
$(Receiver, \mathbf{P}\ NAME, r)$,

and post-solution:

$(Receiver, \mathbf{P}\ NAME, r)$,
$(HeaderRouter, \mathbf{P}\ HEADER, (s, r, n))$
$(HeaderRec, \mathbf{P}\ HEADER, (s, r, n))$

The following schema specifies the sending of data packets.

```
┌─ Transmit_Data ───────────────────────────────────────────
│ ΔGeneric_Router
├────────────────────────────────────────────────────────────
│ ∃ s, r : NAME;  n, d : N;  dat, dat1 : DATA |
│     (s, r, n, dat, d) ∈ ValueSend ∧
│     s ∈ Sender ∧ (s, r, n) ∈ HeaderRouter ∧
│     (s, r, n, dat1, d − 1) ∈ ValueRouter •
│     ValueRouter' = ValueRouter ∪ {(s, r, n, dat, d)} ∧
│     ValueSend' = ValueSend \ {(s, r, n, dat, d)}
└────────────────────────────────────────────────────────────
```

Here is the corresponding rule with pre-solution:

$(ValueSend, \mathbf{P}\ VALUE, (s, r, n, dat, d))$,
$(Sender, NAME, s)$,
$(HeaderRouter, \mathbf{P}\ HEADER, (s, r, n))$,
$(ValueRouter, \mathbf{P}\ VALUE, (s, r, n, dat1, d − 1))$

and post-solution:

$(ValueRouter, \mathbf{P}\ VALUE, (s, r, n, dat, d))$,
$(Sender, NAME, s)$,
$(HeaderRouter, \mathbf{P}\ HEADER, (s, r, n))$

The following schema defines the delivery of value packets:

```
┌─ Deliver_Data ────────────────────────────────────────────
│ ΔGeneric_Router
├────────────────────────────────────────────────────────────
│ ∃ s, r : NAME;  n, d : N;  dat, dat1 : DATA |
│     (s, r, n, dat, d) ∈ ValueRouter ∧
│     r ∈ Receiver • ValueRec' = ValueRec ∪ {(s, r, n, dat, d)}
└────────────────────────────────────────────────────────────
```

Here is the corresponding rule with pre-solution:

> $(Receiver, NAME, r),$
> $(ValueRouter, \mathbf{P}\ VALUE, (s, r, n, dat, d))$

and post-solution:

> $(Receiver, NAME, r),$
> $(ValueRouter, \mathbf{P}\ VALUE, (s, r, n, dat, d)),$
> $(ValueRec, \mathbf{P}\ VALUE, (s, r, n, dat, d))$

We omit the specification of the transmission of Trailer packets.

4.2 Analysis

- **Class Eventual Delivery**
  Theorem 1 in the first refinement version is translated with three theorems depending on the different kind of packets.
  - **Theorem 6.** $(s, r, n) \in HeaderSend$ **ensures** $(s, r, n) \in HeaderRouter$
  - **Theorem 7.** $(s, r, n, dat, d) \in ValueSend$ **ensures** $(s, r, n, dat, d) \in ValueRouter$
  - **Theorem 8.** $(s, r, n) \in TrailerSend$ **ensures** $(s, r, n) \in TrailerRouter$

  These properties simply ensure that for every kind of packets, a packet to be sent will eventually be in the set of routed packets.

  Proof of Theorem 6:
  Let p is $(s, r, n) \in HeaderSend$. First we want to prove p **unless** q. For every operations set applicable to a solution containing molecule
  $(HeaderSend, \mathbf{P}\ Header, (s, r, n)),$
  we have to prove that p still holds after the application. This is true: in fact, all the operations not acting on that header packet do not add that message to the set of routed messages, while the operation $Transmit\_Header$ acting on that header, modifies the solution making q valid. Thus p **unless** q holds. Furthermore, the set of operations that from a state where p holds generates a state where q holds is the set containing operation $Transmit\_Header$ acting on the considered molecule.
  The proofs of theorems 7 and 8 are similar: we omit them.
  - **Theorem 9.** $(s, r, n) \in HeaderRouter$ **ensures** $(s, r, n) \in HeaderRec$
  - **Theorem 10.** $(s, r, n, dat, d) \in ValueRouter$ **ensures** $(s, r, n, dat, d) \in ValueRec$
  - **Theorem 11.** $(s, r, n) \in TrailerRouter$ **ensures** $(s, r, n) \in TrailerRec$
  - **Theorem 12.** $(s, r, n) \in HeaderSend$ **leads_to** $(s, r, n) \in HeaderRec$

  The proofs of these theorems are similar to the one of Theorem 6.
- **Class Order Preserving**
  Theorem 4 is refined here in four properties.
  - **Theorem 13.** $(s, r, n) \in HeaderSend$ **unless** $(s, r, n - 1) \in TrailerRec$
  - **Theorem 14.** $(s, r, n, dat, 1) \in ValueSend$ **unless** $(s, r, n) \in HeaderRec$

- **Theorem 15.** $(s, r, n, dat, d) \in ValueSend$ **unless**
  $(s, r, n, dat, d - 1) \in ValueRec$
- **Theorem 16.** $(s, r, n) \in TrailerSend$ **unless**
  $(s, r, n, dat, d) \in ValueRec$

The meaning of theorem 9 is: a header packet to be sent is not delivered until the trailer packet of the previous message from that sender to that receiver has been delivered as well. The other theorems have similar meanings except that they refer to different kinds of packets.

- **Class Prefix Invariant**
  - **Theorem 17.** invariant$HeaderRec \subseteq PermSH$
  - **Theorem 18.** invariant$ValueRec \subseteq PermSV$
  - **Theorem 19.** invariant$TrailerRec \subseteq PermST$

These invariants are refinements of the invariants of Theorem 4.

4.3 Third version: grid-level routing

In this third version the router is a grid where packets flow until they do reach their destination. Packets first move right on the grid; then, when the column corresponding to the receiver is reached, they switch. *Sender_Struct* and *Receiver_Struct* are the same as in the second version, except that s and r are identified by a name and a number (row for sender and column for receiver).

Here is the refined version for the router: *HeaderTrans*, *ValueTrans* and *TrailerTrans* help to store the passage of variables through the grid nodes.

```
┌─ Router_Struct ─────────────────────────────────
│ HeaderRouter : P(N × N × HEADER)
│ ValueRouter : P(N × N × VALUE)
│ TrailerRouter : P(N × N × TRAILER)
│ HeaderTran : P(N × N × HEADER)
│ ValueTran : P(N × N × VALUE)
│ TrailerTran : P(N × N × TRAILER)
└─────────────────────────────────────────────────
```

The formalization of the architectural style is refined as well: in fact the ports of the components are attached to the external nodes of the grid. The meaning function shows how interactions between components and connectors have been modified:

```
│ M_Conn : Connector ⇸ Router_Struct
│ M_role : RoleInst ⇸ MESSAGE
├──────────────────────────────────────────────────
│ ∀ c : Connector; n : CONNNAME; r : Router_Struct |
│   c ∈ dom M_Conn ∧ r = M_Conn(c)
│   ∧ HeadSub = {∀ n : N; h : HEADER | n < NumReceiver • (n, 1, h)}
│   ∧ ValueSub = {∀ n : N; v : VALUE | n < NumReceiver • (n, 1, v)}
│   ∧ TrailSub = {∀ n : N; t : TRAILER | n < NumReceiver • (n, 1, t)}
│   ∧ HeadSub ⊂ r.HeaderRouter ∧ ValueSub ⊂ r.ValueRouter ∧
│   TrailSub ⊆ TrailerRouter •
│   (HeadSub ∪ ValueSub ∪ TrailSub) = M_role(| {n} × c.roles |)
```

The following schema defines the transmission of header packets toward the right node:

```
┌─ Transmit_Header_Right ─────────────────────────────────
│ ΔGeneric_Router
├──────────────────────────────────────────────────────────
│ ∃ s, r : NAME; i, j, n, nr : N | (i, j, (s, r, n)) ∈ HeaderRouter ∧
│     j ≠ ReceiverNum ∧ (r, nr) ∈ Receiver ∧ j < nr ∧
│     (i, j + 1, (s, r, n − 1)) ∈ TrailerTran •
│     HeaderTran′ = HeaderTran ∪ {(i, j, (s, r, n))} ∧
│     HeaderRouter′ = HeaderRouter \ {(i, j, (s, r, n))}
│     ∪{(i, j + 1, (s, r, n))}
└──────────────────────────────────────────────────────────
```

Here is the corresponding rule with condition:

$j \neq number \wedge nr > j$

pre-solution:

$(HeaderRouter, \mathsf{P}(\mathsf{N} \times \mathsf{N} \times HEADER), (i, j, (s, r, n))),$
$(ReceiverNum, \mathsf{N}, number),$
$(Receiver, \mathsf{P}(NAME, \mathsf{N}), (r, nr)),$
$(TrailerRouter, \mathsf{P}(\mathsf{N} \times \mathsf{N} \times TRAILER), (i, j + 1, (s, r, n − 1)))$

and post-solution:

$(ReceiverNum, \mathsf{N}, number),$
$(Receiver, \mathsf{P}(NAME, \mathsf{N}), (r, nr)),$
$(TrailerRouter, \mathsf{P}(\mathsf{N} \times \mathsf{N} \times TRAILER), (i, j + 1, (s, r, n − 1)))$
$(HeaderTran, \mathsf{P}(\mathsf{N} \times \mathsf{N} \times HEADER), (i, j, (s, r, n))),$
$(HeaderRouter, \mathsf{P}(\mathsf{N} \times \mathsf{N} \times HEADER), (i, j + 1, (s, r, n)))$

The following schema specifies the transmission of header packets toward the down node:

```
┌─ Transmit_Header_Down ──────────────────────────────────
│ ΔGeneric_Router
├──────────────────────────────────────────────────────────
│ ∃ s, r : NAME; i, n, nr : N | (i, nr, (s, r, n)) ∈ HeaderRouter ∧
│     i ≠ SenderNum ∧ (r, nr) ∈ Receiver ∧
│     (i + 1, nr, (s, r, n − 1)) ∈ TrailerTran •
│     HeaderTran′ = HeaderTran ∪ {(i, nr, (s, r, n))} ∧
│     HeaderRouter′ = HeaderRouter \ {(i, nr, (s, r, n))}
│     ∪{(i + 1, nr, (s, r, n))}
└──────────────────────────────────────────────────────────
```

Here is the corresponding rule with condition:

$i \neq number,$

pre-solution:

$(HeaderRouter, \mathsf{P}(\mathsf{N} \times \mathsf{N} \times HEADER), (i, nr, (s, r, n)))$,
$(SenderNum, \mathsf{N}, number)$,
$(Receiver, \mathsf{P}(NAME \times \mathsf{N}), (r, nr))$,
$(TrailerTran, \mathsf{P}(\mathsf{N} \times \mathsf{N} \times TRAILER), (i + 1, nr, (s, r, n - 1)))$

and post-solution:

$(SenderNum, \mathsf{N}, number)$,
$(Receiver, \mathsf{P}(NAME \times \mathsf{N}), (r, nr))$,
$(TrailerTran, \mathsf{P}(\mathsf{N} \times \mathsf{N} \times TRAILER), (i + 1, nr, (s, r, n - 1)))$
$(HeaderTran, \mathsf{P}(\mathsf{N} \times \mathsf{N} \times HEADER), (i, nr, (s, r, n)))$,
$(HeaderRouter, \mathsf{P}(\mathsf{N} \times \mathsf{N} \times HEADER), (i + 1, nr, (s, r, n)))$

The following schema defines the sending of data packets toward right:

$$\begin{array}{l}
\rule{0pt}{1em}\underline{\quad Transmit\_Data\_Right \rule{6cm}{0pt}} \\
\quad \Delta\, Generic\_Router \\
\hline
\quad \exists\, s, r : NAME;\ i, j, n, nr, d : \mathsf{N};\ dat, dat1 : DATA \mid \\
\qquad (i, j, (s, r, n, dat, d)) \in ValueRouter \wedge \\
\qquad (i, j, (s, r, n)) \in HeaderTran \wedge \\
\qquad (i, j, (s, r, n, dat1, d - 1)) \in ValueTran \wedge j \neq ReceiverNum \wedge \\
\qquad (r, nr) \in Receiver \wedge j < nr \bullet \\
\qquad ValueTran' = ValueTran \cup \{(i, j, (s, r, n, dat, d))\} \wedge \\
\qquad ValueRouter' = ValueRouter \setminus \{(i, j, (s, r, n, dat, d))\} \\
\qquad \cup\{(i, j + 1, (s, r, n, dat, d))\}
\end{array}$$

Here is the corresponding rule with condition:
$number \neq i \wedge nr > j$
 pre-solution:

$(ReceiverNum, \mathsf{N}, number)$,
$(Receiver, \mathsf{P}(SENDREC \times \mathsf{N}), (r, m, m1, m2, nr))$,
$(HeaderTran, \mathsf{P}(\mathsf{N} \times \mathsf{N} \times HEADER), (i, j, (s, r, n)))$,
$(ValueTran, \mathsf{P}(\mathsf{N} \times \mathsf{N} \times VALUE), (i, j, (s, r, n, dat1, d - 1)))$,
$(ValueRouter, \mathsf{P}(\mathsf{N} \times \mathsf{N} \times VALUE), (i, j, (s, r, n, dat, d)))$

and post-solution:

$(ReceiverNum, \mathsf{N}, number)$,
$(Receiver, \mathsf{P}(SENDREC \times \mathsf{N}), (r, m, m1, m2, nr))$,
$(HeaderTran, \mathsf{P}(\mathsf{N} \times \mathsf{N} \times HEADER), (i, j, (s, r, n)))$,
$(ValueTran, \mathsf{P}(\mathsf{N} \times \mathsf{N} \times VALUE), (i, j, (s, r, n, dat, d)))$,
$(ValueRouter, \mathsf{P}(\mathsf{N} \times \mathsf{N} \times VALUE), (i, j + 1, (s, r, n, dat, d)))$

Transmission of data packets toward the down node and transmission of the trailer packets is similar: we omit it to save space.

4.4 Analysis

- **Class Eventual Delivery**
  - **Theorem 20.** $(s, r, n) \in HeaderSend \land (s, b) \in Sender$
    ensures $(b, 1, (s, r, n)) \in HeaderRouter$
    We can write similar theorems for every node in the grid. We here write only the one concerning nodes on the last row:
  - **Theorem 21.** $(s, r, n) \in HeaderSend \land (s, b) \in Sender \land (r, m) \in Receiver$
    ensures $(b, m, (s, r, n)) \in HeaderRouter$
  - **Theorem 22.** $(s, r, n) \in HeaderSend$ **leads_to** $(s, r, n) \in HeaderRec$

  Theorems 20, 21 and 22 can be stated and proved also for value and trailer packets.
- **Class Order Preserving**
  - **Theorem 23.** $(s, r, n) \in HeaderSend$ **unless**
    $(ns, 1, (s, r, n - 1)) \in TrailerTran$
  - **Theorem 24.** $(s, r, n, dat, 1) \in ValueSend$ **unless** $(ns, 1, (s, r, n)) \in HeaderTran$
  - **Theorem 25.** $(s, r, n, dat, d) \in ValueSend$ **unless**
    $(ns, 1, (s, r, n, dat, d - 1)) \in ValueTran$
  - **Theorem 26.** $(s, r, n) \in TrailerSend$ **unless** $(ns, 1, (s, r, n, dat, d)) \in ValueTran$

  These properties ensure that each packet can be routed on the first node on the grid only if the previous packet has been transmitted ahead. These properties can be stated for every node of the grid and for any kind of packets.
- **Class Prefix Invariant**
  - **Theorem 27.** **invariant** $HeaderRec \subseteq PermSH$
  - **Theorem 28.** **invariant** $ValueRec \subseteq PermSV$
  - **Theorem 29.** **invariant** $TrailerRec \subseteq PermST$

  These are the usual properties already stated for the other refinements.

5 Conclusions and future work

The introduction of a chemical semantics for Z offers a formal basis to analyze dynamic properties of non sequential systems [8]. In particular, in this paper we have shown how we can analyze the dynamics of a distributed software architecture specified in Z.

Z was suggested quite early as a means for specifying software architectures; for instance, in [9] it was used to study blackboard systems, that are quite common for some classes of AI applications, as those concerning speech recognition. Even the use of Z to specify architectural styles is not new: for instance, in [1] two different styles are described using Z. However, we think that our new chemical semantics helps in specifying better and more precisely behavioral aspects of architectures. The CHAM itself has been used to study software architectures

[16], however such a notation is not as expressive and modular as Z; the CHAM lacks features especially for describing abstract data structures and their properties. We have chosen an integration approach, which combines advantages of both notations to obtain useful architectural styles descriptions.

Our work can be compared with works which integrate Z with other notations. For instance, in [2] Z schemas are combined with annotations written in Hoare's CSP. CSP is used to specify behavior in form of abstract operations, while Z is used for detailing design features of the system architecture and its main data structures. Such an integration is very low level and not formally specified. In [15] Petri Nets are used to formalize control flows, causal relations, and dynamic behavior of systems statically specified using Z; however, there is no formalization of the interaction between the two notations. [13] offers a more formal model of integration of Z with Petri Nets: Petri Nets are mapped on Z specifications so that graphical representation given by Petri Nets can be used to visualize Z specified systems, yet we think this approach is not giving a formal semantics basis for Z but only a visualizing method. A simpler approach has been suggested by Evans himself in [12]. He uses a Unity like logic [7] to formalize properties on the behavior of systems; an interleaving model with atomic operation interpretation is given but not formalized. Our approach shows how the simplicity and conciseness of Unity logic constructs fit quite well with the operational semantics based on CHAM.

The software architecture we have considered, namely the Message Router, has been studied in [11] and [10]. We deal with it as an architectural style because we have found that it can be the basis for designing other software systems like e-mail systems or news systems.

We are developing a model checking framework based on the CHAM semantics and on the logic given in order to make the verification process automatic; we have already built a symbolic animator of specifications based on the chemical semantics [8].

Acknowledgements. Partial support for this work was provided by the Commission of European Union under ESPRIT Programme Basic Research Project 9102 (COORDINATION), and by the Italian MURST 40%- "Progetto Ingegneria del Software".

References

1. G. Abowd, R. Allen, and D. Garlan. Formalizing Style to Understand Descriptions of Software Architecture. *ACM Transactions on Software Engineering and Methodology*, 4(4):319–364, October 1995.
2. M. Benjamin. A Message Passing System. An example of combining Z and CSP. In J. Nicholls, editor, *Proc. 4th Z Users Workshop (ZUM89)*, Workshops in Computing, pages 221–228, Oxford, 1989. Springer-Verlag, Berlin.
3. G. Berry and G. Boudol. The Chemical Abstract Machine. *Theoretical Computer Science*, 96:217–248, 1992.
4. G. Boudol. Some Chemical Abstract Machines. In J. deBakker, W. deRoever, and G. Rozenberg, editors, *A Decade of Concurrency*, volume 803 of *Lecture Notes in Computer Science*, pages 92–123. Springer-Verlag, Berlin, 1993.

5. P. Breuer and J. Bowen. Towards Correct Executable Semantics for Z. In J. Bowen and J. Hall, editors, *Proc. 8th Z Users Workshop (ZUM94)*, Workshops in Computing, pages 185–212, Cambridge, 1994. Springer-Verlag, Berlin.

6. S. Brien and J. Nicholls. Z Base Standard, November 1992. Programming Research Group.

7. K. M. Chandy and J. Misra. *Parallel Programming Design*. Addison-Wesley, 1988.

8. P. Ciancarini, S. Cimato, and C. Mascolo. Engineering Formal Requirements: an Analysis and Testing Method for Z Documents. *Annals of Software Engineering*, (to appear), 1997.

9. I. Craig. *Formal Specification of Advanced AI Architectures*. Ellis Horwood, Chichester, 1991.

10. C. Creveuil and G. Roman. Formal Specification and Design of a Message Router. *ACM Transactions on Software Engineering and Methodology*, 3(4):271–307, October 1994.

11. H. Cunningham and Y. Cai. Specification and Refinement of a Message Router. In *Proc. 7th IEEE Int. Workshop on Sw Specification and Design*, pages 20–29. IEEE Computer Society Press, December 1993.

12. A. Evans. Specifying and Verifying Concurrent Systems Using Z. In M. Bertran, T. Denvir, and M. Naftalin, editors, *Proc. FME'94 Industrial Benefits of Formal Methods*, volume 873 of *Lecture Notes in Computer Science*, pages 366–380. Springer-Verlag, Berlin, 1994.

13. A. Evans. Visualizing Concurrent Z Specifications. In J. Bowen and J. Hall, editors, *Proc. 8th Z Users Workshop (ZUM94)*, Workshops in Computing, pages 269–281, Cambridge, 1994. Springer-Verlag, Berlin.

14. D. Garlan and M. Shaw. An Introduction to Software Architecture. In V. Ambriola and G. Tortora, editors, *Advances in Software Engineering and Knowledge Engineering*, pages 1–40. World Scientific Publishing Co., 1992.

15. X. He. PZ Nets: A Formal Method Integrating Petri Nets with Z. In *Proc. 7th Int. Conf. on Software Engineering and Knowledge Engineering*, pages 173–180, Rockville, Maryland, 1995. Knowledge Systems Institute.

16. P. Inverardi and A. Wolf. Formal Specification and Analysis of Software Architectures Using the Chemical Abstract Machine Model. *IEEE Transactions on Software Engineering*, 21(4):373–386, April 1995.

17. D. Perry and A. Wolf. Foundations for the Study of Software Architecture. *ACM SIGSOFT Software Engineering Notes*, 17(4):40–52, October 1992.

18. M. Shaw and D. Garlan. Formulations and Formalisms in Software Architecture. In J. vanLeeuwen, editor, *Computer Science Today*, volume 1000 of *Lecture Notes in Computer Science*, pages 307–323. Springer-Verlag, Berlin, 1995.

19. J. Spivey. *Understanding Z*. Cambridge Tracts in Theoretical Computer Science. Cambridge University Press, 1988.

Weak Refinement in Z

John Derrick, Eerke Boiten, Howard Bowman and Maarten Steen*

Computing Laboratory, University of Kent, Canterbury, CT2 7NF, UK.
(Phone: + 44 1227 764000, Email: J.Derrick@ukc.ac.uk.)

Abstract. An important aspect in the specification of distributed systems is the role of the internal (or unobservable) operation. Such operations are not part of the user interface (i.e. the user cannot invoke them), however, they are essential to our understanding and correct modelling of the system. Various conventions have been employed to model internal operations when specifying distributed systems in Z. If internal operations are distinguished in the specification notation, then refinement needs to deal with internal operations in appropriate ways. However, in the presence of internal operations, standard Z refinement leads to undesirable implementations.

In this paper we present a generalization of Z refinement, called weak refinement, which treats internal operations differently from observable operations when refining a system. We illustrate some of the properties of weak refinement through a specification of a telecommunications protocol.

Keywords: Refinement; Distributed Systems; Internal Operations; Process Algebras; Concurrency.

1 Introduction

Now the use of Z for the specification of sequential systems is gaining acceptance, attention is being turned to new domains of applicability - one such example is the use of Z for the specification of concurrent and distributed systems [5, 17, 14, 13, 19]. One aspect that is important in the specification of distributed systems is the role of the internal (or unobservable) operation. Such operations are not part of the user interface (i.e. the user cannot invoke them), however, they are essential to our understanding and correct modelling of the system. Internal operations (or actions) arise naturally in distributed systems, either as a result of modelling concurrency or the non-determinism that is inherent in a model of such a system. For example, internal operations can be used to model communication (e.g. as in the language CCS [15]), non-determinism arises as a by-product of this interpretation. Internal operations are also central to abstraction specification through hiding, a particularly important example of

---

* This work was partially funded by British Telecom Research Labs., and the EPSRC under grant number GR/K13035.

this is to enable communication to be internalised - a central facet in the design of distributed systems.

Languages specifically targeted at concurrent systems typically have a notion of internal action or operation built into the language. For example, internal operations form a vital part of the theory of process algebras, and a special symbol is reserved for the occurrence of such an internal event (e.g. i or τ). If an internal operation is distinguished in the specification notation, then refinement and equivalence relations defined over the language need to deal with internal operations in appropriate ways. One way is to treat an internal event no differently from observable events, an example of such a relation is strong bisimulation in a process algebra [15]. However, it is well recognised that this is inappropriate as a refinement relation, and that internal events should typically have a different role within refinement and equivalence relations. Examples of relations in which the observable is differentiated from the internal are weak bisimulation [15], testing equivalence [3], reduction and extension [4], failures refinement [11] and Hennessy's testing pre-orders [10]. Central to these relations is the understanding that internal events are unobservable, and that refinement relations must refine the observable behaviour of a specification differently from its internal behaviour.

A number of authors have adopted conventions for specifying internal operations when modelling systems in Z. In each case the internal operation is specified as normal and either has a distinguished name or informal commentary telling us that it is not part of the user interface. If internal operations appear explicitly in a Z specification, we need to consider the possibility of refining these specifications. How should we treat the refinement of internal operations in Z? We seek here to contribute to the debate by making a proposal called *weak refinement*. This has a similar relation to ordinary Z refinement as weak bisimulation does to strong bisimulation in a process algebra. In particular, we define weak refinement by considering the stand point of an external observer of the system, who manipulates operations in the user interface.

Such an external observer will require that a retrieve relation is still defined between the state spaces of the abstract and concrete specifications and that each abstract observable operation AOp is recast as a concrete observable operation COp. The weak refinement relation is defined to ensure that the observable behaviour of the concrete specification is a refinement of the observable behaviour of the abstract specification.

Throughout the paper we assume the state plus operations style of Z specification, and our discussion takes place within that context.

The structure of the paper is as follows. In Section 2 we review the need for internal operations in Z specifications. Section 3 presents an example of a specification and refinement involving internal operations, the example illustrates that standard Z refinement is too liberal in the presence of internal operations. Section 4 formulates the generalization that we call weak refinement, which is motivated by the treatment of internal events in process algebras. Section 5 revisits the protocol example to show that weak refinement has the required properties of a refinement where internal operations have been specified. Section 6 discusses some properties of this refinement, and we conclude in Section 7.

2 Internal Operations

When modelling sequential systems in Z, the operations represent the user interface. That is, a state change occurs in the system if and only if the user invokes one of the operations. However, when modelling concurrent and distributed systems it is convenient to model *internal* operations. These internal operations represent operations over which the user has no control (hence the name internal). Since they are not part of the user interface they can be invoked by the system (potentially non-deterministically) whenever their pre-conditions hold. They can arise either due to the natural non-determinism of a distributed system [11], or due to communication within the system [15] or due to some aspect of the system being hidden at this level of abstraction [2]. The necessity for the specification of internal events in process algebras is well recognised [15], and a number of researchers have found it convenient or necessary to specify internal operations in Z when specifying distributed systems [7, 16, 19, 21, 9]. In each case the internal operation is specified as normal and either has a distinguished name or informal commentary telling us that it is not part of the user interface. We will see examples of both below. Used in this way, Z is clearly sufficient as a notation for the specification of internal operations or actions.

Is this necessary however, why not leave internal operations to process algebras? Well, Z is particularly suited to the specification of parts of a distributed system which contain large amounts of state information. Typical to this class are managed objects [20] or the information viewpoint of the Open Distributed Processing reference model [12], where the specifications contain a lot of state but there is also a need to model internal operations such as alarms.

Although Z is adequate as a notation for the specification of internal actions/operations, the usual Z refinement rules for operations are inappropriate for specifications containing internal operations. As we shall see (at the end of Section 3.2) they are inappropriate because they allow a refinement to contain more non-determinism than is acceptable. This situation is clearly undesirable, and we must re-formulate refinement for internal operations if they are to be used in Z specifications. This is what we seek to do here.

3 Refinement

A Z specification describes the state space together with a collection of operations. The Z refinement relation [18, 21], defined between two Z specifications, allows both the state space and the individual operations to be refined in a uniform manner.

Operation refinement is the process of recasting each abstract operation AOp into a concrete operation COp, such that (informally) the following holds. The pre-condition of *COp* may be weaker than the pre-condition of *AOp*, and *COp* may have a stronger post-condition than *AOp*. That is, *COp* must be applicable whenever *AOp* is, and if *AOp* is applicable, then every state which *COp* might produce must be one of those which *AOp* might produce. Data refinement extends operation refinement by allowing the state space of the concrete operations to be different from the state space of the abstract operations. Refinement for sequential systems specified in Z is well documented and understood. How does refinement behave in the presence of internal operations?

As an illustration of refinement involving internal operations we consider the specification and refinement of a telecoms protocol (the Signalling System No. 7 standard) adapted from [21]. The first specification defines the external view of the protocol, subsequently we develop a sectional view which specifies the route that messages take through the protocol.

3.1   Specification 1: the external view

Let M be the set of messages that the protocol handles. The state of the system comprises two sequences which represent messages that have arrived in the protocol (*in*), and those that have been forwarded (*out*).

$$
\begin{array}{|l}
\_\_Ext_____ \\
\hline
in, out : \operatorname{seq} M \\
\hline
\exists s : \operatorname{seq} M \bullet in = s \frown out \\
\hline
\end{array}
$$

Incoming messages are added to the left of *in*, and the messages contained in *in* but not in *out* represent those currently inside the protocol. The state invariant specifies that the protocol must not corrupt or re-order. Initially, no messages have been sent:

$$ ExtInit \,\widehat{=}\, [\, Ext' \mid in' = \langle\rangle \,] $$

Two operations then model the transmission (*Transmit*) and reception (*Receive*) of messages into and out of the protocol. Their specification is straightforward. In the *Receive* operation, either no message is available (e.g. they are all on route in the protocol) or the next one is output, this choice is made non-deterministically at this level of abstraction.

$$
\begin{array}{|l}
\_\_Transmit_____ \\
\hline
\Delta Ext \\
m? : M \\
\hline
in' = \langle m? \rangle \frown in \\
out' = out \\
\hline
\end{array}
$$

```
┌─ Receive ─────────────────────────────────────────
│ ΔExt
├────────────────────────────────────
│ in' = in
│ #out' = #out + 1 ∨ out' = out
└─────────────────────────────────────────────────
```

3.2 Specification 2: the sectional view

The sectional view specifies the route the messages take through a number of *sections*. Let N be the number of sections. Each section in the route may receive and send messages, and those which have been received but not yet sent on are in the section. The messages pass through the sections in order. In the state schema, *ins i* represents the messages currently inside section i, *rec i* the messages that have been received by section i, and *sent i* the messages that have been sent onwards from section i. The state and initialization schemas are then given by

```
┌─ Section ──────────────────         ┌─ SectionInit ──────────────
│ rec, ins, sent : seq(seq M)         │ Section'
├────────────────────────────         ├────────────────────────────
│ N = #rec = #ins = #sent             │ ∀ i : 1..N •
│ rec = ins ⌢⌢ sent                   │    rec i = ins i = sent i = ⟨⟩
│ front sent = tail rec               └────────────────────────────
└────────────────────────────
```

where $⌢⌢$ denotes pairwise concatenation of the two sequences (so for every i we have $rec\ i = ins\ i ⌢ sent\ i$). The predicate *front sent = tail rec* ensures that messages that are sent from one section are those that have been received by the next. This specification also has operations to transmit and receive messages, and they are specified as follows:

```
┌─ STransmit ──────────────────────────────────────
│ ΔSection
│ m? : M
├────────────────────────────────────
│ head rec' = ⟨m?⟩ ⌢ (head rec)
│ tail rec' = tail rec
│ sent' = sent
└─────────────────────────────────────────────────
```

```
┌─ SReceive ───────────────────────────────────────
│ ΔSection
├────────────────────────────────────
│ rec' = rec
│ front ins' = front ins
│ last ins' = front(last ins)
│ front sent' = front sent
│ last sent' = ⟨last(last ins)⟩ ⌢ (last sent)
└─────────────────────────────────────────────────
```

Here, the new message received is added to the first section in the route in *STransmit*, and *SReceive* will deliver from the last section in the route. In the first specification, messages arrive non-deterministically, in the sectional view this is represented by the progress of the messages through the sections. Therefore in this more detailed design, we need to specify how the messages progress through the sections, and we do so by defining an operation *Daemon* which non-deterministically selects a section to make progress. The oldest message is then transfered to the following section, and nothing else changes. The important part of this operation is then:

```
┌─ Daemon ──────────────────────────────────────────────
│ ΔSection
├───────────────────────────────────────────────────────
│ ∃ i : 1..N − 1 |
│     ins i ≠ ⟨⟩ •
│     ins′i = front(ins i)
│     ins′(i + 1) = ⟨last(ins i)⟩ ⌢ ins(i + 1)
│     ∀j : 1..N | j ≠ i ∧ j ≠ i + 1 • ins′j = ins j
└───────────────────────────────────────────────────────
```

Daemon is an internal operation (the informal commentary accompanying the specification tells us this), and so can be invoked by the system whenever its pre-condition holds. As noted in [21]: *This operation is not part of the user interface. The user cannot invoke Daemon, but it is essential to our understanding of the system and to its correctness.*

The sectional view is a refinement of the original, where the retrieve relation (which is a total function, i.e. $\forall$ *Section* • $\exists_1$ *Ext* • *Retrieve*) is given by:

```
┌─ Retrieve ────────────────────────────────────────────
│ Ext
│ Section
├───────────────────────────────────────────────────────
│ head rec = in
│ last sent = out
└───────────────────────────────────────────────────────
```

Under this refinement *STransmit* and *SReceive* correspond to *Transmit* and *Receive* respectively, and the internal operation *Daemon* corresponds to the external operation ΞExt (i.e. the identity operation on *Ext*), and we can prove (with appropriate quantification over the states) the refinement by showing that:

SectionInit ∧ *Retrieve* ⇒ *ExtInit*
pre *Transmit* ∧ *Retrieve* ⇒ pre *STransmit*
pre *Transmit* ∧ *Retrieve* ∧ *STransmit* ∧ *Retrieve′* ⇒ *Transmit*
pre *Receive* ∧ *Retrieve* ⇒ pre *SReceive*
pre *Receive* ∧ *Retrieve* ∧ *SReceive* ∧ *Retrieve′* ⇒ *Receive*
pre ΞExt ∧ *Retrieve* ⇒ pre *Daemon*
pre ΞExt ∧ *Retrieve* ∧ *Daemon* ∧ *Retrieve′* ⇒ ΞExt

So far so good. We can specify a system that contains non-determinism in some of the operations in its user interface (e.g. *Receive*), but which doesn't contain any internal operations. We can then refine this specification to one that contains internal operations that correctly model (in the sense of a refinement existing) the abstract specification. Here we have used the standard Z refinement relations, which have been perfectly adequate at this level.

However, we can refine this sectional view further. Consider the *Daemon* operation. This operation is partial (as it does not specify what happens if $ins\ i = \langle\rangle$ for every i), and using standard Z refinement we can weaken its pre-condition, and refine it to the following:

$$
\begin{array}{|l}
\hline
\_\_NDaemon_____ \\
\Delta Section \\
\hline
(\forall i : 1..N - 1 \bullet ins\ i = \langle\rangle \Rightarrow ins'1 = \langle m \rangle \wedge m \in M)\ \vee \\
(\exists i : 1..N - 1\ | \\
\quad ins\ i \neq \langle\rangle\ \bullet \\
\quad ins'i = front(ins\ i) \\
\quad ins'(i + 1) = \langle last(ins\ i)\rangle \frown ins(i + 1) \\
\quad \forall j : 1..N\ |\ j \neq i \wedge j \neq i + 1 \bullet ins'j = ins\ j) \\
\hline
\end{array}
$$

This operation has the same functionality as before, except that in addition the system can invoke it non-deterministically (since it is an internal operation) initially to insert an arbitrary message into the first section. Thus initially there are two possible behaviours of the system: as before the user could invoke *Transmit* to insert a message into the protocol, or now the system could non-deterministically invoke *NDaemon* which corrupts the input stream of the protocol before the user has inserted any messages.

The specification which contains the sectional view operations together with this new *NDaemon* is a refinement of the sectional view. Yet clearly implementations which introduce arbitrary amounts of noise into a stream of protocol messages are unacceptable. But in this situation, using standard Z refinement this has been allowed to happen, what has gone wrong?

We have used standard Z refinement here, and at issue is the refinement of internal operations. Internal operations have behaviour which isn't subject to the normal interpretation of operations (that are in the user interface), so it is not surprising then that using normal refinement brings about unexpected (and undesirable) consequences.

3.3 The firing condition interpretation

One possible solution is described by Strulo in [19], which has the merit of simplicity, but, as we shall see, perhaps constrains refinement too far. Strulo calls internal operations *active*, and operations in the user interface *passive*. The firing condition interpretation is the idea that the pre-condition states the only times the operation can happen at all instead of saying an operation is undefined (but possible) outside its pre-condition.

To define refinement, [19] identifies three regions for an operation (unconstrained, empty and interesting) and the applicability and correctness refinement rules are then re-interpreted for internal operations as:

$\vdash COp \Rightarrow AOp$
$\vdash (\exists \, State' \bullet AOp) \wedge (\exists \, State' \bullet \neg \, AOp) \Rightarrow$
$$(\exists \, State' \bullet COp) \wedge (\exists \, State' \bullet \neg \, COp)$$

The three regions of an operation represent: (1) states where the operation is divergent because no constraints are made on the after state (the unconstrained region), (2) states outside the usual pre-condition but which aren't divergent (the empty region), and (3) the remaining states where some but not all after states are allowed (the interesting region). For a full discussion the reader should consult [19].

In terms of these interpretations and the regions of definition of an operation, the first condition prevents an operation becoming possible where it was impossible, and the second condition ensures that the concrete operation doesn't become impossible where it was defined and possible.

Application of these ideas to the above example shows that with the firing condition interpretation, *NDaemon* is not a refinement of *Daemon*. Thus we successfully stop the pre-condition of an internal operation from being weakened in an unacceptable manner. However, to achieve this a barrier has been placed between observable and unobservable operation refinements. In particular, for hybrid specifications (ones involving both internal and observable operations), the refinement rules used depend on the type of operation - standard refinement for observable operations, and the firing condition interpretation for internal operations.

But the division is not always as simple as that, on occasion we may wish to *introduce* internal operations during a refinement, or we may wish to *remove* internal operations in a refinement. The refinement of the external view to the sectional view is an example of the introduction of internal operations, and we will give an example of their removal shortly.

However, we find that under the firing condition interpretation, the sectional view is not a refinement of the external view of the protocol, because now *Daemon* does not correspond to ΞExt under the firing condition interpretation refinement rules. To overcome this, can we restrict the use of the firing condition interpretation refinement rules to when the abstract operation is internal? The following very simple example will illustrate that we cannot.

Consider an abstract specification with an operation *AOp* in the user interface, and an internal operation *IOp*. The concrete specification consists of a single operation *COp*. Both have state space *State* consisting of a *mode* : $\{0, 1\}$. Initially *mode* is set to 0. The only operations in the specifications are given by:

$$\boxed{\begin{array}{l} AOp \\ \hline \Delta State \\ \hline mode = 0 \wedge mode' = 1 \end{array}} \qquad \boxed{\begin{array}{l} IOp \\ \hline \Delta State, \\ error! : yes \mid no \\ \hline mode = 1 \wedge mode' = 0 \\ error! = yes \end{array}}$$

$$\boxed{\begin{array}{l} COp \\ \hline \Delta State, \\ error! : yes \mid no \\ \hline mode = mode' = 0 \wedge error! = yes \end{array}}$$

It is natural to view the concrete specification as a refinement of the abstract. In the abstract, after invoking AOp an error message will occur (triggered by the internal operation IOp happening non-deterministically, which it eventually always will), in the concrete, after invoking COp an error message will occur. This type of removal of internal events lies at the heart of all treatments of internal operations in process algebras. However, under the firing condition interpretation, the concrete operation is not a refinement of the abstract, because no operation that was possible can become impossible - even if the internal behaviour has moved elsewhere. The fact that IOp has an output here is immaterial to the essence of the example - the aspect of internal operations with output is discussed in Section 5.2.

So, to summarise, standard Z refinement is too liberal in the presence of internal operations. An alternative approach is that suggested in [19], however, this involves a different interpretation of operations, and the refinement of internal behaviour can be too strict as the last example shows. In the next section we will seek an alternative generalization of refinement that steers a middle course by using ideas from process algebras.

4 Weak Refinement

To define weak refinement we will consider the standpoint of an *external* observer. Such an external observer will require that a retrieve relation is still defined between the state spaces of the abstract and concrete specifications and that each observable operation AOp is recast as a concrete operation COp. The refinement relation will ensure that the observable behaviour of the concrete specification is a refinement of the observable behaviour of the abstract specification.

The weak refinement rules have the same form as standard refinement, namely that:

- $\forall I_k; \ Cstate' \bullet Cinit_w \vdash \exists I_l; \ Astate' \bullet Ainit_w \wedge Ret'$

- $\forall I_k \bullet \text{pre}_w AOp \wedge Ret \vdash \exists I_l \bullet \text{pre}_w COp$

$$- \forall I_k; \ I_p; \ I_q \bullet \text{pre}_w \ AOp \wedge Ret \wedge COp_w \vdash \exists I_m; \ I_n; \ Astate' \bullet Ret' \wedge AOp_w$$

except that the subscript w denotes a weak counterpart which we will define below and I_k are sequences of internal operations. The next subsection reviews the treatment of internal events in process algebras, and we use these ideas to motivate our formulation of weak refinement in the following subsection.

4.1 Internal events in Process Algebras

Refinement in a process algebra is defined in terms of the transitions a behaviour can undertake, and we write $P \xrightarrow{a} P'$ if a process (or behaviour) P can perform the action a and then evolve to the process P'. Refinements and equivalences are given in terms of such transitions. For each relation, two versions are possible - a strong relation which treats all actions identically whether observable or not, and a weak version that makes allowances for internal events and is only concerned with observable transitions.

To make allowances for internal actions, consideration is given to what is meant by an observable transition. An observable transition is taken to be any observable action preceded or succeeded by any (finite) number of internal events.

Weak or observable relations now replace transitions $P \xrightarrow{a} P'$ by their observable counterpart: $P \xRightarrow{a} P'$, which means that process P can evolve to process P' by undergoing an unspecified (but finite) number of internal events, followed by the action a, followed by an unspecified number of internal events. Weak bisimulation (or observational equivalence) is an example of a relation defined in such a fashion [15].

4.2 Formulating weak refinement

Throughout this subsection let the state spaces of the abstract and concrete specifications be $Astate$ and $Cstate$ respectively. Let Ret be the retrieve relation defined between the specifications. AOp and COp stand for operations on the abstract and concrete state spaces where COp implements AOp. The initial states are given by $Cinit$ and $Ainit$.

Our formulation of weak refinement will be motivated by the approach taken in process algebras. Application of an operation in Z corresponds to a transition in a process algebra, and in weak refinement in place of the application of an operation Op we allow a finite number of internal operations before and after the occurrence of the operation. This corresponds to the change from $P \xrightarrow{a} P'$ to $P \xRightarrow{a} P'$ in a process algebra when moving from a strong to observable scenario. This can be described in the Z schema calculus by saying there exist internal operations $i_1, \ldots, i_k, j_1, \ldots, j_l$ (for $k, l \geq 0$) and the application of the composition $i_1 \, \S \ldots \S \, i_k \, \S \, Op \, \S \, j_1 \, \S \ldots \S \, j_l$. Throughout this section we abbreviate $i_1 \, \S \ldots \S \, i_k$ to i^k, and we will let I_k denote a sequence of internal actions $\langle i_1, \ldots, i_k \rangle$.

We can now re-formulate each of the three conditions for refinement for a system containing internal operations. We begin with the initialization condition.

Initialization

Without internal operations the relationship required upon initialization is that each possible initial state of the concrete specification must represent a possible initial state of the abstract specification. In the presence of internal operations an initial state might evolve internally to another state. Therefore, "each possible initial state of the concrete specification" now includes all possible evolutions of the initial state under internal operations. Likewise "a possible initial state of the abstract specification" can now include a potential evolution of the initial state due to internal operations.

To formalise this (using the abbreviation $i^k = i_1 \,\S\, \ldots \,\S\, i_k$) we require that:

$$\forall I_k; \; Cstate' \bullet Cinit \,\S\, i^k \vdash \exists I_l; \; Astate' \bullet Ainit \,\S\, i^l \wedge Ret$$

The quantification of the internal operations in $Cinit \,\S\, i^k$ is important. What we wish to ensure is that *every* initial concrete path (including all possible internal operations) can be matched by *some* initial abstract path (possibly involving internal operations). We abbreviate the condition to

$$\forall I_k; \; Cstate' \bullet Cinit_w \vdash \exists I_l; \; Astate' \bullet Ainit_w \wedge Ret'$$

Applicability

Applicability must ensure that if an abstract and concrete state are related by the retrieve relation, then the concrete operation should terminate whenever the abstract operation terminated, where termination is usually expressed in terms of satisfaction of the pre-condition of an operation. In the presence of internal operations we must allow for potential invocation of internal operations, and hence require that: if an abstract and concrete state are related by the retrieve relation, then whenever the abstract operation terminates possibly after any internal evolution then the concrete operation terminates after some internal evolution.

This is described by saying there exists internal operations $i_1, \ldots, i_k$ such that $\mathrm{pre}(i_1 \,\S\, \ldots \,\S\, i_k \,\S\, AOp)$ where $\S$ is schema composition in the Z schema calculus. We abbreviate $\mathrm{pre}(i_1 \,\S\, \ldots \,\S\, i_k \,\S\, AOp)$ to $\mathrm{pre}(i^k \,\S\, AOp)$ or $\mathrm{pre}_w AOp$.

Applicability can then be expressed as

$$\forall I_k \bullet \mathrm{pre}(i^k \,\S\, AOp) \wedge Ret \vdash \exists I_l \bullet \mathrm{pre}(i^l \,\S\, COp)$$

Using the abbreviation $\mathrm{pre}_w AOp$, where we note that we have replaced $\mathrm{pre}\, AOp$ by the condition that AOp is applicable after a number of internal operations, applicability in weak refinement reduces to

$$\forall I_k \bullet \mathrm{pre}_w AOp \wedge Ret \vdash \exists I_l \bullet \mathrm{pre}_w COp$$

Correctness

For correctness, we require the weak analogy to the following: if an abstract state and a concrete state are related by Ret, and both the abstract and concrete operations are guaranteed to terminate, then every possible state after the concrete operation must be related by Ret' to a possible state after the abstract operation [18]. For the weak version pre AOp is replaced by $\text{pre}_w\, AOp$ and we ask that, every possible state after the concrete operation must be related by Ret' to a possible state after the abstract operation, except that now 'after' means an arbitrary number of internal operations may occur before and after the abstract operation. The condition thus becomes, in full,

$$\forall I_k;\; I_p;\; I_q \bullet \text{pre}(i^k \,\S\, AOp) \wedge Ret \wedge i^p \,\S\, COp \,\S\, i^q \vdash$$
$$\exists I_m;\; I_n;\; Astate' \bullet Ret' \wedge i^n \,\S\, AOp \,\S\, i^m$$

which we abbreviate to

$$\forall I_k;\; I_p;\; I_q \bullet \text{pre}_w\, AOp \wedge Ret \wedge COp_w \vdash \exists I_m;\; I_n;\; Astate' \bullet Ret' \wedge AOp_w$$

Again the quantification of the internal operations in COp_w is important. We need to ensure is that every path involving COp and possible internal operations can be matched by some path involving AOp and (possibly) internal operations. Hence the quantification in COp_w is universal over all sequences of internal operations before and after COp.

Rules for Internal operations

We will also apply the applicability and correctness rules to internal operations. For internal operations we don't want applicability to prevent an internal operation becoming impossible where it was previously possible, indeed we want to refine out such internal operations in appropriate fashions. So for an internal operation I (defined on a state space $state$) we define its weak pre-condition (not its pre-condition) by

$$\text{pre}_w\, I = \text{pre}\, \Xi state = state$$

Although this definition of the weak pre-condition for internal operations looks strange, it does not allow us to arbitrarily weaken the pre-condition of an internal operation under weak refinement. The circumstances when we can are governed by what observable operations are present in the abstract specification, and the correctness rules for observable operations prevent the arbitrary weakening of pre-conditions of internal operations.

Applicability for internal operations will reduce to checking that the concrete state is implied by the abstract state (modulo the retrieve relation).

The final piece in the jigsaw is the meaning of correctness for internal operations. We define the weak version of an operation Op by

$$Op_w = \begin{cases} i^k \,\S\, Op \,\S\, i^l & \text{for observable } Op, \\ i^k & \text{for internal operation } Op, k \geq 0 \end{cases}$$

where $i^0 = \Xi state$ and appropriate quantification will be taken over k (and l) according to the context. This ensures that we can match up an occurrence of an internal operation in the abstract specification by zero (using $\Xi state$) or more (using i^k) internal actions in the concrete specification.

Thus to summarise, weak refinement requires that

- $\forall I_k; \ Cstate' \bullet Cinit_w \vdash \exists I_l; \ Astate' \bullet Ainit_w \wedge Ret'$

- $\forall I_k \bullet \mathrm{pre}_w \, AOp \wedge Ret \vdash \exists I_l \bullet \mathrm{pre}_w \, COp$

- $\forall I_k; \ I_p; \ I_q \bullet \mathrm{pre}_w \, AOp \wedge Ret \wedge COp_w \vdash \exists I_m; \ I_n; \ Astate' \bullet Ret' \wedge AOp_w$

where

$$Op_w = \begin{cases} i^k \, \substack{\circ \\ \circ} \, Op \, \substack{\circ \\ \circ} \, i^l & \text{for observable } Op, \\ i^k & \text{for internal operation } Op, k \geq 0 \end{cases}$$

and $i^0 = \Xi state$ and $\mathrm{pre}_w(Op) = \mathrm{pre}(i^k \, \substack{\circ \\ \circ} \, Op)$.

In the next section we show how these rules are applied in practice, and we shall see that although the full generality introduces complexity, in practice the overheads are not large.

5 Examples

We apply the theory we developed above to the examples presented at the start of the paper. In the protocol example, the intuitive behaviour we wish to capture is that the sectional view is a refinement of the external view, but that the third specification is not a refinement of the sectional view. This is indeed the case with weak refinement.

5.1 The Signalling Protocol

First we show the sectional view of the protocol is a weak refinement of the external view. We first prove the initialization is correct, noting that the retrieve relation is total and functional, so that we can use the usual simplification, and we show that:

$$\forall I_k; \ Ext'; \ Section' \bullet SectionInit_w \wedge Retrieve \vdash \exists I_l \bullet ExtInit_w$$

This reduces to $\forall Ext'; \ Section' \bullet SectionInit \wedge Retrieve \vdash ExtInit$, since there are no internal operations in the external specification, and no internal operation is applicable after $SectionInit$ in the sectional view. This can be verified as normal.

To verify applicability, we need to show that

$$\forall I_k \bullet \mathrm{pre}_w \, Transmit \wedge Retrieve \vdash \exists I_l \bullet \mathrm{pre}_w \, STransmit$$
$$\forall I_k \bullet \mathrm{pre}_w \, Receive \wedge Retrieve \vdash \exists I_l \bullet \mathrm{pre}_w \, SReceive$$
$$\forall I_k \bullet \mathrm{pre}_w \, \Xi Ext \wedge Retrieve \vdash \exists I_l \bullet \mathrm{pre}_w \, Daemon$$

The last equation reduces to $Ext \wedge Section \vdash Section$ since $Daemon$ is internal and for internal operations we have defined $pre_w\ Daemon = \Xi Section$. In the case of $Transmit$, the weak pre-condition requirement reduces to

$$\text{pre } Transmit \wedge Retrieve \vdash \exists I_l \bullet pre(i^l \,{}_9^\circ\, STransmit)$$

which is true with $l = 0$. A similar argument holds for the weak pre-condition of $Receive$.

Similarly, to verify correctness, we need to show that

$$\forall I_p;\ I_q \bullet \text{pre } Transmit \wedge Retrieve \wedge STransmit_w \wedge Retrieve' \vdash Transmit$$
$$\forall I_p;\ I_q \bullet \text{pre } Receive \wedge Retrieve \wedge SReceive_w \wedge Retrieve' \vdash Receive$$
$$\forall I_p;\ I_q \bullet \text{pre } \Xi Ext \wedge Retrieve \wedge Daemon_w \wedge Retrieve' \vdash \Xi Ext$$

For the first, we need to check that occurrences of the $Daemon$ operation before and after $STransmit$ in the concrete specification still leave us in a state that is consistent with that produced by $Transmit$ in the abstract. This is found to be true (since $Daemon \Rightarrow \Xi Ext$). The second case is similar. For the third this reduces to showing that

$$\forall k \bullet Ext \wedge Retrieve \wedge Daemon^k \wedge Retrieve' \vdash \Xi Ext$$

where $Daemon^k$ denotes k sequential compositions of $Daemon$. Again this is found to be true.

Therefore the sectional view is indeed a weak refinement of the external view. Moreover, the additional verification requirements imposed by the generality of weak refinement are not large in this example, being confined to the consideration of one internal operation - $Daemon$.

We shall show now that the third specification is *not* a weak refinement of the sectional view. That is, we are not at liberty to weaken the pre-condition of an internal operation arbitrarily. Consider the initialization rule that (for total functional $Retrieve$):

$$\forall I_k;\ Astate;\ Cstate \bullet Cinit_w \wedge Retrieve \vdash \exists I_l \bullet Ainit_w$$

Now in the sectional view it is not possible to apply $Daemon$ initially. However, it is possible to apply $NDaemon$ initially (where it arbitrarily inserts a new element into the protocol). Thus for the third specification to be a weak refinement of the sectional view we require that

$$SectionInit \,{}_9^\circ\, NDaemon \vdash SectionInit$$

This is clearly not true, since after $NDaemon$, ins is no longer empty.

In addition to the initialization requirement failing in this example, the requirement that

$$\forall I_k;\ I_p;\ I_q \bullet pre_w\ STransmit \wedge Retrieve \wedge STransmit_w \wedge Retrieve' \vdash$$
$$\exists I_m;\ I_n \bullet STransmit_w$$

is also violated for the same reasons as the initial condition fails.

5.2 Internal operations with output

In the second example, presented in section 3.3, to show that the concrete is a weak refinement of the abstract, we need to prove that:

$$\forall I_k \bullet \mathrm{pre}_w\ AOp \wedge Ret \vdash \exists I_l \bullet \mathrm{pre}_w\ COp$$
$$\forall I_k;\ I_p;\ I_q \bullet \mathrm{pre}_w\ AOp \wedge Ret \wedge COp_w \vdash \exists I_m;\ I_n;\ Astate' \bullet Ret' \wedge AOp_w$$

In the refinement we will simply link the states for which $mode = 0$ as the state $mode = 1$ was purely an intermediate state for the purposes of specifying the temporal ordering of the operations. Hence the retrieve relation will be:

```
┌─ Ret ──────────────────────────────────
│  State
│ ────────────────────────────
│  mode = 0
└────────────────────────────────────────
```

With this retrieve relation we will in fact show that the concrete operation COp implements both abstract operations AOp and IOp. Since the concrete specification does not have any internal operations we just need to show that:

$$\forall I_k \bullet \mathrm{pre}_w\ AOp \wedge Ret \vdash \mathrm{pre}\ COp$$
$$\forall I_k \bullet \mathrm{pre}_w\ AOp \wedge Ret \wedge COp \wedge Ret' \vdash \exists I_m;\ I_n \bullet AOp_w$$
$$\forall I_k \bullet \mathrm{pre}_w\ IOp \wedge Ret \vdash \mathrm{pre}\ COp$$
$$\forall I_k \bullet \mathrm{pre}_w\ IOp \wedge Ret \wedge COp \wedge Ret' \vdash \exists I_m;\ I_n \bullet IOp_w$$

We can calculate the pre-conditions needed. Note that in the case of $\mathrm{pre}_w\ AOp$ this includes states from which the system can perform an internal operation and then for AOp to successfully terminate.

```
┌─ pre_w AOp ──────────────       ┌─ pre COp ──────────────
│  State                          │  State
│ ──────────────────────          │ ──────────────────────
│  mode = 0 ∨ mode = 1            │  mode = 0
└──────────────────────────       └──────────────────────────
```

The applicability and correctness for the implementation of AOp as COp are then easily verified. Consideration of the internal operation amounts to showing that (because of the way the pre-condition of an internal operation is defined)

$$Ret \vdash \mathrm{pre}\ COp$$
$$Ret \wedge COp \wedge Ret' \vdash \exists k \bullet IOp^k$$

and the latter holds for $k = 0$.

So the concrete specification is indeed a weak refinement of the abstract. This illustrates an interesting aspect of specifying internal operations in Z - they can output data (in fact some interpretations of unobservableness in Z outlaw this possibility e.g. [6], but generally this is the case). This is in contrast to a process algebra where typically internal actions can have no data attributes.

Consider full LOTOS [2], where the internal action is written *i*. Internal actions in LOTOS can arise as a result of direct specification or as a result of hiding observable actions. In the first case, it is syntactically illegal to associate a data attribute with an internal action, e.g. the behaviour

  i!7; *B*

is not well-formed. Here action prefix is represented by ; and a value declaration on an action is given by a !. In the second case, upon hiding an observable action with data, the data is hidden as well as the action. So, for example, in the behaviour

  hide *g* in *g*!5; *stop*

the transition *i* can be performed, but no data is associated with the occurrence of the internal action *i*.

However, it is desirable to be able to specify an internal event which does have data associated with it. Indeed [19] contains an example of such an operation - an alarm notification in a managed object. This is a typical example of the kind of application where it is necessary to be able to specify an atomic internal operation which has output associated with it. Used in this style Z is more expressive than LOTOS in terms of internal events it can specify.

Whether or not such an internal event is unobservable is debatable, and perhaps such events mark the difference between active systems as opposed to reactive systems - the latter often modelled using a process algebra. In an active system events can be under the control of the system but not the environment (e.g. an alarm operation), such events are internal but can have observable effects (such as an alarm notification). This differs from the notion of internal in a process algebra, which equates internal with no observable transition, including output. In such an interpretation the operation *IOp* defined above would not be internal as we can observe its occurrence via its output, and the term *active* used in [19] could be used instead. However, the theory of weak refinement developed here is equally applicable to such a class of events.

6 Properties

6.1 Reducing non-determinism

An important aspect of refinement, in both the sequential and concurrent worlds, is the ability to strengthen an implementation by reducing the non-determinism in the abstract specification. Indeed this is a property of standard Z refinement in the absence of internal operations. Adding internal operations in a specification has introduced an additional form of non-determinism into the language. We shall see that weak-refinement allows us to reduce this type of non-determinism by removing internal operations.

Consider the behaviours described by the following transition diagrams, where *a* and *b* are observable events, and *i* represents an internal operation:

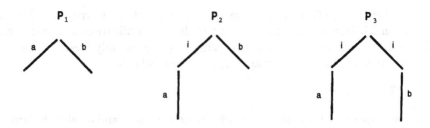

These specifications are not equivalent in any sense, for example in a process algebraic setting none of them are weak bisimulation equivalent. However, we would like a refinement to remove the non-determinism which is present in terms of the internal events, and for P_1 to refine P_2 which in turn refines P_3. Indeed, seen as processes they are related in the sense that, for example, P_1 **red** P_2 **red** P_3, where **red** is the LOTOS reduction relation [2].

Weak refinement, which we denote $\sqsubseteq_w$, also exhibits this property, that is $P_3 \sqsubseteq_w P_2 \sqsubseteq_w P_1$, but $P_1 \not\sqsubseteq_w P_2 \not\sqsubseteq_w P_3$. In terms of Z specifications we are giving these diagrams their obvious interpretation, for example, a Z specification of behaviour P_2 would be given by (the internal operation here is i, all the others are observable):

State
$state : \{0, 1, 2, 3\}$

Init
$\Delta State$
$state' = 0$

_a_
$\Delta State$
$state = 1 \wedge state' = 3$

_b_
$\Delta State$
$state = 0 \wedge state' = 2$

_i_
$\Delta State$
$state = 0 \wedge state' = 1$

A slightly more complex example is given by the two behaviours defined by the following, where again the event i is internal and all others are observable.

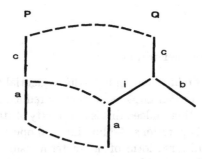

Interpreted as Z specifications we find that P is a weak refinement of Q. This example is interesting because by resolving the non-determinism, the implementation never offers the operation b. The retrieve relation which shows this is a weak refinement is given by the dotted lines in the above diagram. Because pre $b \wedge Ret$ has predicate which is false, b can be implemented by any operation in the concrete specification (e.g. $\Xi State$ will do).

Notice that, as one would hope, Q is not a weak refinement of P, because we have to quantify over *all* paths of internal operations in the concrete specification in the correctness criteria for weak refinement.

6.2 Weak refinement and refinement

In specifications without internal operations, refinement and weak refinement clearly coincide. In the presence of internal operations, neither implies the other. Since our motivation in defining weak refinement was to rule out some "refinements" of internal operations, refinement doesn't imply weak refinement (the protocol specifications provided an example of this).

However, neither does weak refinement imply standard Z refinement. The last example given above exhibits a weak refinement (P is a weak refinement of Q), which does not have a retrieve relation which will define a standard refinement between them.

One desirable property that standard Z refinement possesses is that it is a congruence. That is, if specification S is refined by S', then in any context $C[.]$, $C[S']$ refines $C[S]$. A consequence of this is that operations can be refined individually and the whole specification is then a refinement of the original.

However, weak refinement is not a congruence, due to the presence of internal operations. To see this consider the two specifications given by the following behaviours:

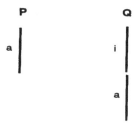

Then under weak refinement these are equivalent, i.e. $P \sqsubseteq_w Q$ and $Q \sqsubseteq_w P$. However, if we add just one further operation to each specification which is applicable at the initial state, i.e. we specify the behaviour

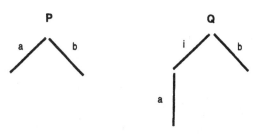

then, as we observed earlier, Q is not a weak refinement of P. So congruence is lost with weak refinement. Incidentally, this counter-example is the same example that shows weak bisimulation is not a congruence in a process algebra, so the result here is not surprising and the ability to find observational relations which *are* congruences can be non-trivial.

7 Conclusions

The motivation for this work arose out of our interest in the use of Z for the specification of distributed systems, and in particular its use within the Open Distributed Processing (ODP) standardization initiative [12]. ODP is a joint standardisation activity of the ISO and ITU. A reference model has been defined which describes an architecture for building *open* distributed systems. Central to this architecture is a *viewpoints* model. This enables distributed systems to be described from a number of different perspectives. There are five viewpoints: enterprise, information, computational, engineering and technology. Requirements and specifications of an ODP system can be made from any of these viewpoints. Z and LOTOS are strong candidates for use in some of the ODP viewpoints, Z for the information viewpoint and LOTOS for the computational and engineering viewpoints.

The use of different viewpoints specified in different languages means we have to have mechanisms to check for the *consistency* of specifications. One aspect of this work has been the development of means to check for the consistency of two Z specifications [1], and a means to translate LOTOS specifications into Z [9]. Development of viewpoints written in different languages will be undertaken using different refinement relations, and this led to the need to develop a notion of weak-refinement in Z which is related to refinements in LOTOS. A full discussion of the relationships between the differing refinement relations is given in [8] (which incidentally assumes the firing condition interpretation discussed above).

Work related to that discussed here is that of Strulo in [19]. His proposal has much greater simplicity than that discussed here, however, it perhaps lacks full generality and involves a different interpretation of the pre-condition of an operation. Our aim here was to generalise standard Z refinement to deal with internal operations in a fully general manner, whilst maintaining the established interpretation of operations in Z.

References

1. E. Boiten, J. Derrick, H. Bowman, and M. Steen. Consistency and refinement for partial specification in Z. In M.-C. Gaudel and J. Woodcock, editors, *FME'96: Industrial Benefit of Formal Methods, Third International Symposium of Formal Methods Europe*, volume 1051 of *Lecture Notes in Computer Science*, pages 287–306. Springer-Verlag, March 1996.
2. T. Bolognesi and E. Brinksma. Introduction to the ISO Specification Language LOTOS. *Computer Networks and ISDN Systems*, 14(1):25–59, 1988.

3. E. Brinksma. A theory for the derivation of tests. In S. Aggarwal and K. Sabnani, editors, *Protocol Specification, Testing and Verification, VIII*, pages 63–74, Atlantic City, USA, June 1988. North-Holland.

4. E. Brinksma, G. Scollo, and C. Steenbergen. Process specification, their implementation and their tests. In B. Sarikaya and G. v. Bochmann, editors, *Protocol Specification, Testing and Verification, VI*, pages 349–360, Montreal, Canada, June 1986. North-Holland.

5. E. Cusack. Object oriented modelling in Z for Open Distributed Systems. In J. de Meer, V. Heymer, and R. Roth, editors, *IFIP TC6 International Workshop on Open Distributed Processing*, pages 167–178, Berlin, Germany, September 1991. North-Holland.

6. E. Cusack and G. H. B. Rafsanjani. ZEST. In S. Stepney, R. Barden, and D. Cooper, editors, *Object Orientation in Z*, Workshops in Computing, pages 113–126. Springer-Verlag, 1992.

7. E. Cusack and C. Wezeman. Deriving tests for objects specified in Z. In J. P. Bowen and J. E. Nicholls, editors, *Seventh Annual Z User Workshop*, pages 180–195, London, December 1992. Springer-Verlag.

8. J. Derrick, H. Bowman, E. Boiten, and M. Steen. Comparing LOTOS and Z refinement relations. In *FORTE/PSTV'96*, pages 501–516, Kaiserslautern, Germany, October 1996. Chapman & Hall.

9. J. Derrick, E.A.Boiten, H. Bowman, and M. Steen. Supporting ODP - translating LOTOS to Z. In *First IFIP International workshop on Formal Methods for Open Object-based Distributed Systems*, Paris, March 1996. Chapman & Hall.

10. M. Hennessy. *Algebraic Theory of Processes*. MIT Press, 1988.

11. C. A. R. Hoare. *Communicating Sequential Processes*. Prentice Hall, 1985.

12. ITU Recommendation X.901-904 — ISO/IEC 10746 1-4. *Open Distributed Processing - Reference Model - Parts 1-4*, July 1995.

13. L. Lamport. TLZ. In J.P. Bowen and J.A. Hall, editors, *ZUM'94, Z User Workshop*, pages 267–268, Cambridge, United Kingdom, June 1994.

14. P. Mataga and P. Zave. Formal specification of telephone features. In J. P. Bowen and J. A. Hall, editors, *Eighth Annual Z User Workshop*, pages 29–50, Cambridge, July 1994. Springer-Verlag.

15. R. Milner. *Communication and Concurrency*. Prentice-Hall, 1989.

16. G. H. B. Rafsanjani. ZEST - Z Extended with Structuring: A users's guide. Technical report, BT, June 1994.

17. S. Rudkin. Modelling information objects in Z. In J. de Meer, V. Heymer, and R. Roth, editors, *IFIP TC6 International Workshop on Open Distributed Processing*, pages 267–280, Berlin, Germany, September 1991. North-Holland.

18. J. M. Spivey. *The Z notation: A reference manual*. Prentice Hall, 1989.

19. B. Strulo. How firing conditions help inheritance. In J. P. Bowen and M. G. Hinchey, editors, *Ninth Annual Z User Workshop*, LNCS 967, pages 264–275, Limerick, September 1995. Springer-Verlag.

20. C. Wezeman and A. J. Judge. Z for managed objects. In J. P. Bowen and J. A. Hall, editors, *Eighth Annual Z User Workshop*, pages 108–119, Cambridge, July 1994. Springer-Verlag.

21. J. Woodcock and J. Davies. *Using Z: Specification, Refinement, and Proof*. Prentice Hall, 1996.

Select Z Bibliography

Jonathan P. Bowen

The University of Reading, Department of Computer Science
Whiteknights, PO Box 225, Reading, Berks RG6 6AY, UK
Email: J.P.Bowen@reading.ac.uk
URL: http://www.cs.reading.ac.uk/people/jpb/

Abstract. This bibliography contains a list of references concerned with the formal Z notation that are either available as published papers, books, selected technical reports, or on-line. The bibliography is in alphabetical order by author name(s).

Introduction

The list of references presented here is maintained in electronic form, in BIBTEX bibliography database format, which is compatible with the widely used LATEX document preparation system [326]. It is intended to keep the bibliography up to date and to issue it to coincide with the regular International Conference of Z User. The latest version of BIBTEX source file used for this bibliography [60] is available as a searchable on-line database on the World Wide Web under the following Uniform Resource Location (URL):

> http://www.comlab.ox.ac.uk/archive/z/bib.html

The actual BIBTEX source is also available as part of the on-line Z archive [579] via anonymous FTP on the Internet under:

> ftp://ftp.comlab.ox.ac.uk/pub/Zforum/z97.bib

Alternatively, if Internet access is difficult, via electronic mail by sending a message containing the command 'send z z97.bib' to the following email address: archive-server@comlab.ox.ac.uk

To add new references concerned with Z to this list, please send details to Jonathan Bowen (contact details above), preferably via electronic mail. It is helpful if you can give as much information as possible so the entry could be included as a reference in future papers concerning Z.

This bibliography has been regularly maintained for Z User Meeting proceedings in the past (e.g., see [65]). For an alternative annotated Z bibliography produced in 1995, see [73].

Acknowledgements

Ruaridh Macdonald of RSRE, Malvern, initiated the idea of a Z bibliography and helped maintain it for several years. Joan Arnold at the Oxford University Computing Laboratory has previously assisted in maintaining the bibliography as part of her work as secretary to the European ESPRIT **ProCoS-WG** Working Group (no. 8694) on 'Provably Correct Systems'. Thank you to everybody who has submitted entries over the years.

References

1. G. D. Abowd. Agents: Communicating interactive processes. In D. Diaper, D. Gilmore, Gilbert Cockton, and Brian Shackel, editors, *Human-Computer Interaction: INTERACT'90*, pages 143–148. Elsevier Science Publishers (North-Holland), 1990.

2. G. D. Abowd. *Formal Aspects of Human-Computer Interaction*. DPhil thesis, Oxford University Computing Laboratory, Wolfson Building, Parks Road, Oxford, UK, 1991.

3. G. D. Abowd, R. Allen, and D. Garlan. Using style to understand descriptions of software architectures. *ACM Software Engineering Notes*, 18(5):9–20, December 1993.

4. G. D. Abowd, R. Allen, and D. Garlan. Formalizing style to understand descriptions of software architecture. *ACM Transactions on Software Engineering and Methodology (TOSEM)*, 4(4):319–364, October 1995.
   The formal model is described using the Z specification language.

5. J.-R. Abrial. Data semantics. In J. W. Klimbie and K. L. Koffeman, editors, *IFIP TC2 Working Conference on Data Base Management*, pages 1–59. North-Holland, April 1974.
   A seminal paper for the formal Z notation, as noted in [237].

6. J.-R. Abrial. The B tool. In Bloomfield et al. [46], pages 86–87.

7. J.-R. Abrial. The B method for large software, specification, design and coding (abstract). In Prehn and Toetenel [427], pages 398–405.

8. J.-R. Abrial. *The B-Book: Assigning Programs to Meanings*. Cambridge University Press, 1996.
   This book is a reference manual for the B-Method developed by Jean-Raymond Abrial, also the originator of the Z notation. B is designed for tool-assisted software development whereas Z is designed mainly for specification.
   Contents: Mathematical reasoning; Set notation; Mathematical objects; Introduction to abstract machines; Formal definition of abstract machines; Theory of abstract machines; Constructing large abstract machines; Example of abstract machines; Sequencing and loop; Programming examples; Refinement; Constructing large software systems; Example of refinement;
   Appendices: Summary of the most current notations; Syntax; Definitions; Visibility rules; Rules and axioms; Proof obligations.

9. J.-R. Abrial, E. Börger, and H. Langmaack, editors. *Formal Methods for Industrial Applications: Specifying and Programming the Steam Boiler*, volume 1165 of *Lecture Notes in Computer Science*. Springer-Verlag, 1996.
   A comparative collection of formal methods case studies. See [108, 453].

10. J.-R. Abrial, S. A. Schuman, and B. Meyer. Specification language. In R. M. McKeag and A. M. Macnaghten, editors, *On the Construction of Programs: An Advanced Course*, pages 343–410. Cambridge University Press, 1980.

11. J.-R. Abrial and I. H. Sørensen. KWIC-index generation. In J. Staunstrup, editor, *Program Specification: Proceedings of a Workshop*, volume 134 of *Lecture Notes in Computer Science*, pages 88–95. Springer-Verlag, 1981.

12. M. Ainsworth, A. H. Cruikchank, P. J. L. Wallis, and L. J. Groves. Viewpoint specification and Z. *Information and Software Technology*, 36(1):43–51, 1994.

13. A. J. Alencar and J. A. Goguen. OOZE: An object-oriented Z environment. In P. America, editor, *Proc. ECOOP'91 European Conference on Object-Oriented Programming*, volume 512 of *Lecture Notes in Computer Science*, pages 180–199. Springer-Verlag, 1991.

14. M. A. Ardis, J. A. Chaves, L. J. Jagadeesan, P. Matega, C. Puchol, M. G. Staskauskas, and J. von Olnhausen. A framework for evaluating specification methods for reactive systems: Experience report. *IEEE Transactions on Software Engineering*, 22(6):378–389, June 1996.

Several different methods, including Modechart, VFSM, ESTEREL, Basic LOTOS, Z, SDL, and C, are applied to a problem encountered in the design of software for AT&T's 5ESS telephone switching system.

15. D. B. Arnold, D. A. Duce, and G. J. Reynolds. An approach to the formal specification of configurable models of graphics systems. In G. Maréchal, editor, *Proc. EUROGRAPH-ICS'87, European Computer Graphics Conference and Exhibition*, pages 439–463. Elsevier Science Publishers (North-Holland), 1987.
    The paper describes a general framework for the formal specification of modular graphics systems, illustrated by an example taken from the Graphical Kernel System (GKS) standard.

16. D. B. Arnold and G. J. Reynolds. Configuring graphics systems components. *IEE/BCS Software Engineering Journal*, 3(6):248–256, November 1988.

17. R. D. Arthan. Formal specification of a proof tool. In Prehn and Toetenel [426], pages 356–370.

18. R. D. Arthan. On free type definitions in Z. In Nicholls [404], pages 40–58.

19. K. Ashoo. The Genesis Z tool – an overview. *BCS-FACS FACTS*, Series II, 3(1):11–13, May 1992.

20. S. Aujla, A. Bryant, and L. Semmens. A rigorous review technique: Using formal notations within conventional development methods. In *Proc. 1993 Software Engineering Standards Symposium*, pages 247–255. IEEE Computer Society Press, 1993.

21. P. B. Austin, K. A. Murray, and A. J. Wellings. File system caching in large point-to-point networks. *IEE/BCS Software Engineering Journal*, 7(1):65–80, January 1992.

22. S. Austin and G. I. Parkin. Formal methods: A survey. Technical report, National Physical Laboratory, Queens Road, Teddington, Middlesex, TW11 0LW, UK, March 1993.

23. C. Bailes and R. Duke. The ecology of class refinement. In Morris and Shaw [388], pages 185–196.

24. M. Bailey. Formal specification using Z. In *Proc. Software Engineering anniversary meeting (SEAS)*, page 99, 1987.

25. J. Bainbridge, R. W. Whitty, and J. B. Wordsworth. Obtaining structural metrics of Z specifications for systems development. In Nicholls [402], pages 269–281.

26. J.-P. Banâtre. About programming environments. In J.-P. Banâtre, S. B. Jones, and D. de Métayer, editors, *Prospects for Functional Programming in Software Engineering*, volume 1 of *Research Reports*, chapter 1, pages 1–22. Springer-Verlag, 1991.

27. P. Bancroft and I. J. Hayes. A formal semantics for a language with type extension. In Bowen and Hinchey [82], pages 299–314.

28. R. Barden and S. Stepney. Support for using Z. In Bowen and Nicholls [84], pages 255–280.

29. R. Barden, S. Stepney, and D. Cooper. The use of Z. In Nicholls [404], pages 99–124.

30. R. Barden, S. Stepney, and D. Cooper. *Z in Practice*. BCS Practitioner Series. Prentice Hall, 1994.

31. G. Barrett. Formal methods applied to a floating-point number system. *IEEE Transactions on Software Engineering*, 15(5):611–621, May 1989.
    A formalization of the IEEE standard for binary floating-point arithmetic in Z is presented. The formal specification is refined into four components. The procedures presented form the basis for the floating-point unit of the Inmos IMS T800 Transputer. This work resulted in a joint UK Queen's Award for Technological Achievement for Inmos Ltd and the Oxford University Computing Laboratory in 1990. It was estimated that the approach saved a year in development time compared to traditional methods.

32. L. M. Barroca, J. S. Fitzgerald, and L. Spencer. The architectural specification of an avionic subsystem. In France and Gerhart [206], pages 17–29.

33. L. M. Barroca and J. A. McDermid. Formal methods: Use and relevance for the development of safety-critical systems. *The Computer Journal*, 35(6):579–599, December 1992.

34. P. Baumann. Z and natural semantics. In Bowen and Hall [74], pages 168–184.

35. P. Baumann and K. Lermer. A framework for the specification of reactive and concurrent systems in Z. In P. S. Thiagarajan, editor, *Foundations of Software Technology and Theoretical Computer Science*, volume 1026 of *Lecture Notes in Computer Science*, pages 62–79. Springer-Verlag, 1995.

36. P. Baumann and K. Lermer. Specifying parallel and distributed real-time systems in Z. In *Proc. 4th International Workshop on Parallel and Distributed Real-Time Systems, Hawaii*, pages 216–222, April 1996.

37. M. Benjamin. A message passing system: An example of combining CSP and Z. In Nicholls [400], pages 221–228.

38. M. Benveniste. Writing operational semantics in Z: A structural approach. In Prehn and Toetenel [426], pages 164–188.

39. S. Bera. Structuring for the VDM specification language. In Bloomfield et al. [46], pages 2–25.

40. P. Bernard and G. Laffitte. The French population census for 1990. In Bowen and Hinchey [82], pages 334–352.

41. J. Bicarregui, J. Dick, and E. Woods. Supporting the length of formal development: From diagrams to VDM to B to C. In Habrias [237], pages 63–75.

42. J. Bicarregui, J. Dick, and E. Woods. Quantitative analysis of formal methods. In Gaudel and Woodcock [217], pages 60–73.

43. J. Bicarregui and B. Ritchie. Invariants, frames and postconditions: A comparison of the VDM and B notations. *IEEE Transactions on Software Engineering*, 21(2):79–89, 1995.

44. P. G. Bishop, editor. *Fault Avoidance*, chapter 3, pages 56–140. Applied Science. Elsevier Science Publishers, 1990.
Section 3.88 (pages 94–96) provides an overview of Z. Other sections describe related techniques.

45. D. Bjørner, C. A. R. Hoare, and H. Langmaack, editors. *VDM and Z – Formal Methods in Software Development*, volume 428 of *Lecture Notes in Computer Science*. VDM-Europe, Springer-Verlag, 1990.
The 3rd VDM-Europe Symposium was held at Kiel, Germany, 17–21 April 1990. A significant number of papers concerned with Z were presented [116, 178, 214, 158, 229, 240, 319, 451, 489, 534, 566].

46. R. Bloomfield, L. Marshall, and R. Jones, editors. *VDM – The Way Ahead*, volume 328 of *Lecture Notes in Computer Science*. VDM-Europe, Springer-Verlag, 1988.
The 2nd VDM-Europe Symposium was held at Dublin, Ireland, 11–16 September 1988. See [6, 39].

47. E. Boiten, J. Derrick, H. Bowman, and M. Steen. Unification and multiple views of data in Z. In J. C. van Vliet, editor, *Proc. Computing Science in the Netherlands*, pages 73–85, November 1995.

48. E. Boiten, J. Derrick, H. Bowman, and M. Steen. Consistency and refinement for partal specifications in Z. In Gaudel and Woodcock [217], pages 287–306.

49. A. Boswell. Specification and validation of a security policy model. In Woodcock and Larsen [563], pages 42–51.
The 1st FME Symposium was held at Odense, Denmark, 19–23 April 1993. Z-related papers include [85, 134, 199, 296, 355, 422].

50. A. Boswell. Specification and validation of a security policy model. *IEEE Transactions on Software Engineering*, 21(2):99–106, 1995.

This paper describes the development of a formal security model in Z for the NATO Air Command and Control System (ACCS): a large, distributed, multilevel-secure system. The model was subject to manual validation, and some of the issues and lessons in both writing and validating the model are discussed.

51. L. Bottaci and J. Jones. *Formal Specification Using Z: A Modelling Approach*. International Thomson Publishing, London, 1995.

52. J. P. Bowen. Formal specification and documentation of microprocessor instruction sets. *Microprocessing and Microprogramming*, 21(1–5):223–230, August 1987.

53. J. P. Bowen. The formal specification of a microprocessor instruction set. Technical Monograph PRG-60, Oxford University Computing Laboratory, Wolfson Building, Parks Road, Oxford, UK, January 1987.
The Z notation is used to define the Motorola M6800 8-bit microprocessor instruction set.

54. J. P. Bowen, editor. *Proc. Z Users Meeting, 1 Wellington Square, Oxford*, Wolfson Building, Parks Road, Oxford, UK, December 1987. Oxford University Computing Laboratory.
The 1987 Z Users Meeting was held on Friday 8 December at the Department of External Studies, Rewley House, 1 Wellington Square, Oxford, UK.

55. J. P. Bowen. Formal specification in Z as a design and documentation tool. In *Proc. Second IEE/BCS Conference on Software Engineering*, number 290 in Conference Publication, pages 164–168. IEE/BCS, July 1988.

56. J. P. Bowen, editor. *Proc. Third Annual Z Users Meeting*, Wolfson Building, Parks Road, Oxford, UK, December 1988. Oxford University Computing Laboratory.
The 1988 Z Users Meeting was held on Friday 16 December at the Department of External Studies, Rewley House, 1 Wellington Square, Oxford, UK. Issued with *A Miscellany of Handy Techniques* by R. Macdonald, *Practical Experience of Formal Specification: A programming interface for communications* by J. B. Wordsworth, and a number of posters.

57. J. P. Bowen. Formal specification of window systems. Technical Monograph PRG-74, Oxford University Computing Laboratory, Wolfson Building, Parks Road, Oxford, UK, June 1989.
Three window systems, X from MIT, WM from Carnegie-Mellon University and the Blit from AT&T Bell Laboratories are covered.

58. J. P. Bowen. POS: Formal specification of a UNIX tool. *IEE/BCS Software Engineering Journal*, 4(1):67–72, January 1989.

59. J. P. Bowen. Formal specification of the ProCoS/safemos instruction set. *Microprocessors and Microsystems*, 14(10):631–643, December 1990.
This article is part of a special feature on *Formal aspects of microprocessor design*, edited by H. S. M. Zedan. See also [461].

60. J. P. Bowen. Z bibliography. Maintained on-line in BIBTEX database format. URL: http://www.comlab.ox.ac.uk/archive/z/bib.html, 1990–1996.
This bibliography is maintained in BIBTEX database source format accessible in searchable form on the World Wide Web. To add entries, please send as complete information as possible to Jonathan Bowen on J.P.Bowen@reading.ac.uk.

61. J. P. Bowen. X: Why Z? *Computer Graphics Forum*, 11(4):221–234, October 1992.
This paper asks whether window management systems would not be better specified through a formal methodology and gives examples in Z of the X Window System.

62. J. P. Bowen. Formal methods in safety-critical standards. In *Proc. 1993 Software Engineering Standards Symposium*, pages 168–177. IEEE Computer Society Press, 1993.

63. J. P. Bowen. Report on Z User Meeting, London 1992. *BCS-FACS FACTS*, Series III, 1(3):7–8, Summer 1993.
Other versions of this report have appeared as follows:

  - Z User Meetings, *Safety Systems: The Safety-Critical Systems Club Newsletter*, 3(1):13, September 1993.
  - Z User Group activities, *JFIT News*, 46:5, September 1993.
  - Report on Z User Meeting, *Information and Software Technology*, 35(10):613, October 1993.
  - Z User Meeting Activities, *High Integrity Systems*, 1(1):93–94, 1994.

64. J. P. Bowen. Comp.specification.z and Z FORUM frequently asked questions. In Bowen and Hinchey [82], pages 561–569.

65. J. P. Bowen. Select Z bibliography. In Bowen and Hinchey [82], pages 527–560.

66. J. P. Bowen. Z glossary. *Information and Software Technology*, 37(5-6):333–334, May–June 1995.

67. J. P. Bowen. *Formal Specification and Documentation using Z: A Case Study Approach*. International Thomson Computer Press, 1996.

68. J. P. Bowen, P. T. Breuer, and K. C. Lano. A compendium of formal techniques for software maintenance. *IEE/BCS Software Engineering Journal*, 8(5):253–262, September 1993.

69. J. P. Bowen, P. T. Breuer, and K. C. Lano. Formal specifications in software maintenance: From code to Z^{++} and back again. *Information and Software Technology*, 35(11/12):679–690, November/December 1993.

70. J. P. Bowen, R. B. Gimson, and S. Topp-Jørgensen. The specification of network services. Technical Monograph PRG-61, Oxford University Computing Laboratory, Wolfson Building, Parks Road, Oxford, UK, August 1987.

71. J. P. Bowen, R. B. Gimson, and S. Topp-Jørgensen. Specifying system implementations in Z. Technical Monograph PRG-63, Oxford University Computing Laboratory, Wolfson Building, Parks Road, Oxford, UK, February 1988.

72. J. P. Bowen and M. J. C. Gordon. Z and HOL. In Bowen and Hall [74], pages 141–167.

73. J. P. Bowen and M. J. C. Gordon. A shallow embedding of Z in HOL. *Information and Software Technology*, 37(5-6):269–276, May–June 1995.
    Revised version of [72].

74. J. P. Bowen and J. A. Hall, editors. *Z User Workshop, Cambridge 1994*, Workshops in Computing. Springer-Verlag, 1994.
    Proceedings of the Eighth Annual Z User Meeting, St. John's College, Cambridge, UK. Published in collaboration with the British Computer Society. For individual papers, see [34, 72, 92, 115, 117, 163, 191, 193, 212, 241, 243, 248, 252, 327, 356, 420, 474, 541, 562, 567]. The proceedings also includes an *Introduction and Opening Remarks*, a *Select Z Bibliography* and a section answering *Frequently Asked Questions*.

75. J. P. Bowen, He Jifeng, R. W. S. Hale, and J. M. J. Herbert. Towards verified systems: The SAFEMOS project. In C. J. Mitchell and V. Stavridou, editors, *The Mathematics of Dependable Systems*, volume 55 of *The Institute of Mathematics and its Applications Conference Series*, pages 23–48. Oxford University Press, 1995.

76. J. P. Bowen and M. G. Hinchey. Formal methods and safety-critical standards. *IEEE Computer*, 27(8):68–71, August 1994.

77. J. P. Bowen and M. G. Hinchey. Seven more myths of formal methods: Dispelling industrial prejudices. In Naftalin et al. [389], pages 105–117.

78. J. P. Bowen and M. G. Hinchey. Editorial. *Information and Software Technology*, 37(5-6):258–259, May–June 1995.
    A special issue on Z. See [66, 79, 73, 87, 213, 340, 354, 358, 529].

79. J. P. Bowen and M. G. Hinchey. Report on Z User Meeting (ZUM'94). *Information and Software Technology*, 37(5-6):335–336, May–June 1995.

80. J. P. Bowen and M. G. Hinchey. Seven more myths of formal methods. *IEEE Software*, 12(4):34–41, July 1995.

This article deals with further myths in addition to those presented in [239]. Previous versions issued as:
- Technical Report PRG-TR-7-94, Oxford University Computing Laboratory, June 1994.
- Technical Report 357, University of Cambridge, Computer Laboratory, January 1995.

81. J. P. Bowen and M. G. Hinchey. Ten commandments of formal methods. *IEEE Computer*, 28(4):56–63, April 1995.
Previously issued as: Technical Report 350, University of Cambridge, Computer Laboratory, September 1994.

82. J. P. Bowen and M. G. Hinchey, editors. *ZUM'95: The Z Formal Specification Notation, 9th International Conference of Z Users, Limerick, Ireland, September 7-9, 1995, Proceedings*, volume 967 of *Lecture Notes in Computer Science*. Springer-Verlag, 1995.
Proceedings of the Ninth Annual Z User Meeting, University of Limerick, Ireland. For individual papers presented at the main conference, see [27, 40, 102, 121, 148, 157, 164, 190, 207, 221, 228, 242, 278, 282, 299, 325, 345, 348, 350, 370, 416, 439, 448, 475, 492, 505, 533, 539]. Some papers formed part of an associated Educational Issues Session organized by Neville Dean [120, 149, 235, 371, 417, 547]. The proceedings also includes as appendices a *Select Z Bibliography* [65] and a section answering *Frequently Asked Questions* [64].

83. J. P. Bowen and M. G. Hinchey. Formal models and the specification process. In A. B. Tucker, Jr., editor, *The Computer Science and Engineering Handbook*, chapter 107, pages 2302–2322. CRC Press, 1997. Section X, Software Engineering.

84. J. P. Bowen and J. E. Nicholls, editors. *Z User Workshop, London 1992*, Workshops in Computing. Springer-Verlag, 1993.
Proceedings of the Seventh Annual Z User Meeting, DTI Offices, London, UK. Published in collaboration with the British Computer Society. For individual papers, see [28, 90, 129, 136, 144, 142, 172, 266, 295, 322, 337, 347, 359, 405, 411, 428, 446, 515, 531]. The proceedings also includes an *Introduction and Opening Remarks*, a *Select Z Bibliography* and a section answering *Frequently Asked Questions*.

85. J. P. Bowen and V. Stavridou. The industrial take-up of formal methods in safety-critical and other areas: A perspective. In Woodcock and Larsen [563], pages 183–195.

86. J. P. Bowen and V. Stavridou. Safety-critical systems, formal methods and standards. *IEE/BCS Software Engineering Journal*, 8(4):189–209, July 1993.
A survey on the use of formal methods, including B and Z, for safety-critical systems. Winner of the 1994 IEE Charles Babbage Premium award. A previous version is also available as Oxford University Computing Laboratory Technical Report PRG-TR-5-92.

87. J. P. Bowen, S. Stepney, and R. Barden. Annotated Z bibliography. *Information and Software Technology*, 37(5-6):317–332, May–June 1995.
Revised version of [493].

88. H. Bowman and J. Derrick. Modelling distributed systems using Z. In K. M. George, editor, *ACM Symposium on Applied Computing*, pages 147–151. ACM Press, February 1995.

89. H. Bowman, J. Derrick, P. Linington, and M. Steen. FDTs for ODP. *Computer Standards & Interfaces*, 17(5–6):457–479, September 1995.

90. A. Bradley. Requirements for Defence Standard 00-55. In Bowen and Nicholls [84], pages 93–94.

91. P. T. Breuer. Z! in progress: Maintaining Z specifications. In Nicholls [402], pages 295–318.

92. P. T. Breuer and J. P. Bowen. Towards correct executable semantics for Z. In Bowen and Hall [74], pages 185–209.

93. S. M. Brien. The development of Z. In D. J. Andrews, J. F. Groote, and C. A. Middelburg, editors, *Semantics of Specification Languages (SoSL)*, Workshops in Computing, pages 1–14. Springer-Verlag, 1994.

94. S. M. Brien and J. E. Nicholls. Z base standard. Technical Monograph PRG-107, Oxford University Computing Laboratory, Wolfson Building, Parks Road, Oxford, UK, November 1992. Accepted for standardization under ISO/IEC JTC1/SC22.
This is the first publicly available version of the proposed ISO Z Standard. See also [486] for the current most widely available Z reference manual.

95. C. Britton, M. Loomes, and R. Mitchell. Formal specification as constructive diagrams. *Microprocessing and Microprogramming*, 37(1–5):175–178, January 1993.

96. M. Brossard-Guerlus and F. Klay. Introducing formal specification in an industrial context: An experiment in Z. In Habrias [237], pages 229–242.

97. D. J. Brown and J. P. Bowen. The Event Queue: An extensible input system for UNIX workstations. In *Proc. European Unix Users Group Conference*, pages 29–52. EUUG, May 1987.
Available from EUUG Secretariat, Owles Hall, Buntingford, Hertfordshire SG9 9PL, UK.

98. D. Brownbridge. Using Z to develop a CASE toolset. In Nicholls [400], pages 142–149.

99. J.-M. Bruel, A. Benzekri, and Y. Raymaud. Z and the specification of real-time systems. In Habrias [237], pages 77–91.

100. A. Bryant. Structured methodologies and formal notations: Developing a framework for synthesis and investigation. In Nicholls [400], pages 229–241.

101. T. Bryant and A. Evans. Formalizing the Object Management Group's Core Object Model. *Computer Standards & Interfaces*, 17(5–6):481–489, September 1995.

102. T. Bryant, A. Evans, L. Semmens, R. Milovanovic, S. Stockman, M. Norris, and C. Selley. Using Z to rigorously review a specification of a network management system. In Bowen and Hinchey [82], pages 423–433.

103. T. Bryant and L. Semmens, editors. *Methods Integration*, Electronic Workshops in Computing. Springer-Verlag, 1996.
Proceedings of the Methods Integration Workshop, University of Leeds, UK, 25–26 March 1996. See [188, 208, 273, 277, 312, 332, 435].

104. G. R. Buckberry. ZED: A Z notation editor and syntax analyser. *BCS-FACS FACTS*, Series II, 2(3):13–23, November 1991.

105. A. Burns and I. W. Morrison. A formal description of the structure attribute model for tool interfacing. *IEE/BCS Software Engineering Journal*, 4(2):74–78, March 1989.

106. A. Burns and A. J. Wellings. Occam's priority model and deadline scheduling. In *Proc. 7th Occam User Group Meeting, Grenoble*, 1987.

107. J. S. Busby and D. Hutchison. The practical integration of manufacturing applications. *Software Practice and Experience*, 22(2):183–207, 1992.

108. R. Büsser and M. Weber. A steam-boiler control specification with Statecharts and Z. In Abrial et al. [9], pages 109–128.

109. P. Butcher. A behavioural semantics for Linda-2. *IEE/BCS Software Engineering Journal*, 6(4):196–204, July 1991.

110. M. J. Butler. Service extension at the specification level. In Nicholls [402], pages 319–336.

111. D. Carrington. ZOOM workshop report. In Nicholls [404], pages 352–364.
This paper records the activities of a workshop on Z and object-oriented methods held in August 1992 at Oxford. A comprehensive bibliography is included.

112. D. Carrington, D. J. Duke, R. Duke, P. King, G. A. Rose, and G. Smith. Object-Z: An object-oriented extension to Z. In S. Vuong, editor, *Formal Description Techniques, II (FORTE'89)*, pages 281–296. Elsevier Science Publishers (North-Holland), 1990.

113. D. Carrington, D. J. Duke, I. J. Hayes, and J. Welsh. Deriving modular designs from formal specifications. *ACM Software Engineering Notes*, 18(5):89–98, December 1993.

114. D. Carrington and G. Smith. Extending Z for object-oriented specifications. In *Proc. 5th Australian Software Engineering Conference (ASWEC'90)*, pages 9–14, 1990.

115. D. Carrington and P. Stocks. A tale of two paradigms: Formal methods and software testing. In Bowen and Hall [74], pages 51–68.
    Also available as Technical Report 94-4, Department of Computer Science, University of Queensland, 1994.

116. P. Chalin and P. Grogono. Z specification of an object manager. In Bjørner et al. [45], pages 41–71.

117. D. K. C. Chan and P. W. Trinder. An object-oriented data model supporting multi-methods, multiple inheritance, and static type checking: A specification in Z. In Bowen and Hall [74], pages 297–315.

118. W. Chantatub and M. Holcombe. Software testing strategies for software requirements and design. In *Proc. EuroSTAR'94*, pages 40/1–40/29, 3000-2 Hartley Road, Jacksonville, Florida 32257, USA, 1994. Software Quality Engineering.
    The paper describes how to construct a detailed Z specification using traditional software engineering techniques (ERDs, DFDs, etc.) in a top down manner. It introduces a number of notational devices to help with the management of large Z specifications. Some issues about proving consistency between levels are also addressed.

119. J. Y. Chauvet. Le cas "legislation viellesse": Etude de cas. In Habrias [237], pages 243–264.

120. P. Ciaccia and P. Ciancarini. A course on formal methods in software engineering: Matching requirements with design. In Bowen and Hinchey [82], pages 482–496.

121. P. Ciaccia, P. Ciancarini, and W. Penzo. A formal approach to software design: The Clepsydra methodology. In Bowen and Hinchey [82], pages 5–24.

122. B. Cohen. Justification of formal methods for system specifications & A rejustification of formal notations. *IEE/BCS Software Engineering Journal*, 4(1):26–38, January 1989.

123. B. Cohen and D. Mannering. The rigorous specification and verification of the safety aspects of a real-time system. In *COMPASS '90*, 1990.

124. B. P. Collins, J. E. Nicholls, and I. H. Sørensen. Introducing formal methods: The CICS experience with Z. In B. Neumann et al., editors, *Mathematical Structures for Software Engineering*. Oxford University Press, 1991.

125. J. Cooke. Editorial – formal methods: What? why? and when? *The Computer Journal*, 35(5):417–418, October 1992.
    An editorial introduction to two special issues on *Formal Methods*. See also [33, 126, 373, 457, 559] for papers relevant to Z.

126. J. Cooke. Formal methods – mathematics, theory, recipes or what? *The Computer Journal*, 35(5):419–423, October 1992.

127. A. C. Coombes, L. Barroca, J. S. Fitzgerald, J. A. McDermid, L. Spencer, and A. Saeed. Formal specification of an aerospace system: The attitude monitor. In Hinchey and Bowen [274], pages 307–332.

128. A. C. Coombes and J. A. McDermid. A tool for defining the architecture of Z specifications. In Nicholls [402], pages 77–92.

129. A. C. Coombes and J. A. McDermid. Using diagrams to give a formal specification of timing constraints in Z. In Bowen and Nicholls [84], pages 119–130.

130. D. Cooper. Educating management in Z. In Nicholls [400], pages 192–194.

131. V. A. O. Cordeiro, A. C. A. Sampaio, and S. L. Meira. From MooZ to Eiffel – a rigorous approach to system development. In Naftalin et al. [389], pages 306–325.

132. S. Craggs and J. B. Wordsworth. Hursley Lab wins another Queen's Award & Hursley and Oxford – a marriage of minds & Z stands for quality. *Developments, IBM Hursley Park*, 8:1–2, 21 April 1992.

133. I. Craig. *The Formal Specification of Advanced AI Architectures*. AI Series. Ellis Horwood, September 1991.

This book contains two rather large (and relatively complete) specifications of Artificial Intelligence (AI) systems using Z. The architectures are the blackboard and Cassandra architectures. As well as showing that formal specification *can* be used in AI at the architecture level, the book is intended as a case-studies book, and also contains introductory material on Z (for AI people). The book assumes a knowledge of Z, so for non-AI people its primary use is for the presentation of the large specifications. The blackboard specification, with explanatory text, is around 100 pages.

134. D. Craigen, S. L. Gerhart, and T. J. Ralston. Formal methods reality check: Industrial usage. In Woodcock and Larsen [563], pages 250–267.
    The 1st FME Symposium was held at Odense, Denmark, 19–23 April 1993. Z-related papers include [85, 134, 199, 296, 355, 422].

135. D. Craigen, S. L. Gerhart, and T. J. Ralston. An international survey of industrial applications of formal methods. Technical Report NIST GCR 93/626-V1 & 2, Atomic Energy Control Board of Canada, US National Institute of Standards and Technology, and US Naval Research Laboratories, 1993.
    Volume 1: Purpose, Approach, Analysis and Conclusions; Volume 2: Case Studies. Order numbers: PB93-178556/AS & PB93-178564/AS; National Technical Information Service, 5285 Port Royal Road, Springfield, VA 22161, USA.

136. D. Craigen, S. L. Gerhart, and T. J. Ralston. An international survey of industrial applications of formal methods. In Bowen and Nicholls [84], pages 1–5.

137. D. Craigen, S. L. Gerhart, and T. J. Ralston. Formal methods reality check: Industrial usage. *IEEE Transactions on Software Engineering*, 21(2):90–98, 1995.
    Revised version of [134].

138. D. Craigen, S. L. Gerhart, and T. J. Ralston. Formal methods technology transfer: Impediments and innovation. In Hinchey and Bowen [274], pages 399–419.

139. D. Craigen, S. Kromodimoeljo, I. Meisels, W. Pase, and M. Saaltink. EVES: An overview. In Prehn and Toetenel [426], pages 389–405.

140. E. Cusack. Inheritance in object oriented Z. In P. America, editor, *Proc. ECOOP'91 European Conference on Object-Oriented Programming*, volume 512 of *Lecture Notes in Computer Science*, pages 167–179. Springer-Verlag, 1991.

141. E. Cusack. Object oriented modelling in Z for open distributed systems. In J. de Meer, editor, *Proc. International Workshop on ODP*. Elsevier Science Publishers (North-Holland), 1992.

142. E. Cusack. Using Z in communications engineering. In Bowen and Nicholls [84], pages 196–202.

143. E. Cusack and M. Lai. Object oriented specification in LOTOS and Z (or my cat really is object oriented!). In J. W. de Bakker, W. P. de Roever, and G. Rozenberg, editors, *REX/FOOL School/Workshop on Foundations of Object-Oriented Languages*, volume 489 of *Lecture Notes in Computer Science*, pages 179–202. Springer-Verlag, 1990.

144. E. Cusack and C. Wezeman. Deriving tests for objects specified in Z. In Bowen and Nicholls [84], pages 180–195.

145. R. S. M. de Barros. Deriving relational database programs from formal specifications. In Naftalin et al. [389], pages 703–723.

146. R. S. M. de Barros and D. J. Harper. Formal development of relational database applications. In D. J. Harper and M. C. Norrie, editors, *Specifications of Database Systems, Glasgow 1991*, Workshops in Computing, pages 21–43. Springer-Verlag, 1992.

147. R. S. M. de Barros and D. J. Harper. A method for the specification of relational database applications. In Nicholls [404], pages 261–286.

148. P. D. de Lima Machado and S. L. Meira. On the use of formal specifications in the design and simulation of artificial neural nets. In Bowen and Hinchey [82], pages 63–82.

149. N. Dean. Mental models of Z: I – sets and logic. In Bowen and Hinchey [82], pages 498–507.

150. N. Dean. *The Essence of Discrete Mathematics*. The Essence of Computing Series. Prentice Hall, 1997.
An introductory book using a Z-like notation.

151. N. Dean and M. G. Hinchey. Introducing formal methods through rôle-playing. *ACM SIGCSE Bulletin*, 27(1):302–306, March 1995.

152. B. Dehbonei and F. Mejia. Formal methods in the railways signalling industry. In Naftalin et al. [389], pages 26–34.

153. N. Delisle and D. Garlan. Formally specifying electronic instruments. In *Proc. 5th International Workshop on Software Specification and Design*. IEEE Computer Society, May 1989. Also published in ACM SIGSOFT Software Engineering Notes 14(3).

154. N. Delisle and D. Garlan. A formal specification of an oscilloscope. *IEEE Software*, 7(5):29–36, September 1990.
Unlike most work on the application of formal methods, this research uses formal methods to gain insight into system architecture. The context for this case study is electronic instrument design.

155. J. Derrick, E. Boiten, H. Bowman, and M. Steen. Supporting ODP – translating LOTOS to Z. In E. Najm and J.-B. Stefani, editors, *First IFIP International Workshop on Formal Methods for Open Object-based Distributed Systems*. Chapman & Hall, March 1996.

156. J. Derrick, H. Bowman, and M. Steen. Maintaining cross viewpoint consistency using Z. In K. Raymond and L. Armstrong, editors, *IFIP TC6 International Conference on Open Distributed Processing*, pages 413–424. Chapman & Hall, February 1995.

157. J. Derrick, H. Bowman, and M. Steen. Viewpoints and objects. In Bowen and Hinchey [82], pages 449–468.

158. R. Di Giovanni and P. L. Iachini. HOOD and Z for the development of complex systems. In Bjørner et al. [45], pages 262–289.

159. A. J. J. Dick, P. J. Krause, and J. Cozens. Computer aided transformation of Z into Prolog. In Nicholls [400], pages 71–85.

160. A. Diller. *Z: An Introduction to Formal Methods*. John Wiley & Sons, 1990.
This book offers a comprehensive tutorial to Z from the practical viewpoint. Many natural deduction style proofs are presented and exercises are included. A second edition is now available [162].

161. A. Diller. Z and Hoare logics. In Nicholls [404], pages 59–76.

162. A. Diller. *Z: An Introduction to Formal Methods*. John Wiley & Sons, 2nd edition, 1994.
This book offers a comprehensive tutorial to Z from the practical viewpoint. Many natural deduction style proofs are presented and exercises are included. Z as defined in the 2nd edition of *The Z Notation* [486] is used throughout.
Contents: Tutorial introduction; Methods of reasoning; Case studies; Specification animation; Reference manual; Answers to exercises; Glossaries of terms and symbols; Bibliography.

163. A. Diller and R. Docherty. Z and abstract machine notation: A comparison. In Bowen and Hall [74], pages 250–263.

164. M. d'Inverno and M. Priestley. Structuring specification in Z to build a unifying framework for hypertext systems. In Bowen and Hinchey [82], pages 83–102.

165. A. J. Dix. *Formal Methods for Interactive Systems*. Computers and People Series. Academic Press, 1991.

166. A. J. Dix, J. Finlay, G. D. Abowd, and R. Beale. *Human-Computer Interaction*. Prentice Hall International, 1993.

167. R. F. Docherty. Translation from Z to AMN. In Habrias [237], pages 205–228.

168. C. J. Dodge. *A Fast Fourier Transform Accelerator for a Transputer System*. PhD thesis, University of Aberdeen, Department of Biomedical Physics, Foresterhill, Aberdeen AB9 2ZD, UK, 1993.
    The design includes a detailed Z specification.

169. C. J. Dodge, P. G. B. Ross, A. R. Allen, and P. E. Undrill. Formal methods in the design of an FFT accelerator for a Transputer based image processing system. *Medical and Biological Engineering and Computing*, 29:91, 1991. Supplement.

170. C. J. Dodge, P. E. Undrill, A. R. Allen, and P. G. B. Ross. Application of Z in digital hardware design. *IEE Proceedings: Computers and Digital Techniques*, 143(1):79–86, 1996.

171. V. Doma and R. Nicholl. EZ: A system for automatic prototyping of Z specifications. In Prehn and Toetenel [426], pages 189–203.

172. C. Draper. Practical experiences of Z and SSADM. In Bowen and Nicholls [84], pages 240–251.

173. D. A. Duce, D. J. Duke, P. J. W. ten Hagen, I. Herman, and G. J. Reynolds. Formal methods in the development of PREMO. *Computer Standards & Interfaces*, 17(5–6):491–509, September 1995.

174. D. A. Duce, D. J. Duke, P. J. W. ten Hagen, and G. J. Reynolds. PREMO - an initial approach to a formal definition. *Computer Graphics Forum*, 13(3):C–393–C–406, 1994.
    PREMO (Presentation Environments for Multimedia Objects) is a work item proposal by the ISO/IEC JTC11/SC24 committee, which is responsible for international standardization in the area of computer graphics and image processing.

175. D. J. Duke. Structuring Z specifications. In *Proc. 14th Australian Computer Science Conference*, 1991.

176. D. J. Duke. Enhancing the structures of Z specifications. In Nicholls [404], pages 329–351.

177. D. J. Duke. *Object-Oriented Formal Specification*. PhD thesis, Department of Computer Science, University of Queensland, St. Lucia 4072, Australia, 1992.

178. D. J. Duke and R. Duke. Towards a semantics for Object-Z. In Bjørner et al. [45], pages 244–261.

179. D. J. Duke and M. D. Harrison. Event model of human-system interaction. *IEE/BCS Software Engineering Journal*, 10(1):3–12, January 1995.

180. D. J. Duke and M. D. Harrison. Mapping user requirements to implementations. *IEE/BCS Software Engineering Journal*, 10(1):13–20, January 1995.

181. R. Duke and D. J. Duke. Aspects of object-oriented formal specification. In *Proc. 5th Australian Software Engineering Conference (ASWEC'90)*, pages 21–26, 1990.

182. R. Duke, I. J. Hayes, P. King, and G. A. Rose. Protocol specification and verification using Z. In S. Aggarwal and K. Sabnani, editors, *Protocol Specification, Testing, and Verification VIII*, pages 33–46. Elsevier Science Publishers (North-Holland), 1988.

183. R. Duke, P. King, G. A. Rose, and G. Smith. The Object-Z specification language. In T. Korson, V. Vaishnavi, and B. Meyer, editors, *Technology of Object-Oriented Languages and Systems: TOOLS 5*, pages 465–483. Prentice Hall, 1991.

184. R. Duke, P. King, G. A. Rose, and G. Smith. The Object-Z specification language: Version 1. Technical Report 91-1, Department of Computer Science, University of Queensland, St. Lucia 4072, Australia, April 1991.
    The most complete (and currently the standard) reference on Object-Z. It has been reprinted by ISO JTC1 WG7 as document number 372. A condensed version of this report was published as [183].

185. R. Duke, G. Rose, and G. Smith. Object-Z: A specification language advocated for the description of standards. *Computer Standards & Interfaces*, 17(5–6):511–533, September 1995.

186. R. Duke, G. A. Rose, and A. Lee. Object-oriented protocol specification. In L. Logrippo, R. L. Probert, and H. Ural, editors, *Protocol Specification, Testing, and Verification X*, pages 325–338. Elsevier Science Publishers (North-Holland), 1990.

187. R. Duke and G. Smith. Temporal logic and Z specifications. *Australian Computer Journal*, 21(2):62–69, May 1989.

188. L. Dunckley and A. Smith. Improving access of the commercial software developer to formal methods: Integrating MERISE with Z. In Bryant and Semmens [103].

189. D. Edmond. *Information Modeling: Specification and Implementation*. Prentice Hall, 1992.

190. D. Edmond. Refining database systems. In Bowen and Hinchey [82], pages 25–44.

191. M. Engel. Specifying real-time systems with Z and the Duration Calculus. In Bowen and Hall [74], pages 282–294.

192. A. S. Evans. Specifying & verifying concurrent systems using Z. In Naftalin et al. [389], pages 366–400.

193. A. S. Evans. Visualising concurrent Z specifications. In Bowen and Hall [74], pages 269–281.

194. P. C. Fencott, A. J. Galloway, M. A. Lockyer, S. J. O'Brien, and S. Pearson. Formalising the semantics of Ward/Mellor SA/RT essential models using a process algebra. In Naftalin et al. [389], pages 681–702.

195. N. E. Fenton and D. Mole. A note on the use of Z for flowgraph transformation. *Information and Software Technology*, 30(7):432–437, 1988.

196. E. Fergus and D. C. Ince. Z specifications and modal logic. In P. A. V. Hall, editor, *Proc. Software Engineering 90*, volume 1 of *British Computer Society Conference Series*. Cambridge University Press, 1990.

197. C. Fidge, M. Utting, P. Kearney, and I. J. Hayes. Integrating real-time scheduling theory and program refinement. In Gaudel and Woodcock [217], pages 327–346.

198. C. J. Fidge. Specification and verification of real-time behaviour using Z and RTL. In J. Vytopil, editor, *Formal Techniques in Real-Time and Fault-Tolerant Systems*, Lecture Notes in Computer Science, pages 393–410. Springer-Verlag, 1992.

199. C. J. Fidge. Real-time refinement. In Woodcock and Larsen [563], pages 314–331.

200. C. J. Fidge. Adding real time to formal program development. In Naftalin et al. [389], pages 618–638.

201. C. J. Fidge. Proof obligations for real-time refinement. In Till [522], pages 279–305.

202. K. Finney. Mathematical notation in formal specification: Too difficult for the masses? *IEEE Transactions on Software Engineering*, 22(2):158–159, February 1996.

203. M. Flynn, T. Hoverd, and D. Brazier. Formaliser – an interactive support tool for Z. In Nicholls [400], pages 128–141.

204. I. Fogg, B. Hicks, A. Lister, T. Mansfield, and K. Raymond. A comparison of LOTOS and Z for specifying distributed systems. *Australian Computer Science Communications*, 12(1):88–96, February 1990.

205. D. C. Fowler, P. A. Swatman, and P. M. C. Swatman. Implementing EDI in the public sector: Including formality for enhanced control. In *Proc. 7th International Conference on Electronic Data Interchange*, June 1993.

206. R. B. France and S. L. Gerhart, editors. *Proc. Workshop on Industrial-strength Formal Specification Techniques*. IEEE Computer Society Press, 1995.

207. R. B. France and M. M. Larrondo-Petrie. A two-dimensional view of integrated formal and informal specification techniques. In Bowen and Hinchey [82], pages 434–448.

208. R. B. France, J. Wu, M. M. Larondo-Petrie, and J.-M. Bruel. A tale of two case studies: Using integrated methods to support rigorous requirements specification. In Bryant and Semmens [103].

Includes a study of an integrated Object-Oriented Analysis (OOA) method (Fusion) and formal specification technique (Z) used to create requirements models that are graphical and analyzable.

209. N. E. Fuchs. Specifications are (preferably) executable. *IEE/BCS Software Engineering Journal*, 7(5):323–334, September 1992.

210. P. H. B. Gardiner, P. J. Lupton, and J. C. P. Woodcock. A simpler semantics for Z. In Nicholls [402], pages 3–11.

211. D. Garlan. The role of reusable frameworks. *ACM SIGSOFT Software Engineering Notes*, 15(4):42–44, September 1990.

212. D. Garlan. Integrating formal methods into a professional master of software engineering program. In Bowen and Hall [74], pages 71–85.

213. D. Garlan. Making formal methods effective for professional software engineers. *Information and Software Technology*, 37(5-6):261–268, May–June 1995.
    Revised version of [212].

214. D. Garlan and N. Delisle. Formal specifications as reusable frameworks. In Bjørner et al. [45], pages 150–163.

215. D. Garlan and N. Delisle. Formal specification of an architecture for a family of instrumentation systems. In Hinchey and Bowen [274], pages 55–72.

216. D. Garlan and D. Notkin. Formalizing design spaces: Implicit invocation mechanisms. In Prehn and Toetenel [426], pages 31–45.

217. M.-C. Gaudel and J. C. P. Woodcock, editors. *FME'96: Industrial Benefit and Advances in Formal Methods*, volume 1051 of *Lecture Notes in Computer Science*. Formal Methods Europe, Springer-Verlag, 1996.
    The 3rd FME Symposium was held at Oxford, UK, 18–22 October 1996. The proceedings includes Z-related papers [48, 197, 313, 538] and B-related papers [42, 276, 537].

218. S. L. Gerhart. Applications of formal methods: Developing virtuoso software. *IEEE Software*, 7(5):6–10, September 1990.
    This is an introduction to a special issue on Formal Methods with an emphasis on Z in particular. It was published in conjunction with special Formal Methods issues of *IEEE Transactions on Software Engineering* and *IEEE Computer*. See also [154, 239, 390, 484, 546].

219. S. L. Gerhart, D. Craigen, and T. J. Ralston. Observations on industrial practice using formal methods. In *Proc. 15th International Conference on Software Engineering (ICSE), Baltimore, Maryland, USA*, May 1993.

220. S. L. Gerhart, D. Craigen, and T. J. Ralston. Experience with formal methods in critical systems. *IEEE Software*, 11(1):21–28, January 1994.
    Several commercial and exploratory cases in which Z features heavily are briefly presented on page 24. See also [323].

221. D. M. German and D. D. Cowan. Experiments with the Z Interchange Format and SGML. In Bowen and Hinchey [82], pages 224–233.

222. S. Gilmore. Correctness-oriented approaches to software development. Technical Report ECS-LFCS-91-147 (also CST-76-91), Department of Computer Science, University of Edinburgh, Edinburgh EH9 3JZ, UK, 1991.
    This PhD thesis provides a critical evaluation of Z, VDM and algebraic specifications.

223. R. B. Gimson. The formal documentation of a Block Storage Service. Technical Monograph PRG-62, Oxford University Computing Laboratory, Wolfson Building, Parks Road, Oxford, UK, August 1987.

224. R. B. Gimson and C. C. Morgan. Ease of use through proper specification. In D. A. Duce, editor, *Distributed Computing Systems Programme*. Peter Peregrinus, London, 1984.

225. R. B. Gimson and C. C. Morgan. The Distributed Computing Software project. Technical Monograph PRG-50, Oxford University Computing Laboratory, Wolfson Building, Parks Road, Oxford, UK, July 1985.

226. J. Ginbayashi. Analysis of business processes specified in Z against an E-R data model. Technical Monograph PRG-103, Oxford University Computing Laboratory, Wolfson Building, Parks Road, Oxford, UK, December 1992.

227. H. S. Goodman. From Z specifications to Haskell porgrams: A three-pronged approach. In Habrias [237], pages 167–182.

228. H. S. Goodman. The Z-into-Haskell tool-kit: An illustrative case study. In Bowen and Hinchey [82], pages 374–388.

229. R. Gotzhein. Specifying open distributed systems with Z. In Bjørner et al. [45], pages 319–339.

230. W. K. Grassmann and J.-P. Tremblay. *The Formal Specification of Requirements in Z*, chapter 8, pages 441–480. Prentice Hall, 1996.

231. A. M. Gravell. Minimisation in formal specification and design. In Nicholls [400], pages 32–45.

232. A. M. Gravell. What is a good formal specification? In Nicholls [402], pages 137–150.

233. A. M. Gravell and P. Henderson. Executing formal specifications need not be harmful. *IEE/BCS Software Engineering Journal*, 11(2):104–110, March 1996.

234. A. M. Gravell and C. H. Pratten. Formal methods and open systems. *Software—Concepts and Tools*, 16(4):183–188, 1995.

235. D. Gries. Equational logic: A great pedagogical tool for teaching a skill in logic. In Bowen and Hinchey [82], pages 508–509.

236. H. Habrias. Z, chapter 10, pages 267–290. Méthodologies du Logiciel. Masson, Paris, 1993.

237. H. Habrias, editor. *Z Twenty Years on – What is its Future?*, Université de Nantes, France, 1995. IRIN (Institut de Recherche en Informatique de Nantes).
    Proceedings of the 7th International Conference on *Putting into Practice Methods and Tools for Information System Design*, Nantes, France, 10–12 October 1995. This conference considered the future of Z, about twenty years after a seminal paper relating to Z [5]. See [41, 96, 99, 119, 167, 227, 244, 263, 342, 398, 415, 425, 466, 502, 532].

238. F. Halasz and M. Schwartz. The Dexter hypertext reference model. In *NIST Hypertext Standardization Workshop*, January 1990.

239. J. A. Hall. Seven myths of formal methods. *IEEE Software*, 7(5):11–19, September 1990.
    Formal methods are difficult, expensive, and not widely useful, detractors say. Using a case study and other real-world examples, this article challenges such common myths. See also [80].

240. J. A. Hall. Using Z as a specification calculus for object-oriented systems. In Bjørner et al. [45], pages 290–318.

241. J. A. Hall. Specifying and interpreting class hierarchies in Z. In Bowen and Hall [74], pages 120–138.

242. J. A. Hall, D. L. Parnas, N. Plat, J. Rushby, and C. T. Sennett. The future of industrial formal methods. In Bowen and Hinchey [82], pages 238–242.
    Position statements for a panel session moderated by T. King.

243. J. G. Hall and J. A. McDermid. Towards a Z method: Axiomatic specification in Z. In Bowen and Hall [74], pages 213–229.

244. J. G. Hall, J. A. McDermid, and I. Toyn. Model conjectures for Z specifications. In Habrias [237], pages 41–51.

245. P. A. V. Hall. Towards testing with respect to formal specification. In *Proc. Second IEE/BCS Conference on Software Engineering*, number 290 in Conference Publication, pages 159–163. IEE/BCS, July 1988.

246. U. Hamer and J. Peleska. Z applied to the A330/340 CICS cabin communication system. In Hinchey and Bowen [274], pages 253–284.

247. V. Hamilton. The use of Z within a safety-critical software system. In Hinchey and Bowen [274], pages 357–374.

248. J. A. R. Hammond. Producing Z specifications from object-oriented analysis. In Bowen and Hall [74], pages 316–336.

249. J. A. R. Hammond. Z. In J. J. Marciniak, editor, *Encyclopedia of Software Engineering*, volume 2, pages 1452–1453. John Wiley & Sons, 1994.

250. M. D. Harrison. Engineering human-error tolerant software. In Nicholls [404], pages 191–204.

251. A. Harry. *Formal Methods Fact File: VDM and Z*. John Wiley & Sons, 1996.

252. W. Hasselbring. Animation of Object-Z specifications with a set-oriented prototyping language. In Bowen and Hall [74], pages 337–356.

253. W. Hasselbring. Prototyping parallel algorithms in a set-oriented language. Dissertation, Department of Computer Science, University of Dortmund, Hamburg, Germany, 1994.
    This dissertation presents the design and implementation of an approach to prototyping parallel algorithms with ProSet-Linda. The presented approach to designing and implementing ProSet-Linda relies on the use of the formal specification language Object-Z and the prototyping language ProSet itself.

254. H. P. Haughton. Using Z to model and analyse safety and liveness properties of communication protocols. *Information and Software Technology*, 33(8):575–580, October 1991.

255. I. J. Hayes. Applying formal specification to software development in industry. *IEEE Transactions on Software Engineering*, 11(2):169–178, February 1985.

256. I. J. Hayes. Specification directed module testing. *IEEE Transactions on Software Engineering*, 12(1):124–133, January 1986.

257. I. J. Hayes. Using mathematics to specify software. In *Proc. First Australian Software Engineering Conference*. Institution of Engineers, Australia, May 1986.

258. I. J. Hayes. A generalisation of bags in Z. In Nicholls [400], pages 113–127.

259. I. J. Hayes. Interpretations of Z schema operators. In Nicholls [402], pages 12–26.

260. I. J. Hayes. Multi-relations in Z: A cross between multi-sets and binary relations. *Acta Informatica*, 29(1):33–62, February 1992.

261. I. J. Hayes. VDM and Z: A comparative case study. *Formal Aspects of Computing*, 4(1):76–99, 1992.

262. I. J. Hayes, editor. *Specification Case Studies*. Prentice Hall International Series in Computer Science, 2nd edition, 1993.
    This is a revised edition of the first ever book on Z, originally published in 1987; it contains substantial changes to every chapter. The notation has been revised to be consistent with *The Z Notation: A Reference Manual* by Mike Spivey [486]. The CAVIAR chapter has been extensively changed to make use of a form of modularization.
    Divided into four sections, the first provides tutorial examples of specifications, the second is devoted to the area of software engineering, the third covers distributed computing, analyzing the role of mathematical specification, and the fourth part covers the IBM CICS transaction processing system. Appendices include comprehensive glossaries of the Z mathematical and schema notation. The book will be of interest to the professional software engineer involved in designing and specifying large software projects.
    The other contributors are W. Flinn, R. B. Gimson, S. King, C. C. Morgan, I. H. Sørensen and B. A. Sufrin.

263. I. J. Hayes. Specification models. In Habrias [237], pages 1–10.

264. I. J. Hayes and C. B. Jones. Specifications are not (necessarily) executable. *IEE/BCS Software Engineering Journal*, 4(6):330–338, November 1989.

265. I. J. Hayes, C. B. Jones, and J. E. Nicholls. Understanding the differences between VDM and Z. *FACS Europe*, Series I, 1(1):7–30, Autumn 1993.
    Also available as Technical Report UMCS-93-8-1, Department of Computer Science, University of Manchester, UK, 1993.

266. I. J. Hayes and L. Wildman. Towards libraries for Z. In Bowen and Nicholls [84], pages 9–36.

267. He Jifeng, C. A. R. Hoare, M. Fränzle, M. Müller-Ulm, E.-R. Olderog, M. Schenke, A. P. Ravn, and H. Rischel. Provably correct systems. In H. Langmaack, W.-P. de Roever, and J. Vytopil, editors, *Formal Techniques in Real Time and Fault Tolerant Systems*, volume 863 of *Lecture Notes in Computer Science*, pages 288–335. Springer-Verlag, 1994.

268. He Jifeng, C. A. R. Hoare, and J. W. Sanders. Data refinement refined. In B. Robinet and R. Wilhelm, editors, *Proc. ESOP 86*, volume 213 of *Lecture Notes in Computer Science*, pages 187–196. Springer-Verlag, 1986.

269. D. Heath, D. Allum, and L. Dunckley. *Introductory Logic and Formal Methods*. A. Waller, Henley-on-Thames, UK, 1994.

270. B. Hepworth. ZIP: A unification initiative for Z standards, methods and tools. In Nicholls [400], pages 253–259.

271. B. Hepworth and D. Simpson. The ZIP project. In Nicholls [402], pages 129–133.

272. M. G. Hinchey. Formal methods for system specification: An ounce of prevention is worth a pound of cure. *IEEE Potentials Magazine*, 12(3):50–52, October 1993.

273. M. G. Hinchey. JSD $\mathrel{\hat=}$ ΔCSP $\oplus$ TLZ – a case study. In Bryant and Semmens [103].

274. M. G. Hinchey and J. P. Bowen, editors. *Applications of Formal Methods*. Prentice Hall International Series in Computer Science, 1995.
    A collection on industrial examples of the use of formal methods. Chapters relevant to Z include [127, 138, 215, 246, 247, 275, 357].

275. M. G. Hinchey and J. P. Bowen. Applications of formal methods FAQ. In *Applications of Formal Methods* [274], pages 1–15.

276. J. Hoare, J. Dick, D. Neilson, and I. H. Sørensen. Applying the B technologies to CICS. In Gaudel and Woodcock [217], pages 74–84.

277. S. Hooker, M. A. Lockyer, and P. C. Fencott. CASE support for methods integration: Implementation of a translation from a structured to a formal notation. In Bryant and Semmens [103].
    The work presented takes the Z specification of the Semantic Function and implements it in the functional programming language, ML.

278. H.-M. Hörcher. Improving software tests using Z specifications. In Bowen and Hinchey [82], pages 152–166.

279. I. S. C. Houston and M. Josephs. Specifying distributed CICS in Z: Accessing local and remote resources (short communication). *Formal Aspects of Computing*, 6(6):569–579, 1994.

280. I. S. C. Houston and M. B. Jospehs. A formal description of the OMG's Core Object Model and the meaning of compatible extension. *Computer Standards & Interfaces*, 17(5–6):553–558, September 1995.

281. I. S. C. Houston and S. King. CICS project report: Experiences and results from the use of Z in IBM. In Prehn and Toetenel [426], pages 588–596.

282. A. P. Hughes and A. A. Donnelly. An algebraic proof in VDM♣. In Bowen and Hinchey [82], pages 114–133.

283. A. D. Hutcheon and A. J. Wellings. Specifying restrictions on imperative programming languages for use in a distributed embedded environment. *IEE/BCS Software Engineering Journal*, 5(2):93–104, March 1990.

284. P. L. Iachini. Operation schema iterations. In Nicholls [402], pages 50–57.

285. M. Imperato. *An Introduction to Z*. Chartwell-Bratt, 1991.
    Contents: Introduction; Set theory; Logic; Building Z specifications; Relations; Functions; Sequences; Bags; Advanced Z; Case study: a simple banking system.

286. D. C. Ince. Z and system specification. In D. C. Ince and D. Andrews, editors, *The Software Life Cycle*, chapter 12, pages 260–277. Butterworths, 1990.

287. D. C. Ince. *An Introduction to Discrete Mathematics, Formal System Specification and Z*. Oxford Applied Mathematics and Computing Science Series. Oxford University Press, 2nd edition, 1993.

288. INMOS Limited. Specification of instruction set & Specification of floating point unit instructions. In *Transputer Instruction Set – A compiler writer's guide*, pages 127–161. Prentice Hall, 1988.
    Appendices F and G use a Z-like notation to give a specification of the instruction set of the IMS T212 and T414 Transputers, and the T800 floating-point Transputer.

289. A. Jack. It's hard to explain, but Z is much clearer than English. *Financial Times*, page 22, 21 April 1992.

290. D. Jackson. Abstract model checking of infinite specifications. In Naftalin et al. [389], pages 519–531.

291. D. Jackson. Structuring Z specifications with views. *ACM Transactions on Software Engineering and Methodology (TOSEM)*, 4(4):365–389, October 1995.

292. D. Jackson and C. A. Damon. Elements of style: Analyzing a software design feature with a counterexample detector. *IEEE Transactions on Software Engineering*, 22(7):484–495, July 1996.
    Nitpick, a specification checker, is applied to the design of a style mechanism for a word processor, using a subset of Z.

293. D. Jackson and M. Jackson. Problem decomposition for reuse. *IEE/BCS Software Engineering Journal*, 11(1):19–30, January 1996.
    An approach to software development based on the idea of *problem frames* and of structuring Z specifications as *views*.

294. J. Jacky. Formal specifications for a clinical cyclotron control system. *ACM SIGSOFT Software Engineering Notes*, 15(4):45–54, September 1990.

295. J. Jacky. Formal specification and development of control system input/output. In Bowen and Nicholls [84], pages 95–108.

296. J. Jacky. Specifying a safety-critical control system in Z. In Woodcock and Larsen [563], pages 388–402.
    The 1st FME Symposium was held at Odense, Denmark, 19–23 April 1993. Z-related papers include [85, 134, 199, 296, 355, 422].

297. J. Jacky. Specifying a safety-critical control system in Z. *IEEE Transactions on Software Engineering*, 21(2):99–106, 1995.
    Revised version of [296].

298. J. Jacky. *The Way of Z: Practical Programming with Formal Methods*. Cambridge University Press, 1997.

299. J. Jacky and J. Unger. From Z to code: A graphical user interface for a radiation therapy machine. In Bowen and Hinchey [82], pages 315–333.

300. J. Jacob. The varieties of refinements. In Morris and Shaw [388], pages 441–455.

301. Jin Song Dong and R. Duke. An object-oriented approach to the formal specification of ODP trader. In *Proc. IFIP TC6/WG6.1 International Conference on Open Distributed Processing*, pages 341–352, September 1993.

302. Jin Song Dong, R. Duke, and G. A. Rose. An object-oriented approach to the semantics of programming languages. In *Proc. 17th Australian Computer Science Conference (ACSC-17)*, pages 767–775, January 1994.

303. C. W. Johnson. Using Z to support the design of interactive safety-critical systems. *IEE/BCS Software Engineering Journal*, 10(2):49–60, March 1995.

304. M. Johnson and P. Sanders. From Z specifications to functional implementations. In Nicholls [400], pages 86–112.

305. P. Johnson. Using Z to specify CICS. In *Proc. Software Engineering anniversary meeting (SEAS)*, page 303, 1987.

306. C. B. Jones. Interference revisited. In Nicholls [402], pages 58–73.

307. C. B. Jones, R. C. Shaw, and T. Denvir, editors. *5th Refinement Workshop*, Workshop in Computing. Springer-Verlag, 1992.
   The workshop was held at Lloyd's Register, London, UK, 8–10 January 1992. See [460].

308. R. B. Jones. ICL ProofPower. *BCS-FACS FACTS*, Series III, 1(1):10–13, Winter 1992.

309. D. Jordan, J. A. McDermid, and I. Toyn. CADiZ – computer aided design in Z. In Nicholls [402], pages 93–104.

310. M. B. Josephs. The data refinement calculator for Z specifications. *Information Processing Letters*, 27(1):29–33, 1988.

311. M. B. Josephs. A state-based approach to communicating processes. *Distributed Computing*, 3:9–18, 1988.
   A theoretical paper on combining features of CSP and Z.

312. V. Kasurinen and K. Sere. Data modelling in ZIM. In Bryant and Semmens [103].

313. V. Kasurinen and K. Sere. Integrating action systems and Z in a medical system specification. In Gaudel and Woodcock [217], pages 105–119.

314. H. Kilov. Information modeling and Object Z: Specifying generic reusable associations. In O. Etzion and A. Segev, editors, *Proc. NGITS-93 (Next Generation Information Technology and Systems)*, pages 182–191, June 1993.

315. H. Kilov and J. Ross. Declarative specifications of collective behavior: Generic reusable frameworks. In H. Kilov and W. Harvey, editors, *Proc. Workshop on Specification of Behavioral Semantics in Object-Oriented Information Modeling*, pages 71–75, Institute for Information Management and Department of Computer and Information Systems, Robert Morris College, Coraopolos and Pittsburgh, Pennsylvania, USA, 1993. OOPSLA.

316. H. Kilov and J. Ross. Appendix A: A more formal approach. In *Information Modeling: An Object-Oriented Approach*, Object-Oriented Series, pages 199–207. Prentice Hall, 1994.

317. H. Kilov and J. Ross. *Information Modeling: An Object-Oriented Approach*. Object-Oriented Series. Prentice Hall, 1994.

318. P. King. Printing Z and Object-Z LaTeX documents. Department of Computer Science, University of Queensland, May 1990.
   A description of a Z style option 'oz.sty', an extended version of Mike Spivey's 'zed.sty' [483], for use with the LaTeX document preparation system [326]. It is particularly useful for printing Object-Z documents [112, 178].

319. S. King. Z and the refinement calculus. In Bjørner et al. [45], pages 164–188.
   Also published as Technical Monograph PRG-79, Oxford University Computing Laboratory, February 1990.

320. S. King and I. H. Sørensen. Specification and design of a library system. In McDermid [364].

321. S. King, I. H. Sørensen, and J. C. P. Woodcock. Z: Grammar and concrete and abstract syntaxes. Technical Monograph PRG-68, Oxford University Computing Laboratory, Wolfson Building, Parks Road, Oxford, UK, 1988.

322. J. C. Knight and D. M. Kienzle. Preliminary experience using Z to specify a safety-critical system. In Bowen and Nicholls [84], pages 109–118.

323. J. C. Knight and B. Littlewood. Critical task of writing dependable software. *IEEE Software*, 11(1):16–20, January 1994.
    Guest editors' introduction to a special issue of *IEEE Software* on *Safety-Critical Systems*. A short section on formal methods mentions several Z books on page 18. See also [220].

324. R. D. Knott and P. J. Krause. The implementation of Z specifications using program transformation systems: The SuZan project. In C. Rattray and R. G. Clark, editors, *The Unified Computation Laboratory*, volume 35 of *IMA Conference Series*, pages 207–220, Oxford, UK, 1992. Clarendon Press.

325. I. Kraan and P. Baumann. Implementing Z in Isabelle. In Bowen and Hinchey [82], pages 355–373.

326. L. Lamport. LaTeX *User's Guide & Reference Manual: A document preparation system*. Addison-Wesley Publishing Company, 2nd edition, 1993.
    Z specifications may be produced using the document preparation system LaTeX together with a special LaTeX style option. The most widely used style files are fuzz.sty [485], zed.sty [483] and oz.sty [318].

327. L. Lamport. TLZ. In Bowen and Hall [74], pages 267–268. Abstract.

328. K. C. Lano. Z^{++}, an object-orientated extension to Z. In Nicholls [402], pages 151–172.

329. K. C. Lano. Refinement in object-oriented specification languages. In Till [522], pages 236–259.

330. K. C. Lano and P. T. Breuer. From programs to Z specifications. In Nicholls [400], pages 46–70.

331. K. C. Lano, P. T. Breuer, and H. P. Haughton. Reverse engineering COBOL via formal methods. *Software Maintenance: Research and Practice*, 5:13–35, 1993.
    Also published in a shortened form as Chapter 16 in [536].

332. K. C. Lano and S. Goldsack. Integrated formal and object-oriented methods: The VDM^{++} approach. In Bryant and Semmens [103].
    Structure Diagrams are formalized in terms of TLZ [327], a combination of the Z notation and the simple temporal logic of Lamport's TLA, with changes in state being a function of the events in the RPT+ formalization.

333. K. C. Lano and H. P. Haughton. An algebraic semantics for the specification language Z^{++}. In *Proc. Algebraic Methodology and Software Technology Conference (AMAST '91)*. Springer-Verlag, 1992.

334. K. C. Lano and H. P. Haughton. Reasoning and refinement in object-oriented specification languages. In O. L. Madsen, editor, *ECOOP '92: European Conference on Object-Oriented Programming*, volume 615 of *Lecture Notes in Computer Science*, pages 78–97. Springer-Verlag, 1992.

335. K. C. Lano and H. P. Haughton. *The Z^{++} Manual*. Lloyd's Register of Shipping, 29 Wellesley Road, Croydon CRO 2AJ, UK, 1992.

336. K. C. Lano and H. P. Haughton, editors. *Object Oriented Specification Case Studies*. Object Oriented Series. Prentice Hall International, 1993.
    Contents: Chapters introducing object oriented methods, object oriented formal specification and the links between formal and structured object-oriented techniques; seven case studies in particular object oriented formal methods, including:
    The Unix Filing System: A MooZ Specification; An Object-Z Specification of a Mobile Phone System; Object-oriented Specification in VDM^{++}; Specifying

a Concept-recognition System in Z^{++}; Specification in OOZE; Refinement in Fresco; SmallVDM: An Environment for Formal Specification and Prototyping in Smalltalk.

A glossary, index and bibliography are also included. The contributors are some of the leading figures in the area, including the developers of the above methods and languages: Silvio Meira, Gordon Rose, Roger Duke, Antonio Alencar, Joseph Goguen, Alan Wills, Cassio Souza dos Santos, Ana Cavalcanti.

337. K. C. Lano and H. P. Haughton. Reuse and adaptation of Z specifications. In Bowen and Nicholls [84], pages 62–90.

338. K. C. Lano and H. P. Haughton. *Reverse Engineering and Software Maintenance: A Practical Approach*. International Series in Software Engineering. McGraw Hill, 1993.

339. K. C. Lano and H. P. Haughton. Standards and techniques for object-oriented formal specification. In *Proc. 1993 Software Engineering Standards Symposium*, pages 237–246. IEEE Computer Society Press, 1993.

340. K. C. Lano and H. P. Haughton. Formal development in B Abstract Machine Notation. *Information and Software Technology*, 37(5-6):303–316, May–June 1995.

341. G. Laycock. Formal specification and testing: A case study. *Software Testing, Verification and Reliability*, 2(1):7–23, May 1992.

342. Y. Ledru and Y. Chiaramella. Integrating and teaching Z and CSP. In Habrias [237], pages 131–147.

343. D. Lightfoot. *Formal Specification using Z*. Macmillan, 1991.
Contents: Introduction; Sets in Z; Using sets to describe a system – a simple example; Logic: propositional calculus; Example of a Z specification document; Logic: predicate calculus; Relations; Functions; A seat allocation system; Sequences; An example of sequences – the aircraft example again; Extending a specification; Collected notation; Books on formal specification; Hints on creating specifications; Solutions to exercises. Also available in French.

344. P. A. Lindsay. On transferring VDM verification techniques to Z. In Naftalin et al. [389], pages 190–213.
Also available as Technical Report 94-10, Department of Computer Science, University of Queensland, 1994.

345. B. Liskov and J. M. Wing. Specifications and their use in defining subtypes. In Bowen and Hinchey [82], pages 246–263.

346. R. L. London and K. R. Milsted. Specifying reusable components using Z: Realistic sets and dictionaries. *ACM SIGSOFT Software Engineering Notes*, 14(3):120–127, May 1989.

347. M. Love. Animating Z specifications in SQL*Forms3.0. In Bowen and Nicholls [84], pages 294–306.

348. M. Luck and M. d'Inverno. Structuring a Z specification to provide a formal framework for autonomous agent systems. In Bowen and Hinchey [82], pages 47–62.

349. P. J. Lupton. Promoting forward simulation. In Nicholls [402], pages 27–49.

350. A. MacDonald and D. Carrington. Structuring Z specifications: Some choices. In Bowen and Hinchey [82], pages 203–223.

351. R. Macdonald. Z usage and abusage. Report no. 91003, RSRE, Ministry of Defence, Malvern, Worcestershire, UK, February 1991.
This paper presents a miscellany of observations drawn from experience of using Z, shows a variety of techniques for expressing certain class of idea concisely and clearly, and alerts the reader to certain pitfalls which may trap the unwary.

352. B. P. Mahony and I. J. Hayes. A case-study in timed refinement: A central heater. In Morris and Shaw [388], pages 138–149.

353. B. P. Mahony and I. J. Hayes. A case-study in timed refinement: A mine pump. *IEEE Transactions on Software Engineering*, 18(9):817–826, September 1992.

354. K. C. Mander and F. Polack. Rigorous specification using structured systems analysis and Z. *Information and Software Technology*, 37(5-6):285–291, May–June 1995.
Revised version of [420].

355. A. Martin. Encoding W: A logic for Z in 2OBJ. In Woodcock and Larsen [563], pages 462–481.

356. P. Mataga and P. Zave. Formal specification of telephone features. In Bowen and Hall [74], pages 29–50.

357. P. Mataga and P. Zave. Multiparadigm specification of an AT&T switching system. In Hinchey and Bowen [274], pages 375–398.

358. P. Mataga and P. Zave. Using Z to specify telephone features. *Information and Software Technology*, 37(5-6):277–283, May–June 1995.
Revised version of [356].

359. I. Maung and J. R. Howse. Introducing Hyper-Z – a new approach to object orientation in Z. In Bowen and Nicholls [84], pages 149–165.

360. M. D. May. Use of formal methods by a silicon manufacturer. In C. A. R. Hoare, editor, *Developments in Concurrency and Communication*, University of Texas at Austin Year of Programming Series, chapter 4, pages 107–129. Addison-Wesley Publishing Company, 1990.

361. M. D. May, G. Barrett, and D. E. Shepherd. Designing chips that work. In C. A. R. Hoare and M. J. C. Gordon, editors, *Mechanized Reasoning and Hardware Design*, pages 3–19. Prentice Hall International Series in Computer Science, 1992.

362. M. D. May and D. E. Shepherd. Verification of the IMS T800 microprocessor. In *Proc. Electronic Design Automation*, pages 605–615, London, UK, September 1987.

363. J. A. McDermid. Special section on Z. *IEE/BCS Software Engineering Journal*, 4(1):25–72, January 1989.
A special issue on Z, introduced and edited by Prof. J. A. McDermid. See also [58, 122, 482, 555].

364. J. A. McDermid, editor. *The Theory and Practice of Refinement: Approaches to the Formal Development of Large-Scale Software Systems*. Butterworth Scientific, 1989.
This book contains papers from the 1st Refinement Workshop held at the University of York, UK, 7–8 January 1988. Z-related papers include [320, 394].

365. J. A. McDermid. Formal methods: Use and relevance for the development of safety critical systems. In P. A. Bennett, editor, *Safety Aspects of Computer Control*. Butterworth-Heinemann, Oxford, UK, 1993.
This paper discusses a number of formal methods and summarizes strengths and weaknesses in safety critical applications; a major safety-related example is presented in Z.

366. M. A. McMorran and J. E. Nicholls. Z user manual. Technical Report TR12.274, IBM United Kingdom Laboratories Ltd, Hursley Park, Winchester, Hampshire SO21 2JN, UK, July 1989.

367. M. A. McMorran and S. Powell. *Z Guide for Beginners*. Blackwell Scientific, 1993.

368. S. L. Meira and A. L. C. Cavalcanti. Modular object-oriented Z specifications. In Nicholls [402], pages 173–192.

369. B. Meyer. On formalism in specifications. *IEEE Software*, 2(1):6–26, January 1985.

370. E. Mikk. Compilation of Z specifications into C for automatic test result evaluation. In Bowen and Hinchey [82], pages 167–180.

371. L. Mikuišiak, V. Vojtek, J. Hasaralejko, and J. Hanzelová. Z browser – tool for visualization of Z specifications. In Bowen and Hinchey [82], pages 510–523.

372. C. Minkowitz, D. Rann, and J. H. Turner. A C++ library for implementing specifications. In France and Gerhart [206], pages 61–75.

373. V. Mišić, D. Velašević, and B. Lazarević. Formal specification of a data dictionary for an extended ER data model. *The Computer Journal*, 35(6):611–622, December 1992.

374. J. D. Moffett and M. S. Sloman. A case study representing a model: To Z or not to Z? In Nicholls [402], pages 254–268.

375. B. Q. Monahan. Book review. *Formal Aspects of Computing*, 1(1):137–142, January–March 1989.
    A review of *Understanding Z: A Specification Language and Its Formal Semantics* by Mike Spivey [481].

376. B. Q. Monahan and R. C. Shaw. Model-based specifications. In J. A. McDermid, editor, *Software Engineer's Reference Book*, chapter 21. Butterworth-Heinemann, Oxford, UK, 1991.
    This chapter contains a case study in Z, followed by a discussion of the respective trade-offs in specification between Z and VDM.

377. C. C. Morgan. Data refinement using miracles. *Information Processing Letters*, 26(5):243–246, January 1988.

378. C. C. Morgan. Procedures, parameters, and abstraction: Separate concerns. *Science of Computer Programming*, 11(1), October 1988.

379. C. C. Morgan. The specification statement. *ACM Transactions on Programming Languages and Systems (TOPLAS)*, 10(3), July 1988.

380. C. C. Morgan. Types and invariants in the refinement calculus. In *Proc. Mathematics of Program Construction Conference*, Twente, June 1989.

381. C. C. Morgan. *Programming from Specifications*. Prentice Hall International Series in Computer Science, 2nd edition, 1994.
    This book presents a rigorous treatment of most elementary program development techniques, including iteration, recursion, procedures, parameters, modules and data refinement.

382. C. C. Morgan and K. A. Robinson. Specification statements and refinement. *IBM Journal of Research and Development*, 31(5), September 1987.

383. C. C. Morgan and J. W. Sanders. Laws of the logical calculi. Technical Monograph PRG-78, Oxford University Computing Laboratory, Wolfson Building, Parks Road, Oxford, UK, September 1989.
    This document records some important laws of classical predicate logic. It is designed as a reservoir to be tapped by *users* of logic, in system development.

384. C. C. Morgan and B. A. Sufrin. Specification of the Unix filing system. *IEEE Transactions on Software Engineering*, 10(2):128–142, March 1984.

385. C. C. Morgan and T. Vickers, editors. *On the Refinement Calculus*. Formal Approaches to Computing and Information Technology series (FACIT). Springer-Verlag, 1994.
    This book collects together the work accomplished at Oxford on the refinement calculus: the rigorous development, from state-based assertional specification, of executable imperative code.

386. C. C. Morgan and J. C. P. Woodcock. What is a specification? In D. Craigen and K. Summerskill, editors, *Formal Methods for Trustworthy Computer Systems (FM89)*, Workshops in Computing, pages 38–43. Springer-Verlag, 1990.

387. C. C. Morgan and J. C. P. Woodcock, editors. *3rd Refinement Workshop*, Workshops in Computing. Springer-Verlag, 1991.
    The workshop was held at the IBM Laboratories, Hursley Park, UK, 9–11 January 1990. See [459].

388. J. M. Morris and R. C. Shaw, editors. *4th Refinement Workshop*, Workshops in Computing. Springer-Verlag, 1991.

The workshop was held at Cambridge, UK, 9–11 January 1991. For Z related papers, see [23, 300, 352, 549, 557, 545].

389. M. Naftalin, T. Denvir, and M. Bertran, editors. *FME'94: Industrial Benefit of Formal Methods*, volume 873 of *Lecture Notes in Computer Science*. Formal Methods Europe, Springer-Verlag, 1994.

The 2nd FME Symposium was held at Barcelona, Spain, 24–28 October 1994. Z-related papers include [77, 131, 145, 192, 194, 200, 290, 344]. B-related papers include [152, 442, 504].

390. K. T. Narayana and S. Dharap. Formal specification of a look manager. *IEEE Transactions on Software Engineering*, 16(9):1089–1103, September 1990.

A formal specification of the look manager of a dialog system is presented in Z. This deals with the presentation of visual aspects of objects and the editing of those visual aspects.

391. K. T. Narayana and S. Dharap. Invariant properties in a dialog system. *ACM SIGSOFT Software Engineering Notes*, 15(4):67–79, September 1990.

392. T. C. Nash. Using Z to describe large systems. In Nicholls [400], pages 150–178.

393. Ph. W. Nehlig and D. A. Duce. GKS-9x: The design output primitive, an approach to specification. *Computer Graphics Forum*, 13(3):C–381–C–392, 1994.

394. D. S. Neilson. Hierarchical refinement of a Z specification. In McDermid [364].

395. D. S. Neilson. From Z to C: Illustration of a rigorous development method. Technical Monograph PRG-101, Oxford University Computing Laboratory, Wolfson Building, Parks Road, Oxford, UK, 1990.

396. D. S. Neilson. Machine support for Z: The zedB tool. In Nicholls [402], pages 105–128.

397. D. S. Neilson and D. Prasad. zedB: A proof tool for Z built on B. In Nicholls [404], pages 243–258.

398. K. Nguyen and R. Duke. A formal analysis method for conceptual modelling of information systems. In Habrias [237], pages 93–110.

399. J. E. Nicholls. Working with formal methods. *Journal of Information Technology*, 2(2):67–71, June 1987.

400. J. E. Nicholls, editor. *Z User Workshop, Oxford 1989*, Workshops in Computing. Springer-Verlag, 1990.

Proceedings of the Fourth Annual Z User Meeting, Wolfson College & Rewley House, Oxford, UK, 14–15 December 1989. Published in collaboration with the British Computer Society. For the opening address see [413]. For individual papers, see [37, 98, 100, 130, 159, 203, 231, 258, 270, 304, 330, 392, 418, 471, 490, 543].

401. J. E. Nicholls. A survey of Z courses in the UK. In *Z User Workshop, Oxford 1990* [402], pages 343–350.

402. J. E. Nicholls, editor. *Z User Workshop, Oxford 1990*, Workshops in Computing. Springer-Verlag, 1991.

Proceedings of the Fifth Annual Z User Meeting, Lady Margaret Hall, Oxford, UK, 17–18 December 1990. Published in collaboration with the British Computer Society. For individual papers, see [25, 91, 110, 128, 210, 232, 259, 271, 284, 306, 309, 328, 368, 374, 396, 401, 410, 430, 456, 544, 570]. The proceedings also includes an *Introduction and Opening Remarks*, a *Selected Z Bibliography*, a selection of posters and information on Z tools.

403. J. E. Nicholls. Domains of application for formal methods. In *Z User Workshop, York 1991* [404], pages 145–156.

404. J. E. Nicholls, editor. *Z User Workshop, York 1991*, Workshops in Computing. Springer-Verlag, 1992.

Proceedings of the Sixth Annual Z User Meeting, York, UK. Published in collaboration with the British Computer Society. For individual papers, see [18, 29, 147, 111, 161, 176, 250, 397, 403, 421, 449, 472, 516, 530, 560, 575].

405. J. E. Nicholls. Plain guide to the Z base standard. In Bowen and Nicholls [84], pages 52–61.

406. J. E. Nicholls et al. Z in the development process. Technical Report PRG-TR-1-89, Oxford University Computing Laboratory, Wolfson Building, Parks Road, Oxford, UK, June 1989. Proceedings of a discussion workshop held on 15 December 1988 in Oxford, UK, with contributions by Peter Collins, David Cooper, Anthony Hall, Patrick Hall, Brian Hepworth, Ben Potter and Andrew Ricketts.

407. C. J. Nix and B. P. Collins. The use of software engineering, including the Z notation, in the development of CICS. *Quality Assurance*, 14(3):103–110, September 1988.

408. A. Norcliffe and G. Slater. *Mathematics for Software Construction*. Series in Mathematics and its Applications. Ellis Horwood, 1991.
Contents: Why mathematics; Getting started: sets and logic; Developing ideas: schemas; Functions; Functions in action; A real problem from start to finish: a drinks machine; Sequences; Relations; Generating programs from specifications: refinement; The role of proof; More examples of specifications; Concluding remarks; Answers to exercises.

409. A. Norcliffe and S. Valentine. Z readers video course. PAVIC Publications, 1992. Sheffield Hallam University, 33 Collegiate Crescent, Sheffield S10 2BP, UK.
Video-based Training Course on the Z Specification Language. The course consists of 5 videos, each of approximately one hour duration, together with supporting texts and case studies.

410. A. Norcliffe and S. H. Valentine. A video-based training course in reading Z specifications. In Nicholls [402], pages 337–342.

411. G. Normington. Cleanroom and Z. In Bowen and Nicholls [84], pages 281–293.

412. C. O' Halloran. Evaluation semantics in Z. In Naftalin et al. [389], pages 502–518.

413. B. Oakley. The state of use of formal methods. In Nicholls [400], pages 1–5.
A record of the opening address at ZUM'89.

414. C. E. Parker. Z tools catalogue. ZIP project report ZIP/BAe/90/020, British Aerospace, Software Technology Department, Warton PR4 1AX, UK, May 1991.

415. H. Parker, F. Polack, and K. C. Mander. The industrial trial of SAZ: Reflections on the use of an integrated specification method. In Habrias [237], pages 111–129.

416. D. L. Parnas. Language-free mathematical models for software design. In Bowen and Hinchey [82], pages 3–4. Extended abstract.

417. D. L. Parnas. Teaching programming as engineering. In Bowen and Hinchey [82], pages 471–481.

418. M. Phillips. CICS/ESA 3.1 experiences. In Nicholls [400], pages 179–185.
Z was used to specify 37,000 lines out of 268,000 lines of code in the IBM CICS/ESA 3.1 release. The initial development benefit from using Z was assessed as being a 9% improvement in the *total development cost* of the release, based on the reduction of programmer days fixing problems.

419. M. Pilling, A. Burns, and K. Raymond. Formal specifications and proofs of inheritance protocols for real-time scheduling. *IEE/BCS Software Engineering Journal*, 5(5):263–279, September 1990.

420. F. Polack and K. C. Mander. Software quality assurance using the SAZ method. In Bowen and Hall [74], pages 230–249.

421. F. Polack, M. Whiston, and P. Hitchcock. Structured analysis – a draft method for writing Z specifications. In Nicholls [404], pages 261–286.

422. F. Polack, M. Whiston, and K. C. Mander. The SAZ project: Integrating SSADM and Z. In Woodcock and Larsen [563], pages 541–557.
423. B. F. Potter, J. E. Sinclair, and D. Till. *An Introduction to Formal Specification and Z*. Prentice Hall International Series in Computer Science, 2nd edition, 1996.
Contents: Formal specification in the context of software engineering; An informal introduction to logic and set theory; A first specification; The Z notation: the mathematical language, relations and functions, schemas and specification structure; A first specification; Formal reasoning; From specification to program: data and operation refinement, operation decomposition; From theory to practice.
424. B. F. Potter and D. Till. The specification in Z of gateway functions within a communications network. In *Proc. IFIP WG10.3 Conference on Distributed Processing*. Elsevier Science Publishers (North-Holland), October 1987.
425. C. H. Pratten. An introduction to proving AMN specifications with PVS and the AMN-PROOF tool. In Habrias [237], pages 149–165.
426. S. Prehn and W. J. Toetenel, editors. *VDM'91: Formal Software Development Methods*, volume 551 of *Lecture Notes in Computer Science*. Springer-Verlag, 1991. Volume 1: Conference Contributions.
The 4th VDM-Europe Symposium was held at Noordwijkerhout, The Netherlands, 21–25 October 1991. Papers with relevance to Z include [17, 38, 139, 171, 216, 281, 535, 548, 574]. See also [427].
427. S. Prehn and W. J. Toetenel, editors. *VDM'91: Formal Software Development Methods*, volume 552 of *Lecture Notes in Computer Science*. Springer-Verlag, 1991. Volume 2: Tutorials.
Papers with relevance to Z include [7, 558]. See also [426].
428. G.-H. B. Rafsanjani and S. J. Colwill. From Object-Z to C^{++}: A structural mapping. In Bowen and Nicholls [84], pages 166–179.
429. RAISE Language Group. *The RAISE Specification Language*. BCS Practitioner Series. Prentice Hall International, 1992.
430. G. P. Randell. Data flow diagrams and Z. In Nicholls [402], pages 216–227.
431. D. Rann, J. Turner, and J. Whitworth. *Z: A Beginner's Guide*. Chapman & Hall, London, 1994.
432. B. Ratcliff. *Introducing Specification Using Z: A Practical Case Study Approach*. International Series in Software Engineering. McGraw-Hill, 1994.
433. A. P. Ravn, H. Rischel, and V. Stavridou. Provably correct safety critical software. In *Proc. IFAC Safety of Computer Controlled Systems 1990 (SAFECOMP'90)*. Pergamon Press, 1990.
Also available as Technical Report CSD-TR-625 from Department of Computer Science, Royal Holloway, University of London, Egham, Surrey TW20 0EX, UK.
434. M. Rawson. OOPSLA'93: Workshop on formal specification of object-oriented systems – position paper. In H. Kilov and W. Harvey, editors, *Proc. Workshop on Specification of Behavioral Semantics in Object-Oriented Information Modeling*, pages 125–135, Institute for Information Management and Department of Computer and Information Systems, Robert Morris College, Coraopolos and Pittsburgh, Pennsylvania, USA, 1993. OOPSLA.
435. M. Rawson and P. Allen. Synthesis – an integrated, object-oriented method and tool for requirements specification in Z. In Bryant and Semmens [103].
436. T. J. Read. Formal specification of reusable Ada software packages. In A. Burns, editor, *Towards Ada 9X Conference Proceedings*, pages 98–117, 1991.
437. J. N. Reed. Semantics-based tools for a specification support environment. In *Mathematical Foundations of Programming Language Semantics*, volume 298 of *Lecture Notes in Computer Science*. Springer-Verlag, 1988.

438. J. N. Reed and J. E. Sinclair. An algorithm for type-checking Z: A Z specification. Technical Monograph PRG-81, Oxford University Computing Laboratory, Wolfson Building, Parks Road, Oxford, UK, March 1990.

439. C. Reilly. Exploring specifications with Mathematica. In Bowen and Hinchey [82], pages 408–420.

440. N. R. Reizer, G. D. Abowd, B. C. Meyers, and P. R. H. Place. Using formal methods for requirements specification of a proposed POSIX standard. In *IEEE International Conference on Requirements Engineering (ICRE'94)*, April 1994.

441. G. J. Reynolds. Yet another approach to the formal specification of a configurable graphics system. In *Proc. Eurographics Association Formal Methods in Computer Graphics*, June 1991.

442. B. Ritchie, J. Bicarregui, and H. P. Haughton. Experiences in using the abstract machine notation in a GKS case study. In Naftalin et al. [389], pages 93–104.

443. K. A. Robinson. Refining Z specifications to programs. In *Proc. Australian Software Engineering Conference*, pages 87–97, 1987.

444. G. A. Rose. Object-Z. In Stepney et al. [494], pages 59–77.

445. G. A. Rose and P. Robinson. A case study in formal specifications. In *Proc. First Australian Software Engineering Conference*, May 1986.

446. A. R. Ruddle. Formal methods in the specification of real-time, safety-critical control systems. In Bowen and Nicholls [84], pages 131–146.

447. P. Rudkin. Modelling information objects in Z. In J. de Meer, editor, *Proc. International Workshop on ODP*. Elsevier Science Publishers (North-Holland), 1992.

448. J. Rushby. Mechanizing formal methods: Challenges and opportunities. In Bowen and Hinchey [82], pages 105–113.

449. M. Saaltink. Z and Eves. In Nicholls [404], pages 223–242.

450. H. Saiedian. The mathematics of computing. *Journal of Computer Science Education*, 3(3):203–221, 1992.

451. A. C. A. Sampaio and S. L. Meira. Modular extensions to Z. In Bjørner et al. [45], pages 211–232.

452. P. Sanders, M. Johnson, and R. Tinker. From Z specifications to functional implementations. *British Telecom Technology Journal*, 7(4), October 1989.

453. M. Schenke and A. P. Ravn. Refinement from a control problem to programs. In Abrial et al. [9], pages 403–427.

454. S. A. Schuman and D. H. Pitt. Object-oriented subsystem specification. In L. G. L. T. Meertens, editor, *Program Specification and Transformation*, pages 313–341. Elsevier Science Publishers (North-Holland), 1987.

455. S. A. Schuman, D. H. Pitt, and P. J. Byers. Object-oriented process specification. In C. Rattray, editor, *Specification and Verification of Concurrent Systems*, pages 21–70. Springer-Verlag, 1990.

456. L. T. Semmens and P. M. Allen. Using Yourdon and Z: An approach to formal specification. In Nicholls [402], pages 228–253.

457. L. T. Semmens, R. B. France, and T. W. G. Docker. Integrated structured analysis and formal specification techniques. *The Computer Journal*, 35(6):600–610, December 1992.

458. C. T. Sennett. Formal specification and implementation. In C. T. Sennett, editor, *High-Integrity Software*, Computer Systems Series. Pitman, 1989.

459. C. T. Sennett. Using refinement to convince: Lessons learned from a case study. In Morgan and Woodcock [387], pages 172–197.

460. C. T. Sennett. Demonstrating the compliance of Ada programs with Z specifications. In Jones et al. [307].

461. D. E. Shepherd. Verified microcode design. *Microprocessors and Microsystems*, 14(10):623–630, December 1990.

This article is part of a special feature on *Formal aspects of microprocessor design*, edited by H. S. M. Zedan. See also [59].

462. D. E. Shepherd and G. Wilson. Making chips that work. *New Scientist*, 1664:61–64, May 1989.

A general article containing information on the formal development of the T800 floating-point unit for the Transputer including the use of Z.

463. D. Sheppard. *An Introduction to Formal Specification with Z and VDM*. International Series in Software Engineering. McGraw Hill, 1995.

464. L. B. Sherrell and D. L. Carver. Z meets Haskell: A case study. In *COMPSAC '93: 17th Annual International Computer Software and Applications Conference*, pages 320–326. IEEE Computer Society Press, November 1993.

The paper traces the development of a simple system, the class manager's assistant, from an existing Z specification, through design in Z, to a Haskell implementation.

465. L. B. Sherrell and D. L. Carver. FunZ: An intermediate specification language. *The Computer Journal*, 38(3):193–206, 1995.

466. J. E. Sinclair and D. C. Ince. The use of Z in specifying security properties. In Habrias [237], pages 27–39.

467. R. Sinnott and K. J. Turner. Modeling ODP viewpoints. In H. Kilov, W. Harvey, and H. Mili, editors, *Proc. Workshop on Precise Behavioral Specifications in Object-Oriented Information Modeling, OOPSLA 1994*, pages 121–128, Robert Morris College, Coraopolos and Pittsburgh, Pennsylvania 15108-1189, USA, 1994. OOPSLA.

468. R. O. Sinnott and K. J. Turner. Specifying multimedia binding objects in Z. In O. Spaniol, C. Linnhoff-Popien, and B. Meyer, editors, *Trends in Distributed Systems*, volume 1161 of *Lecture Notes in Computer Science*, pages 244–258. Springer-Verlag, 1996.

469. R. O. Sinnott and K. J. Turner. Specifying ODP computational objects in Z. In E. Najm and J.-B. Stefani, editors, *Proc. Formal Methods for Open Object-based Distributed Systems*, March 1996.

470. R. O. Sinnott and K. J. Turner. Type checking in distributed systems: A complete model and its Z specification. In J. Rolia, editor, *International Conference on Open Distributed Processing (ICODP) and Distributed Platforms (ICDP)*, May 1997. To appear.

471. A. Smith. The Knuth-Bendix completion algorithm and its specification in Z. In Nicholls [400], pages 195–220.

472. A. Smith. On recursive free types in Z. In Nicholls [404], pages 3–39.

473. G. Smith. *An Object-Oriented Approach to Formal Specification*. PhD thesis, Department of Computer Science, University of Queensland, St. Lucia 4072, Australia, October 1992.

A detailed description of a version of Object-Z similar to (but not identical to) that in [184]. The thesis also includes a formalization of temporal logic history invariants and a fully-abstract model of classes in Object-Z.

474. G. Smith. A object-oriented development framework for Z. In Bowen and Hall [74], pages 89–107.

475. G. Smith. Extending $\mathcal{W}$ for Object-Z. In Bowen and Hinchey [82], pages 276–295.

476. G. Smith. A fully abstract semantics of classes for Object-Z. *Formal Aspects of Computing*, 7(3):289–313, 1995.

477. G. Smith and R. Duke. Modelling a cache coherence protocol using Object-Z. In *Proc. 13th Australian Computer Science Conference (ACSC-13)*, pages 352–361, 1990.

478. P. Smith and R. Keighley. The formal development of a secure transaction mechanism. In Prehn and Toetenel [426], pages 457–476.

479. I. Sommerville. *Software Engineering*, chapter 9, pages 153–168. Addison-Wesley Publishing Company, 4th edition, 1992.
    A chapter entitled *Model-Based Specification* including examples using Z.

480. I. H. Sørensen. A specification language. In J. Staunstrup, editor, *Program Specification: Proceedings of a Workshop*, volume 134 of *Lecture Notes in Computer Science*, pages 381–401. Springer-Verlag, 1981.

481. J. M. Spivey. *Understanding Z: A Specification Language and its Formal Semantics*, volume 3 of *Cambridge Tracts in Theoretical Computer Science*. Cambridge University Press, January 1988.
    Published version of 1985 DPhil thesis.

482. J. M. Spivey. An introduction to Z and formal specifications. *IEE/BCS Software Engineering Journal*, 4(1):40–50, January 1989.

483. J. M. Spivey. A guide to the zed style option. Oxford University Computing Laboratory, December 1990.
    A description of the Z style option 'zed.sty' for use with the LaTeX document preparation system [326].

484. J. M. Spivey. Specifying a real-time kernel. *IEEE Software*, 7(5):21–28, September 1990.
    This case study of an embedded real-time kernel shows that mathematical techniques have an important role to play in documenting systems and avoiding design flaws.

485. J. M. Spivey. *The ƒUZZ Manual*. Computing Science Consultancy, 34 Westlands Grove, Stockton Lane, York YO3 0EF, UK, 2nd edition, July 1992.
    The manual describes a Z type-checker and 'fuzz.sty' style option for LaTeX documents [326]. The package is compatible with the book, *The Z Notation: A Reference Manual* by the same author [486].

486. J. M. Spivey. *The Z Notation: A Reference Manual*. Prentice Hall International Series in Computer Science, 2nd edition, 1992.
    This is a revised edition of the first widely available reference manual on Z originally published in 1989. The book provides a complete and definitive guide to the use of Z in specifying information systems, writing specifications and designing implementations. See also the draft Z standard [94].
    Contents: Tutorial introduction; Background; The Z language; The mathematical tool-kit; Sequential systems; Syntax summary; Changes from the first edition; Glossary.

487. J. M. Spivey. The consistency theorem for free type definitions in Z. *Formal Aspects of Computing*, 8:369–375, 1996.

488. J. M. Spivey. Richer types for Z. *Formal Aspects of Computing*, 8:565–584, 1996.

489. J. M. Spivey and B. A. Sufrin. Type inference in Z. In Bjørner et al. [45], pages 426–438.

490. J. M. Spivey and B. A. Sufrin. Type inference in Z. In Nicholls [400], pages 6–31.
    Also published as [489].

491. S. Stepney. *High Integrity Compilation: A Case Study*. Prentice Hall, 1993.

492. S. Stepney. Testing as abstraction. In Bowen and Hinchey [82], pages 137–151.

493. S. Stepney and R. Barden. Annotated Z bibliography. *Bulletin of the European Association of Theoretical Computer Science*, 50:280–313, June 1993.

494. S. Stepney, R. Barden, and D. Cooper, editors. *Object Orientation in Z*. Workshops in Computing. Springer-Verlag, 1992.
    This is a collection of papers describing various OOZ approaches – Hall, ZERO, MooZ, Object-Z, OOZE, Schuman & Pitt, Z^{++}, ZEST and Fresco (an object-oriented VDM method) – in the main written by the methods' inventors, and all specifying the same two examples. The collection is a revised and expanded version of a ZIP report distributed at the 1991 Z User Meeting at York.

495. S. Stepney, R. Barden, and D. Cooper. A survey of object orientation in Z. *IEE/BCS Software Engineering Journal*, 7(2):150–160, March 1992.

496. S. Stepney and S. P. Lord. Formal specification of an access control system. *Software— Practice and Experience*, 17(9):575–593, September 1987.

497. P. Stocks. *Applying formal methods to software testing*. PhD thesis, Department of Computer Science, University of Queensland, St. Lucia 4072, Australia, 1993.

498. P. Stocks and D. A. Carrington. Deriving software test cases from formal specifications. In *6th Australian Software Engineering Conference*, pages 327–340, July 1991.

499. P. Stocks and D. A. Carrington. Test template framework: A specification-based testing case study. In *Proc. International Symposium on Software Testing and Analysis (ISSTA '93)*, pages 11–18, June 1993.
    Also available in a longer form as Technical Report UQCS-255, Department of Computer Science, University of Queensland.

500. P. Stocks and D. A. Carrington. Test templates: A specification-based testing framework. In *Proc. 15th International Conference on Software Engineering*, pages 405–414, May 1993.
    Also available in a longer form as Technical Report UQCS-243, Department of Computer Science, University of Queensland.

501. P. Stocks, K. Raymond, D. Carrington, and A. Lister. Modelling open distributed systems in Z. *Computer Communications*, 15(2):103–113, March 1992.
    In a special issue on the practical use of FDTs (Formal Description Techniques) in communications and distributed systems, edited by Dr. Gordon S. Blair.

502. B. Stoddart, C. Fencott, and S. Dunne. Modelling hypbrid systems in Z. In Habrias [237], pages 11–25.

503. B. Stoddart and P. Knaggs. The event calculus (formal specification of real time systems by means of diagrams and Z schemas). In *5th International Conference on putting into practice methods and tools for information system design*, University of Nantes, Institute Universitaire de Technologie, 3 Rue du Maréchal Joffre, 44041 Nantes Cedex 01, France, September 1992.

504. A. C. Storey and H. P. Haughton. A strategy for the production of verifiable code using the B method. In Naftalin et al. [389], pages 346–365.

505. B. Strulo. How firing conditions help inheritance. In Bowen and Hinchey [82], pages 264–275.

506. B. A. Sufrin. Formal system specification: Notation and examples. In D. Neel, editor, *Tools and Notations for Program Construction*. Cambridge University Press, 1982.
    An example of a filing system specification, this was the first published use of the schema notation to put together states.

507. B. A. Sufrin. Towards formal specification of the ICL data dictionary. *ICL Technical Journal*, August 1984.

508. B. A. Sufrin. Formal methods and the design of effective user interfaces. In M. D. Harrison and A. F. Monk, editors, *People and Computers: Designing for Usability*. Cambridge University Press, 1986.

509. B. A. Sufrin. Formal specification of a display-oriented editor. In N. Gehani and A. D. McGettrick, editors, *Software Specification Techniques*, International Computer Science Series, pages 223–267. Addison-Wesley Publishing Company, 1986.
    Originally published in Science of Computer Programming, 1:157–202, 1982.

510. B. A. Sufrin. A formal framework for classifying interactive information systems. In *IEE Colloquium on Formal Methods and Human-Computer Interaction*, number 09 in IEE Digest, pages 4/1–14, London, UK, 1987. The Institution of Electrical Engineers.

511. B. A. Sufrin. Effective industrial application of formal methods. In G. X. Ritter, editor, *Information Processing 89, Proc. 11th IFIP Computer Congress*, pages 61–69. Elsevier Science Publishers (North-Holland), 1989.
This paper presents a Z model of the Unix *make* utility.

512. B. A. Sufrin and He Jifeng. Specification, analysis and refinement of interactive processes. In M. D. Harrison and H. Thimbleby, editors, *Formal Methods in Human-Computer Interaction*, volume 2 of *Cambridge Series on Human-Computer Interaction*, chapter 6, pages 153–200. Cambridge University Press, 1990.
A case study on using Z for process modelling.

513. B. A. Sufrin and J. C. P. Woodcock. Towards the formal specification of a simple programming support environment. *IEE/BCS Software Engineering Journal*, 2(4):86–94, July 1987.

514. P. A. Swatman. *Increasing Formality in the Specification of High-Quality Information Systems in a Commercial Context*. Phd thesis, Curtin University of Technology, School of Computing, Perth, Western Australia, July 1992.

515. P. A. Swatman. Using formal specification in the acquisition of information systems: Educating information systems professionals. In Bowen and Nicholls [84], pages 205–239.

516. P. A. Swatman, D. Fowler, and C. Y. M. Gan. Extending the useful application domain for formal methods. In Nicholls [404], pages 125–144.

517. P. A. Swatman and P. M. C. Swatman. Formal specification: An analytic tool for (management) information systems. *Journal of Information Systems*, 2(2):121–160, April 1992.

518. P. A. Swatman and P. M. C. Swatman. Is the information systems community wrong to ignore formal specification methods? In R. Clarke and J. Cameron, editors, *Managing Information Technology's Organisational Impact*. Elsevier Science Publishers (North-Holland), October 1992.

519. P. A. Swatman and P. M. C. Swatman. Managing the formal specification of information systems. In *Proc. International Conference on Organization and Information Systems*, September 1992.

520. P. A. Swatman, P. M. C. Swatman, and R. Duke. Electronic data interchange: A high-level formal specification in Object-Z. In *Proc. 6th Australian Software Engineering Conference (ASWEC'91)*, 1991.

521. S. Thompson. Specification techniques [9004-0316]. *ACM Computing Reviews*, 31(4):213, April 1990.
A review of *Formal methods applied to a floating-point number system* [31].

522. D. Till, editor. *6th Refinement Workshop*, Workshop in Computing. Springer-Verlag, 1994. The workshop was held at City University, London, UK, 5–7 January 1994. See [201, 329].

523. B. S. Todd. A model-based diagnostic program. *IEE/BCS Software Engineering Journal*, 2(3):54–63, May 1987.

524. R. Took. The presenter – a formal design for an autonomous display manager. In I. Sommerville, editor, *Software Engineering Environments*, pages 151–169. Peter Peregrinus, London, 1986.

525. I. Toyn. *CADiZ Quick Reference Guide*. York Software Engineering Ltd, University of York, York YO1 5DD, UK, 1990.
A guide to the CADiZ (Computer Aided Design in Z) toolkit. This makes use of the Unix *troff* family of text formatting tools. Contact David Jordan at the address above or on yse@minster.york.ac.uk via e-mail for further information on CADiZ. See also [309] for a paper introducing CADiZ. Support for LaTeX [326] is now available.

526. I. Toyn and J. A. McDermid. CADiZ: An architecture for Z tools and its implementation. Technical document, Department of Computer Science, University of York, York YO1 5DD, UK, November 1993.

527. I. Toyn and J. A. McDermid. CADiZ: An architecture for Z tools and its implementation. *Software—Practice and Experience*, 25(3):305–330, March 1995.

528. O. Traynor, P. Kearney, E. Kazmierczak, Li Wang, and E. Karlsen. Extending Z with modules. *Australian Computer Science Communications*, 17(1), 1995. Proc. ACSC'95.

529. S. Valentine. The programming language Z^{--}. *Information and Software Technology*, 37(5-6):293–301, May–June 1995.

530. S. H. Valentine. Z^{--}, an executable subset of Z. In Nicholls [404], pages 157–187.

531. S. H. Valentine. Putting numbers into the mathematical toolkit. In Bowen and Nicholls [84], pages 9–36.

532. S. H. Valentine. An algebraic introduction of real numbers into Z. In Habrias [237], pages 183–204.

533. S. H. Valentine. Equal rights for schemas in Z. In Bowen and Hinchey [82], pages 183–202.

534. M. J. van Diepen and K. M. van Hee. A formal semantics for Z and the link between Z and the relational algebra. In Bjørner et al. [45], pages 526–551.

535. K. M. van Hee, L. J. Somers, and M. Voorhoeve. Z and high level Petri nets. In Prehn and Toetenel [426], pages 204–219.

536. H. J. van Zuylen, editor. *The REDO Compendium: Reverse Engineering for Software Maintenance*. John Wiley & Sons, 1993.
An overview of the results of the ESPRIT REDO project, including the use of Z and Z^{++}. See in particular Chapter 16, also published in a longer form as [331].

537. M. Waldén and K. Sere. Refining action systems with B-tool. In Gaudel and Woodcock [217], pages 85–104.

538. M. Weber. Combining Statecharts and Z for the design of safety-critical control systems. In Gaudel and Woodcock [217], pages 307–326.

539. M. M. West. Types and sets in gödel and Z. In Bowen and Hinchey [82], pages 389–407.

540. M. M. West and B. M. Eaglestone. Software development: Two approaches to animation of Z specifications using Prolog. *IEE/BCS Software Engineering Journal*, 7(4):264–276, July 1992.

541. C. Wezeman and A. Judge. Z for managed objects. In Bowen and Hall [74], pages 108–119.

542. C. D. Wezeman. Using Z for network modelling: An industrial experience report. *Computer Standards & Interfaces*, 17(5–6):631–638, September 1995.

543. R. W. Whitty. Structural metrics for Z specifications. In Nicholls [400], pages 186–191.

544. P. J. Whysall and J. A. McDermid. An approach to object-oriented specification using Z. In Nicholls [402], pages 193–215.

545. P. J. Whysall and J. A. McDermid. Object-oriented specification and refinement. In Morris and Shaw [388], pages 151–184.

546. J. M. Wing. A specifier's introduction to formal methods. *IEEE Computer*, 23(9):8–24, September 1990.

547. J. M. Wing. Hints for writing specifications. In Bowen and Hinchey [82], page 497.

548. J. M. Wing and A. M. Zaremski. Unintrusive ways to integrate formal specifications in practice. In Prehn and Toetenel [426], pages 545–570.

549. K. R. Wood. The elusive software refinery: a case study in program development. In Morris and Shaw [388], pages 281–325.

550. K. R. Wood. A practical approach to software engineering using Z and the refinement calculus. *ACM Software Engineering Notes*, 18(5):79–88, December 1993.

551. W. G. Wood. Application of formal methods to system and software specification. *ACM SIGSOFT Software Engineering Notes*, 15(4):144–146, September 1990.

552. J. C. P. Woodcock. Teaching how to use mathematics for large-scale software development. *Bulletin of BCS-FACS*, July 1988.

553. J. C. P. Woodcock. Calculating properties of Z specifications. *ACM SIGSOFT Software Engineering Notes*, 14(4):43–54, 1989.

554. J. C. P. Woodcock. Mathematics as a management tool: Proof rules for promotion. In *Proc. 6th Annual CSR Conference on Large Software Systems*, Bristol, UK, September 1989.

555. J. C. P. Woodcock. Structuring specifications in Z. *IEE/BCS Software Engineering Journal*, 4(1):51–66, January 1989.

556. J. C. P. Woodcock. Z. In D. Craigen and K. Summerskill, editors, *Formal Methods for Trustworthy Computer Systems (FM89)*, Workshops in Computing, pages 57–62. Springer-Verlag, 1990.

557. J. C. P. Woodcock. Implementing promoted operations in Z. In Morris and Shaw [388], pages 366–378.

558. J. C. P. Woodcock. A tutorial on the refinement calculus. In Prehn and Toetenel [427], pages 79–140.

559. J. C. P. Woodcock. The rudiments of algorithm design. *The Computer Journal*, 35(5):441–450, October 1992.

560. J. C. P. Woodcock and S. M. Brien. $\mathcal{W}$: A logic for Z. In Nicholls [404], pages 77–96.

561. J. C. P. Woodcock and J. Davies. *Using Z: Specification, Proof and Refinement*. Prentice Hall International Series in Computer Science, 1996.

    This book contains enough material for three complete courses of study. It provides an introduction to the world of logic, sets and relations. It explains the use of the Z notation in the specification of realistic systems. It shows how Z specifications may be refined to produce executable code; this is demonstrated in a selection of case studies.

    The book strikes a balance between the formality of mathematics and the practical needs of industrial software development, following to the draft ISO standard for Z. It is based upon the experience of the authors in teaching Z to a wide variety of audiences. A set of exercises, solutions, and transparency masters is available on-line to complement the book.

562. J. C. P. Woodcock, P. H. B. Gardiner, and J. R. Hulance. The formal specification in Z of Defence Standard 00-56. In Bowen and Hall [74], pages 9–28.

563. J. C. P. Woodcock and P. G. Larsen, editors. *FME'93: Industrial-Strength Formal Methods*, volume 670 of *Lecture Notes in Computer Science*. Formal Methods Europe, Springer-Verlag, 1993.

    The 1st FME Symposium was held at Odense, Denmark, 19–23 April 1993. Z-related papers include [85, 134, 199, 296, 355, 422].

564. J. C. P. Woodcock and P. G. Larsen. Guest editorial. *IEEE Transactions on Software Engineering*, 21(2):61–62, 1995.

    Best papers from the FME'93 Symposium [563]. See [43, 50, 137, 297].

565. J. C. P. Woodcock and M. Loomes. *Software Engineering Mathematics: Formal Methods Demystified*. Pitman, 1988.

    Also published as: *Software Engineering Mathematics*, Addison-Wesley, 1989.

566. J. C. P. Woodcock and C. C. Morgan. Refinement of state-based concurrent systems. In Bjørner et al. [45], pages 340–351.

    Work on combining Z and CSP.

567. R. Worden. Fermenting and distilling. In Bowen and Hall [74], pages 1–6.

568. J. B. Wordsworth. Teaching formal specification methods in an industrial environment. In *Proc. Software Engineering '86*, London, 1986. IEE/BCS, Peter Peregrinus.

569. J. B. Wordsworth. Specifying and refining programs with Z. In *Proc. Second IEE/BCS Conference on Software Engineering*, number 290 in Conference Publication, pages 8–16. IEE/BCS, July 1988.

    A tutorial summary.

570. J. B. Wordsworth. The CICS application programming interface definition. In Nicholls [402], pages 285–294.

571. J. B. Wordsworth. *Software Development with Z: A A Practical Approach to Formal Methods in Software Engineering*. Addison-Wesley Publishing Company, 1993.

This book provides a guide to developing software from specification to code, and is based in part on work done at IBM's UK Laboratory that won the UK Queen's Award for Technological Achievement in 1992.

Contents: Introduction; A simple Z specification; Sets and predicates; Relations and functions; Schemas and specifications; Data design; Algorithm design; Specification of an oil terminal control system.

572. Xiaoping Jia. *ZTC: A Type Checker for Z – User's Guide*. Institute for Software Engineering, Department of Computer Science and Information Systems, DePaul University, Chicago, IL 60604, USA, 1994.

ZTC is a type checker for the Z specification language. ZTC accepts two forms of input: LaTeX [326] with the zed.sty style option [483] and ZSL, an ASCII version of Z. ZTC can also perform translations between the two input forms. This document is intended to serve as both a user's guide and a reference manual for ZTC.

573. W. D. Young. Comparing specifications paradigms: Gypsy and Z. Technical Report 45, Computational Logic Inc., 1717 W. 6th St., Suite 290, Austin, Texas 78703, USA, 1989.

574. P. Zave and M. Jackson. Techniques for partial specification and specification of switching systems. In Prehn and Toetenel [426], pages 511–525.

Also published as [575].

575. P. Zave and M. Jackson. Techniques for partial specification and specification of switching systems. In Nicholls [404], pages 205–219.

576. P. Zave and M. Jackson. Conjunction as composition. *ACM Transactions on Software Engineering and Methodology (TOSEM)*, 2(4):379–411, October 1993.

Partial specifications written in many different specification languages can be composed if they are all given semantics in the same domain, or alternatively, all translated into a common style of predicate logic. A Z specification is used as an example.

577. P. Zave and M. Jackson. Where do operations come from? A multiparadigm technique. *IEEE Transactions on Software Engineering*, 22(7):508–528, July 1996.

Z is supplemented, primarily with automata and grammars, to provide a rigorous and systematic mapping from input stimuli to convenient operations and arguments for the Z specification.

578. Y. Zhang and P. Hitchcock. EMS: Case study in methodology for designing knowledge-based systems and information systems. *Information and Software Technology*, 33(7):518–526, September 1991.

579. Z archive. URL: http://www.comlab.ox.ac.uk/archive/z.html, 1996.

A computer-based archive server is available for use by anyone with World Wide Web (WWW) access, anonymous FTP access or an electronic mail address. This allows people interested in Z (and other information) to access various archived files. In particular a Z bibliography [60] is available. The preferred method of access to the on-line Z archive is via the Web under the 'URL' (Uniform Resource Locator) given above. Some of the archive is accessible via anonymous FTP under the ftp://ftp.comlab.ox.ac.uk/pub/Zforum/ directory. For information on access via electronic mail, send a message to archive-server@comlab.ox.ac.uk containing the command 'help'.

Comp.specification.z and Z FORUM
Frequently Asked Questions

Jonathan P. Bowen

The University of Reading, Department of Computer Science
Whiteknights, PO Box 225, Reading, Berks RG6 6AY, UK
Email: J.P.Bowen@reading.ac.uk
URL: http://www.cs.reading.ac.uk/people/jpb/

Abstract. This appendix provides some details on how to access information on Z, particularly electronically. It has been generated from a message that is updated and sent out monthly on international computer networks.

This on-line information is available on-line on the following World Wide Web (WWW) hypertext page where it is split into convenient sections and updated each month:

   http://www.cis.ohio-state.edu/hypertext/faq/usenet/z-faq/faq.html

1   Subject: What is it?

Z (pronounced "zed") is a formal specification notation based on set theory and first order predicate logic. It has been developed at the Programming Research Group at the Oxford University Computing Laboratory (OUCL) and elsewhere since the late 1970s. It is used by industry as part of the software (and hardware) development process in Europe, USA and elsewhere. Currently it is undergoing international ISO standardization.

The comp.specification.z electronic USENET newsgroup was established in June 1991 and is intended to handle messages concerned with Z. It has an estimated readership of tens of thousands of people worldwide. Comp.specification.z provides a convenient forum for messages concerned with recent developments and the use of Z. Pointers to and reviews of recent books and articles are particularly encouraged. These may be included in the Z bibliography (see below) if they appear in comp.specification.z.

2   What if I do not have access to USENET news?

There is an associated Z FORUM electronic mailing list that was initiated in January 1986 by Ruaridh Macdonald, RSRE, UK. Articles are now automatically cross-posted between comp.specification.z and the mailing list for those whose do not have access to USENET news. This may apply especially to industrial Z users who are particularly encouraged to subscribe and post their experiences to the list. Please contact zforum-request@comlab.ox.ac.uk with your name, address and email address to join the mailing list (or if you change your email address or wish to be removed from the list). Readers are strongly urged to read the `comp.specification.z` newsgroup rather than the Z FORUM mailing list if possible. Messages for submission to the Z FORUM mailing list and the `comp.specification.z` newsgroup may be emailed to `zforum@comlab.ox.ac.uk`. This method of posting is particularly recommended for important messages like announcements of meetings since not all messages posted on `comp.specification.z` reach the OUCL.

A mailing list for the Z User Meeting educational issues session has been set by Neville Dean, Anglia Polytechnic University, UK. Anyone interested may join by emailing `zugeis-request@comlab.ox.ac.uk` with your contact details.

3 What if I do not have access to email?

If you wish to join the postal Z mailing list, please send your address to Amanda Kingscote, Praxis plc, 20 Manvers Street, Bath BA1 1PX, UK (tel +44-1225-444700, fax +44-1225-465205, email `ark@praxis.co.uk`). This will ensure you receive details of Z meetings, etc., particularly for people without access to electronic mail.

4 How can I join in?

If you are currently using Z, you are welcome to introduce yourself to the newsgroup and Z FORUM list by describing your work with Z or raising any questions you might have about Z which are not answered here. You may also advertize publications concerning Z which you or your colleagues produce. These may then be added to the master Z bibliography maintained at the OUCL (see below).

5 Where are Z-related files archived?

Information on the World Wide Web (WWW) is available under the

        `http://www.comlab.ox.ac.uk/archive/z.html`

page. See also the

    `http://www.comlab.ox.ac.uk/archive/formal-methods.html`

page on formal methods in general. The WWW global hypermedia system is accessible using the 'netscape', 'mosaic' or 'lynx' programs for example. Contact your system manager if WWW access is not available on your system.

Some of the Z archive is also available via anonymous FTP under

        `ftp://ftp.comlab.ox.ac.uk/pub/Zforum`

(IP address 163.1.27.2 if you have access problems). The README file provides some general information and `00index` gives a list of the files. If you cannot access the Internet directly, there is an automatic electronic mail-based electronic archive server which allows access to some of the Z FTP archive. Send an email message containing the command 'help' to `archive-server@comlab.ox.ac.uk` for further information on how to use the server. If you have serious trouble accessing the archive server, contact `archive-management@comlab.ox.ac.uk`.

6 What tools are available?

Various tools for formatting, type-checking and aiding proofs in Z are available. A free LaTeX style file and documentation can be obtained from the OUCL archive:

```
ftp://ftp.comlab.ox.ac.uk/pub/Zforum/zed.sty
```

```
ftp://ftp.comlab.ox.ac.uk/pub/Zforum/zguide.tex
```

A newer style 'csp_zed.sty' is available in the same location, which uses the new font selection scheme and covers CSP and Z symbols. A style for Object-Z 'oz.sty' with a guide 'oz.tex' is also accessible. LaTeX2e users may find 'zed-csp.sty' and 'zed2e.tex' useful.

The *f*UZZ package, a syntax and type-checker with a LaTeX style option and fonts, is available from the Spivey Partnership, 10 Warneford Road, Oxford OX4 1LU, UK. It is compatible with the 2nd edition of Spivey's Z Reference Manual. Access

```
ftp://ftp.comlab.ox.ac.uk/pub/Zforum/fuzz
```

for brief information and an order form. For further information, contact Mike Spivey (email Mike.Spivey@comlab.oxford.ac.uk).

CADiZ is a suite of integrated tools for preparing and type-checking Z specifications as professional quality typeset documents. The Z dialect it recognizes is evolving in line with the standard. The typesetting can be performed by either *troff* or LaTeX for UNIX or Word for Windows. The mouse can be used to interact with a view of the typeset specification to inspect properties deduced by the type-checker, to see the expansion of schema calculus expressions, and to reason about conjectures such as proof obligations. The PC version is integrated with MS Word using OLE2, providing WYSIWYG editing of Z paragraphs directly in Word documents. (The *troff* and LaTeX versions use ordinary text editors on ASCII mark-up.) Further development of the tools is ongoing. CADiZ is a BCS Award winning product available for Sun, SGI and PC machines from York Software Engineering Ltd, The Innovation Centre, York Science Park, Heslington, York, YO1 5DG, UK (email yse@minster.york.ac.uk, tel +44-1904-435206, fax +44-1904-435135).

ProofPower is a suite of tools supporting specification and proof in Higher Order Logic (HOL) and in Z. Short courses on ProofPower-Z are available as demand arises. Information about ProofPower can be obtained automatically by sending email to ProofPower-server@win.icl.co.uk. Contact Roger Jones, International Computers Ltd, Eskdale Road, Winnersh, Wokingham, Berkshire RG11 5TT, UK (tel +44-118-969-3131 ext 6536, fax +44-118-969-7636, email rbj@win.icl.co.uk) for further details.

Zola is a commercial integrated support tool for Z on Sun workstations, for automated assistance at all stages of the specification construction, proving and maintenance process. It is intended for system developers and includes a WYSIWYG editor, type-checker and tactical theorem prover suitable for the creation and maintenance of large specifications. For further information, contact Chris Paine or Will Harwood, Imperial Software Technology, 62–74 Burleigh Street, Cambridge CB1 1DJ, UK (tel +44-1223-462400, fax +44-1223-462500, email fms@ist.co.uk).

ZTC is a Z type-checker available free of charge for educational and non-profit uses. It is intended to be compliant with the 2nd edition of Spivey's Z Reference

Manual. It accepts LATEX with 'zed' or 'oz' styles, and ZSL – an ASCII version of Z. ZANS is a Z animator. It is a research prototype that is still very crude. Both ZTC and ZANS run on Linux, SunOS 4.x, Solaris 2.x, HP-UX 9.0, DOS, and extended DOS. They are available via anonymous FTP under `ftp://ise.cs.depaul.edu/pub` in the directories ZANS-x.xx and ZTC-x.xx, where x.xx are version numbers. Contact Xiaoping Jia `jia@cs.depaul.edu` for further information.

Formaliser is a syntax-directed WYSIWYG Z editor and interactive type checker, running under Microsoft Windows, available from Logica. For further information, contact Susan Stepney, Logica UK Limited, Cambridge Division, Betjeman House, 104 Hills Road, Cambridge CB2 1LQ, UK (tel +44-1223-366343, fax +44-1223-251001, email `stepneys@logica.com`) or see under

> `http://public.logica.com/~formaliser/formlsr/formlsr.htm`

DST-fuzz is a set of tools based on the *f*UZZ package by Mike Spivey, supplying a Motif based user interface for LATEX based pretty printing, syntax and type-checking. A CASE tool interface allows basic functionality for combined application of Z together with structured specifications. The tools are integrated into SoftBench. For further information contact Hans-Martin Hoercher, DST Deutsche System-Techik GmbH, Edisonstr. 3, D-24145 Kiel, Germany (tel +49-431-7109-478, fax +49-431-7109-503, email `hmh@informatik.uni-kiel.d400.de`).

The B-Toolkit is a set of integrated tools which fully supports the B-Method for formal software development and is available from B-Core (UK) Limited, Magdalen Centre, The Oxford Science Park, Oxford OX4 4GA, UK. Contact Ib Sørensen (tel +44-1865-784520, fax +44-1865-784518, email `Ib.Sorensen@comlab.ox.ac.uk`) or see

> `http://www.b-core.com/`

Nitpick is a freely available tool for fully automatically analyzing software specifications in (roughly) a subset of Z. See

> `http://www.cs.cmu.edu/~nitpick/`

Z fonts for MS Windows and Macintosh are available on-line. For hyperlinks to these and other Z tool resources see the WWW Z page:

> `http://www.comlab.ox.ac.uk/archive/z.html#tools`

7   How can I learn about Z?

There are a number of courses on Z run by industry and academia. Oxford University offers industrial short courses in the use Z. As well as introductory courses, recent newly developed material includes advanced Z-based courses on proof and refinement, partly based around the B-Tool. Courses are held in Oxford, or elsewhere (e.g., on a company's premises) if there is enough demand. For further information, contact Jim Woodcock (tel +44-1865-283514, fax +44-1865-273839, email `Jim.Woodcock@comlab.ox.ac.uk`).

Logica offer a five day Z course at company sites. Contact Rosalind Barden (tel +44-1223-366343 ext 4860, fax +44-1223-322315, email `rosalind@logcam.co.uk`) at Logica UK Limited, Betjeman House, 104 Hills Road, Cambridge CB2 1LQ, UK.

Praxis Systems plc runs a range of Z (and other formal methods) courses. For details contact Anthony Hall on +44-1225-444700 or jah@praxis.co.uk.

Formal Systems (Europe) Ltd run a range of Z, CSP and other formal methods courses, primarily in the US and with such lecturers as Jim Woodcock and Bill Roscoe (both lecturers at the OUCL). For dates and prices contact Kate Pearson (tel +44-1865-728460, fax +44-1865-201114) at Formal Systems (Europe) Limited, 3 Alfred Street, Oxford OX1 4EH, UK.

DST Deutsche System-Technik runs a collection of courses for either Z or CSP, mainly in Germany. These courses range from half day introductions to formal methods and Z to one week introductory or advanced courses, held either at DST, or elsewhere. For further information contact Hans-Martin Hoercher, DST Deutsche System-Techik GmbH, Edisonstr. 3, D-24145 Kiel, Germany (tel +49-431-7109-478, fax +49-431-7109-503, email hmh@informatik.uni-kiel.d400.de).

8 What has been published about Z?

A searchable on-line Z bibliography is available on the World Wide Web under

http://www.comlab.ox.ac.uk/archive/z/bib.html

in BIBTEX format.

The following books largely concerning Z have been or are due to be published (in approximate chronological order):

• I. Hayes (ed.), Specification Case Studies, Prentice Hall International Series in Computer Science, 1987. (2nd ed., 1993) URL:

http://www.prenhall.com/013/832543/ptr/83254-3.html

• J. M. Spivey, Understanding Z: A specification language and its formal semantics, Cambridge University Press, 1988.

• D. Ince, An Introduction to Discrete Mathematics, Formal System Specification and Z, Oxford University Press, 1988. (2nd ed., 1993)

• J. C. P. Woodcock & M. Loomes, Software Engineering Mathematics: Formal Methods Demystified, Pitman, 1998. (Also Addison-Wesley, 1989)

• J. M. Spivey, The Z Notation: A reference manual, Prentice Hall International Series in Computer Science, 1989. (2nd ed., 1992) [Widely used as a de facto standard for Z. Often known as ZRM2.]

• A. Diller, Z: An introduction to formal methods, Wiley, 1990.

• J. E. Nicholls (ed.), Z user workshop, Oxford 1989, Springer-Verlag, Workshops in Computing, 1990.

• B. Potter, J. Sinclair & D. Till, An Introduction to Formal Specification and Z, Prentice Hall International Series in Computer Science, 1991. (2nd ed., 1996) URL:

http://www.prenhall.com/013/242206/ptr/24220-6.html

• D. Lightfoot, Formal Specification using Z, MacMillan, 1991.

• A. Norcliffe & G. Slater, Mathematics for Software Construction, Ellis Horwood, 1991.

• J. E. Nicholls (ed.), Z User Workshop, Oxford 1990, Springer-Verlag, Workshops in Computing, 1991.

- I. Craig, The Formal Specification of Advanced AI Architectures, Ellis Horwood, 1991.
- M. Imperato, An Introduction to Z, Chartwell-Bratt, 1991.
- J. B. Wordsworth, Software Development with Z, Addison-Wesley, 1992.
- S. Stepney, R. Barden & D. Cooper (eds.), Object Orientation in Z, Springer-Verlag, Workshops in Computing, August 1992. URL:
  http://public.logica.com/~stepneys/bib/ss/ooz.htm
- J. E. Nicholls (ed.), Z User Workshop, York 1991, Springer-Verlag, Workshops in Computing, 1992. URL:
  http://www.imi.gla.ac.uk/springer/eWiC/Abstracts/9.html
- D. Edmond, Information Modeling: Specification and implementation, Prentice Hall, 1992.
- J. P. Bowen & J. E. Nicholls (eds.), Z User Workshop, London 1992, Springer-Verlag, Workshops in Computing, 1993. URL:
  http://www.comlab.ox.ac.uk/archive/z/zum92.html
- S. Stepney, High Integrity Compilation: A case study, Prentice Hall, 1993. URL:
  http://public.logica.com/~stepneys/bib/ss/hic.htm
- M. McMorran & S. Powell, Z Guide for Beginners, Blackwell Scientific, 1993.
- K. C. Lano & H. Haughton (eds.), Object-oriented Specification Case Studies, Prentice Hall International Object-Oriented Series, 1993.
- B. Ratcliff, Introducing Specification using Z: A practical case study approach, McGraw-Hill, 1994.
- A. Diller, Z: An introduction to formal methods, 2nd ed., Wiley, 1994.
- J. P. Bowen & J. A. Hall (eds.), Z User Workshop, Cambridge 1994, Springer-Verlag, Workshops in Computing, 1994. URL:
  http://www.comlab.ox.ac.uk/archive/z/zum94.html
- R. Barden, S. Stepney & D. Cooper, Z in Practice, Prentice Hall BCS Practitioner Series, 1994. URL:
  http://public.logica.com/~stepneys/bib/ss/zip.htm
- D. Rann, J. Turner & J. Whitworth, Z: A beginner's guide. Chapman & Hall, 1994.
- D. Heath, D. Allum & L. Dunckley, Introductory Logic and Formal Methods. A. Waller, Henley-on-Thames, 1994.
- L. Bottaci and J. Jones, Formal Specification using Z: A modelling approach. International Thomson Publishing, 1995.
- D. Sheppard, An Introduction to Formal Specification with Z and VDM. McGraw Hill International Series in Software Engineering, 1995.
- J. P. Bowen & M. G. Hinchey (eds.), ZUM'95: The Z Formal Specification Notation, Springer-Verlag, Lecture Notes in Computer Science, volume 967, 1995. URL:
  http://www.comlab.ox.ac.uk/archive/z/zum95.html
- J. P. Bowen, Formal Specification and Documentation using Z: A Case Study Approach, International Thomson Compress Press, 1996. URL:
  http://www.comlab.ox.ac.uk/oucl/users/jonathan.bowen/zbook.html
- J. C. P. Woodcock & J. Davies, Using Z: Specification, proof and refinement, Prentice Hall International Series in Computer Science, 1996. URL:
  http://www.comlab.ox.ac.uk/usingz.html

- A. Harry, Formal Methods Fact File: VDM and Z, Wiley, 1996.
- J. Jacky, The Way of Z: Practical Programming with Formal Methods, Cambridge University Press, 1997. URL:

```
http://www.radonc.washington.edu/prostaff/jon/z-book/
```

See also an on-line list of Z books from Blackwells Bookshop under:

```
http://www.blackwell.co.uk/cgi-bin/bb_catsel?09_IBY
```

Formal Methods: A Survey by S. Austin & G. I. Parkin, March 1993 includes information on the use and teaching of Z in industry and academia. Contact DITC Office, Formal Methods Survey, National Laboratory, Teddington, Middlesex TW11 0LW, UK (tel +44-181-943-7002, fax +44-181-977-7091) for a copy.

Oxford University Computing Laboratory Technical Monographs and Reports, including many on Z, is available from the OUCL librarian (tel +44-1865-273837, fax +44-1865-273839, email library@comlab.ox.ac.uk).

For information on formal methods publications in general, see:

```
http://www.comlab.ox.ac.uk/archive/formal-methods/pubs.html
```

9 What is object-oriented Z?

Several object-oriented extensions to or versions of Z have been proposed. The book *Object orientation in Z*, listed above, is a collection of papers describing various OOZ approaches – Hall, ZERO, MooZ, Object-Z, OOZE, Schuman&Pitt, Z^{++}, ZEST and Fresco (an OO VDM method) – in the main written by the methods' inventors, and all specifying the same two examples. A more recent book entitled *Object-oriented specification case studies* surveys the principal methods and languages for formal object-oriented specification, including Z-based approaches.

10 How can I run Z?

Z is a (non-executable in general) specification language, so there is no such thing as a Z compiler/linker/etc. as you would expect for a programming language. Some people have looked at animating subsets of Z for rapid prototyping purposes, using logic and functional programming for example, but this is not really the major point of Z, which is to increase human understandability of the specified system and allow the possibility of formal reasoning and development. However, Prolog seems to be the main favoured language for Z prototyping and some references may be found in the Z bibliography (see above).

11 Where can I meet other Z people?

The 10th International Conference of Z Users (ZUM'97) will be held 3–4 April 1997 at the University of Reading, UK. For general enquiries, contact the Conference Chair, Jonathan Bowen (tel +44-118-931-6544, fax +44-118-975-1994, email J.P.Bowen@reading.ac.uk). Information on ZUM'97 is available under

```
http://www.cs.reading.ac.uk/zum97/
```

Announcements concerning the International Conference of Z Users are issued on electronic newsgroups including `comp.specification.z`, the Z postal mailing list and various specialist electronic mailing lists.

The 9th International Conference of Z Users (ZUM'95) was held 7–9 September 1995 in Limerick, Ireland. The proceedings appeared in the Springer-Verlag LNCS series (volume 967). Previous proceedings for Z User Meetings have been published in the Springer-Verlag Workshops in Computing series since the 4th meeting in 1989. See the URL

`http://www.comlab.ox.ac.uk/archive/z/zum.html`

for further on-line information on previous meetings.

For a list of meetings with a formal methods content, see:

`http://www.comlab.ox.ac.uk/archive/formal-methods/meetings.html`

12   What is the Z User Group?

The Z User Group was set up in 1992 to oversee Z-related activities, and the Z User Meetings in particular. As a subscriber to either `comp.specification.z`, ZFORUM or the postal mailing list, you may consider yourself a member of the Z User Group. There are currently no charges for membership, although this is subject to review if necessary. Contact `zforum-request@comlab.ox.ac.uk` for further information.

13   How can I obtain the draft Z standard?

The proposed Z standard under ISO/IEC JTC1/SC22 is available on-line. See under

`http://www.comlab.ox.ac.uk/oucl/groups/zstandards/`

for the latest information and locations. An early version is also available in printed form from the OUCL librarian (tel +44-1865-273837, fax +44-1865-273839, email `library@comlab.ox.ac.uk`) by requesting Technical Monograph PRG-107.

14   Where else is Z discussed?

The BCS-FACS (British Computer Society Formal Aspects of Computer Science special interest group) and FME (Formal Methods Europe) are two organizations interested in formal methods in general. Contact BCS FACS, Dept of Computer Studies, Loughborough University of Technology, Loughborough, Leicester LE11 3TU, UK (tel +44-1509-222676, fax +44-1509-211586, email `FACS@lut.ac.uk`).

A *FACS Europe* newsletter is issued to members of FACS and FME. Please send suitable Z-related material to the Z column editor, David Till, Dept of Computer Science, City University, Northampton Square, London, EC1V 0HB, UK (tel +44-171-477-8552, email `till@cs.city.ac.uk`) for possible publication. Material from articles appearing on the `comp.specification.z` newsgroup may be included if considered of sufficient interest (with permission from the originator if possible). It would be helpful for posters of articles on `comp.specification.z` to indicate if they do not want further distribution for any reason.

15 How does VDM compare with Z?

See I. J. Hayes, C. B. Jones & J. E. Nicholls, Understanding the differences between VDM and Z, FACS Europe, series I, 1(1):7–30, Autumn 1993 available as an on-line Technical Report from Manchester under

```
ftp://ftp.cs.man.ac.uk/pub/TR/UMCS-93-8-1.ps.Z
```

and I. J. Hayes, VDM and Z: A comparative case study, Formal Aspects of Computing, 4(1):76–99, 1992. VDM is discussed on the (unmoderated) VDM FORUM mailing list. Send a message containing the command 'join vdm-forum name' where name is your real name to mailbase@mailbase.ac.uk. To contact the list administrator, email John Fitzgerald on vdm-forum-request@mailbase.ac.uk.

16 How does the B-Method compare with Z?

B is a tool-based formal method for software development, conceived by the originator of Z, Jean-Raymond Abrial, whereas Z is designed mainly for specification. See

```
http://www.b-core.com/ZVdmB.html
```

for a comparison. See also

```
http://www.comlab.ox.ac.uk/archive/formal-methods/b.html
```

for further information on B.

17 What if I have spotted a mistake or an omission?

Please send corrections or new relevant information about meetings, books, tools, etc., to J.P.Bowen@reading.ac.uk. New questions and model answers are also gratefully received!

Author Index

Lecture Notes in Computer Science

For information about Vols. 1–1133

please contact your bookseller or Springer-Verlag